Challenges in Granular Physics

Challenges in Granular Physics

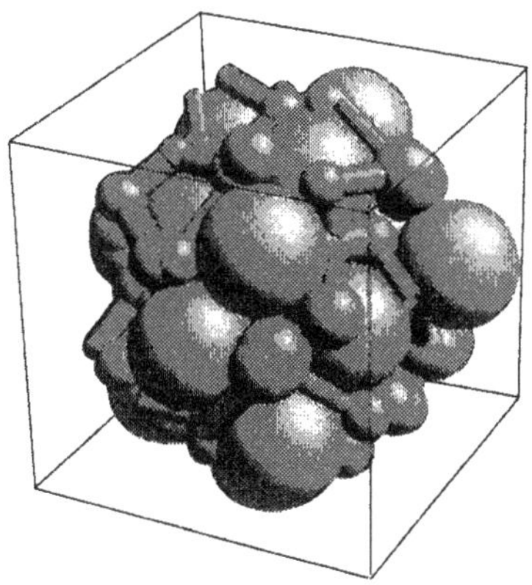

Thomas Halsey

Corporate Strategic Research, ExxonMobil Research and Engineering, USA

Anita Mehta

S N Bose National Centre for Basic Sciences, Calcutta, India

World Scientific

New Jersey • London • Singapore • Hong Kong

Published by

World Scientific Publishing Co Pte Ltd
5 Toh Tuck Link, Singapore 596224
USA office: 27 Warren Street, Suite 401-402, Hackensack, NJ 07601
UK office: 57 Shelton Street, Covent Garden, London WC2H 9HE

British Library Cataloguing-in-Publication Data
A catalogue record for this book is available from the British Library.

CHALLENGES IN GRANULAR PHYSICS

ISBN-13 978-981-238-239-9
ISBN-10 981-238-239-9

Preface

While the study of granular materials has an old and distinguished history, this history has only recently included large numbers of physicists. What explains the upsurge in the interest in the physics of granular materials over the past 15 years? While this is perhaps a question best left to historians or sociologists of science, some possible answers immediately suggest themselves.

- In the last two decades, condensed matter physics has become increasingly mature as a science, and ambitious practitioners, particularly of statistical condensed matter physics, have launched into subjects seemingly far removed from the core areas of physics. With greater and lesser success, this has led to new intellectual movement in areas as diverse as biological physics, finance, neuroscience and engineering. In this view, the vogue for granular physics is a result of statistical physicists in search of new horizons, and increasingly open to working in other scientific fields.
- Condensed matter physics has traditionally been led by experimentalists. Various pioneers have developed the instruments, materials and experimental protocols required for precision experiments to be made on granular systems. Today the experimental study of granular systems has become a sizable international sub-field of its own. Theorists have followed in the wake of this development, puzzled and stimulated by the many beautiful and often counter-intuitive results obtained.
- Granular systems provide instances of many non-equilibrium phenomena in vivid and relatively macroscopic form. One of the many mysterious properties of glasses is their ability to age, or slowly change state, over long periods of time; this property is expressed in the possibly apocryphal downward flow of medieval stained glass over the centuries intervening between that time and the modern age. Granular media show the same property in processes as diverse as settling under tapping (familiar to anyone pouring coffee or sugar into a container) or in the notoriously fickle response of sand dunes to sound waves.
- Another great problem in non-equilibrium physics is pattern formation; the means by which spatial and/or temporal structures are created in systems driven out of equilibrium. Here again granular materials exhibit these types of phenomena, but often in exaggerated and easy to visualize form.
- Fifteen years ago, Bak, Tang and Wiesenfeld used sandpiles as a paradigm in a new theory of why so many natural systems exhibit power-law behavior. This

theory of self-organized criticality has been widely influential, and has attracted considerable attention to avalanches and sandpiles; ironically, the authors were probably more interested in power-law behavior as a general phenomenon of nature than in the specific physics of granular materials.

- The behavior of granular materials is also of great importance in applied science: powder technology, fluid catalytic cracking in oil refining, and the geophysics of landslides and sand dunes are obvious examples.

Although all of these are certainly excellent reasons to be interested in granular materials, something is still missing from these explanations. Granular materials are fascinating in and of themselves, not just because they exhibit phenomena known from other areas of statistical or condensed matter physics. The combined limit of rigid, athermal particles leads us into a new and strange world of physics. While some of the physics therein is reminiscent of relaxation in glasses, kinetics of gases, pattern formation in liquids, or sound propagation in disordered solids; much is novel as well. For in combining properties of gases, liquids and solids, granular materials succeed in creating a new domain for physics, with its own practical rules and aesthetic principles.

In this volume, the reader will see many examples of this special physics of granular systems. What are the properties of static packings of particles? The number of nearest neighbors seems to depend upon the coefficient of friction, or perhaps not. What determines the rheology of dense granular flows? Are they fundamentally a surface phenomenon, or is there a characteristic bulk rheology? What is the role of gravity in this rheology? Are "force chains" of strongly interacting particles fundamental to granular statics and dynamics, or are they an optical illusion with no special relevance? Is the concept of entropy fundamental to understanding the relaxation of granular materials, or is it an equilibrium idea out of place in the non-equilibrium world? The reader will have an opportunity to examine these and other questions; perhaps some readers will succeed in resolving them.

This volume is the tangible result of a conference, "Challenges in Granular Physics," held as an Adriatico Research Conference at the International Centre for Theoretical Physics in Trieste, Italy, in August 2001. We are grateful to the European Union and the Conselho Nacional de Desenvolvimento Cientifico e Tecnolgico, Brazil, for their generous support of this workshop. We are also grateful to the ICTP, Trieste, its director, Prof. Miguel Virasoro, our local co-organizer, Prof. Silvio Franz, and to the superb professional staff of the ICTP, especially Lisa Iannitti, for their untiring work in support of this meeting.

A few meters from our conference facilities, the waves of the Adriatic lap onto the shore; grains of sand are stirred up, settle and avalanche along the beach and under the surf. What, exactly, is going on under the gaze of tourists, waiters, boaters, and a few curious physicists and engineers? Read on and discover.

Thomas C. Halsey

CONTENTS

CHAPTER 1

Multi-Particle Structures in Non-Sequentially Reorganized Hard Sphere Deposits

LUIS A. PUGNALONI
Instituto de Fisica de Liquidos y Sistemas Biologicos, UNLP-CONICET, Casilla de correo 565, 1900 La Plata, Argentina

G. C. BARKER
Institute of Food Research, Norwich Research Park, Colney, Norwich NR4 7UA, UK

ANITA MEHTA
S N Bose National Centre for Basic Sciences, Block JD Sector III Salt Lake, Calcutta 700 098, India

We have examined extended structures, bridges and arches, in computer generated, non-sequentially stabilized, hard sphere deposits. The bridges and arches have well-defined distributions of sizes and shapes. The distribution functions reflect the contraints associated with hard particle packing and the details of the restructuring process. A subpopulation of string-like bridges has been identified. Bridges are fundamental micro-structural elements in real granular systems and their sizes and shapes dominate considerations of structural properties and flow instabilities such as jamming.

Keywords: Granular materials; shaking; bridge structures.

1. Introduction

There has always been a fascination, amongst physicists, with the structures and configurations that exist within disordered packings of hard particles (see, for example, Ref. 7). One interest stems from the fundamental, frustrated geometries that exist within sphere packings, e.g. Refs. 4 and 16; another comes from the parallels between random packings and the structures of real disordered materials like liquids, glasses and granular solids [3, 14]. In particular, it is clear that the mechanical and transport properties of mesoscale disordered materials, like powders and deposits, are strongly dependent on the relative positions and connectivity of the constituent particles. A striking example of this interplay follows when several

First published in *Advances in Complex Systems*, Vol. 4, No. 4 (2001), pp. 289–297.

particles combine to form an 'arch' or 'bridge' near to the outlet of a gravity flow container and cause the flow to stop. This blocking phenomenon has an enormous impact on a wide range of technological and industrial processes; there have been many attempts to quantify the effect and to optimize operational parameters like the outlet size and the internal granular flow pattern, e.g. Ref. 10. However there is little information on the statistical details of the particulate configurations that are the underlying cause of the blocking.

In two dimensions arches and bridges can be observed throughout dense, random packings of hard disks, e.g. Ref. 5, and they appear to be ubiquitous elements of stable granular structures. In a recent report, To *et al.* [15], described experiments in which a jamming arch of monosized disks was repeatedly formed, in gravitational flow, across the outlet of a conical, two-dimensional hopper. These experiments indicated that the jamming arches had configurations that were similar to those of self-avoiding random walks. This statistical appreciation was used to obtain predictions of crucial macroscopic parameters like the jamming probability.

Below we give some details of bridge structures formed in models of hard sphere deposits and explore the role of bridges in three-dimensional disordered packings which are stable under gravity. We do not find 'diffusive' bridge configurations, but we have identified a special, chain-like subpopulation of bridges.

In a stable packing of hard particles each particle rests on three others in such a way that its weight vector passes through the triangle formed from the three contact points (we do not consider situations involving non-point contacts). A bridge is a configuration in which the three point stability conditions of two or more particles are linked, i.e. there are mutual stabilizations. In a simple example, with two particles A and B, particle A would rest on particle B and two other particles whereas particle B would rest on particle A and two further particles. Neither particle A or B could rest, i.e. be part of the stable structure, without the other. Bridge configurations, therefore, are the result of non-sequential stabilizations; they cannot be formed by the sequential placement of individual particles. In practice, almost all processing operations involving granular materials such as pouring and shaking, etc., are non-sequential processes. Two examples of two particle bridges are shown in Fig. 1. Each of these configurations is part of a large, dense, packing of spheres; all those spheres not involved in the bridge have been deleted so that it is clearly visible. The configuration on the left uses only three particles as the base of the bridge whereas that on the right uses four (base particles whilst ensuring the stability of the configuration do not, themselves, involve mutual stabilizations with other particles in the bridge structure). The details of the bridge configurations, in terms of sizes and shapes, are a manifestation of the volume and angular constraints that exist in dense hard particle assemblies. In turn these structures reflect the nature of the processing operations that precede the formation of a stable packing. In this respect bridges can be seen as part of the 'memory' of a granular system.

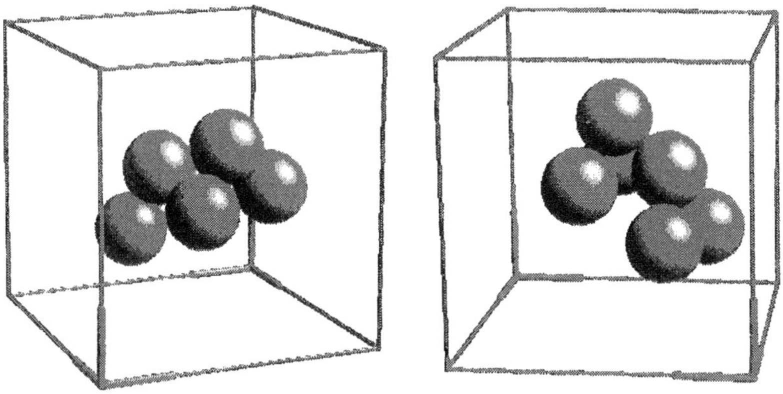

Fig. 1. Simple two particle bridge with three and four base particles.

2. Model Deposits

We have examined bridge structures in hard sphere assemblies that are generated by an established, non-sequential, restructuring algorithm [1, 8]. This algorithm restructures a stable, hard sphere deposit in three distinct stages. Firstly, free volume is introduced homogeneously throughout the system and the particles are given small, random, displacements. Secondly, the packing is compressed in a uniaxial external field using a low-temperature Monte Carlo process. Thirdly, the spheres are stabilized using a steepest decent 'drop and roll' dynamics to find a local minimum of the potential energy. Crucially, during the third phase of the restructuring, the spheres, although moved in sequence, are able to roll in contact with spheres that are in either stable or unstable positions; thus, mutual stabilizations may arise. The final configuration has a well-defined network of contacts and each sphere has a uniquely defined three point stability (in practice, the final configuration may include a few 'rattlers' [16]).

Restructuring simulations are performed in a rectangular cell (a square prism) with periodic boundaries in the lateral directions and a hard, disordered base perpendicular to the compression (external field) direction. Our previous investigations [8], have shown that this restructuring process does not depend strongly on the simulation parameters and that, after many cycles, restructured packings have a steady state described by particular values for the structural descriptors, such as the mean packing fraction and the mean coordination number. Typically, the steady state mean volume fraction is in the range $\phi \sim 0.55$–0.61 and the mean coordination number is $Z \sim 5.6 \pm 0.1$. The nature of the steady state is determined by the size of the expansion phase or the 'amplitude' of the process [8]. We have shown [1, 2] that the random packings generated in this way have many features in common with the states generated in vibrated granular media. In particular, we have shown that by varying the driving amplitude systematically we can explore 'irreversible' and 'reversible' branches of a density versus driving amplitude relationship analogous to the experimentally observed behaviour [11]. We have used monosized spheres in order to avoid any problems with induced size segregation and a disordered base prevents ordering. In the packings we have considered here the Q_6 order parameter, e.g. from Ref. 16, has a value $Q_6/Q_6^{\rm fcc} \sim 0.05$. This small value indicates only a very limited amount of face centered cubic crystalite formation in the system ($Q_6^{\rm fcc}$ is the value of the order parameter for a face-centered cubic crystal structure).

3. Statistics of Bridge Structures

We have identified clusters of mutually stabilized particles in computer generated packings of hard spheres. Each configuration includes approximately $N_{\rm tot} = 2500$ particles and we have examined approximately 100 configurations from each of two steady states, with $\phi = 0.56$ and 0.58, of the reorganization process.

Figure 2 illustrates a mutually stabilized cluster of five particles that is part of a large, stable packing; this figure also shows six particles which form a base (all other particles in the packing are hidden to make the diagram clearer). Also shown in Fig. 2 is the network of contacts for the particles in the bridge. This bridge is quite complex and includes a set of three particles (lower and to the right) that each have two mutual stabilizations.

Figure 3 illustrates a seven-particle bridge with nine base particles. The contact network shows that, although this bridge is larger than that in Fig. 2, it has a simpler topology because all of the mutually stabilized particles are in sequence — the bridge is string-like. The right-hand configuration in Fig. 1, with four base particles, is a string-like bridge. In practice, string-like bridges are common; bridges such as the one illustrated on the left-hand side of Fig. 1 are very rare in our packings.

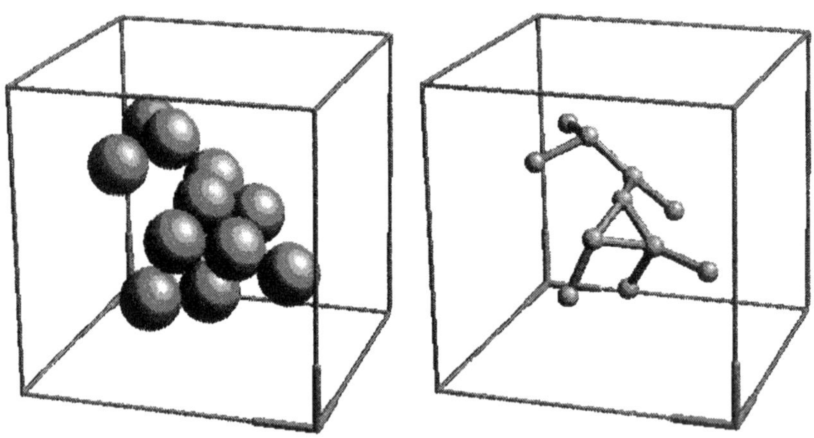

Fig. 2. A five-particle bridge with six base particles and the corresponding contact network.

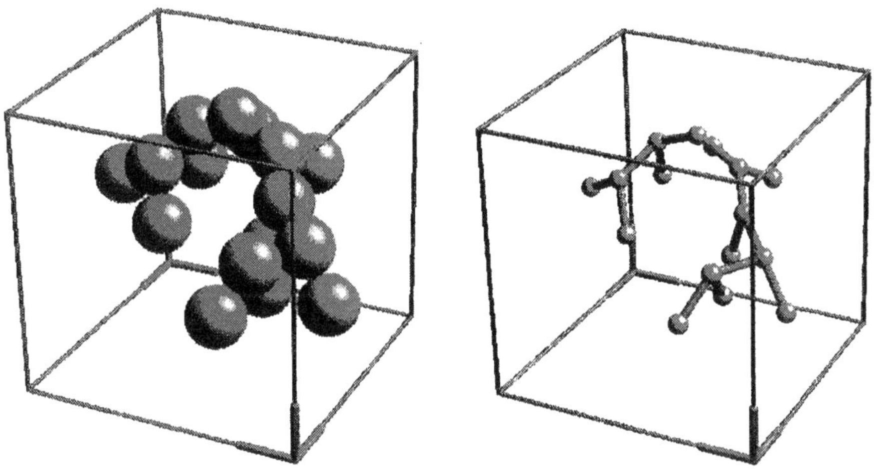

Fig. 3. A seven-particle string-like bridge with nine base particles and the corresponding contact network.

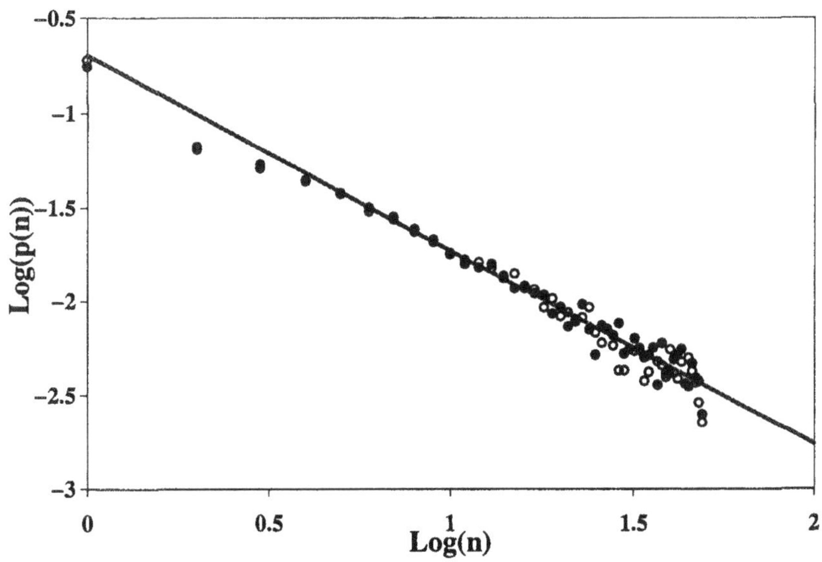

Fig. 4. The size distribution of bridges in non-sequentially reorganized hard sphere deposits; the full circles correspond to packings with $\langle\phi\rangle = 0.58$, the open circles correspond to packings with $\langle\phi\rangle = 0.56$ and the line is a fitted scaling $p(n) \sim n^{-\alpha}$.

Each packing contains a large variety of bridge sizes and shapes. Approximately 80 percent of particles are in mutually stabilized locations. In Fig. 4 we have plotted the size distribution of the bridges as $\text{Log}(p(n))$ against $\text{Log}(n)$ where $p(n) = \langle nN(n)/N_{\text{tot}}\rangle$ and $N(n)$ is the number of bridges which contain n mutually stabilized particles. Angular brackets indicate an average over configurations in the

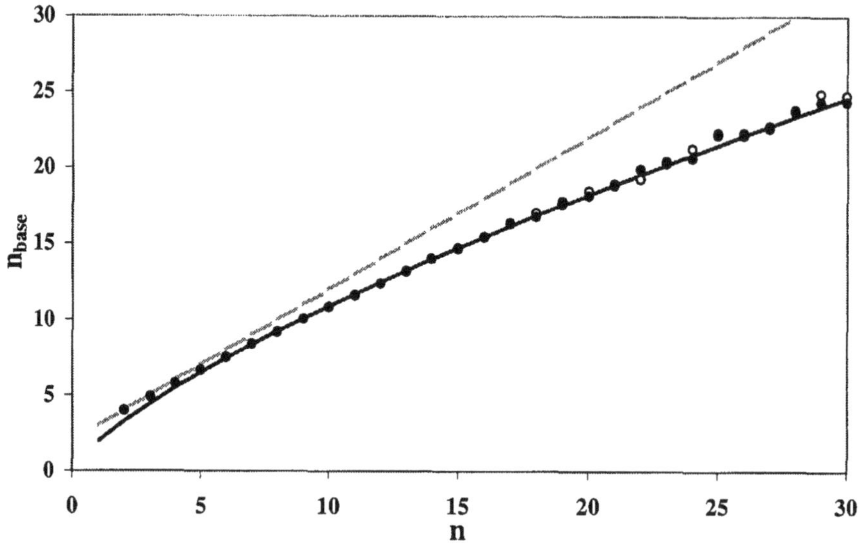

Fig. 5. The mean base size for bridges with size n; the dashed line indicates behaviour for string-like bridges $n_{\text{base}} = n + 2$ and the full line is a scaling fitted to the behaviour of the larger bridges.

steady state. We can consider $p(n)$ as the probability that a particular particle is included in a bridge with size n. Over a wide range of bridge sizes the distribution function has a scaling behaviour of the form $p(n) \sim n^{-\alpha}$ with $\alpha \sim 1.0 \pm 0.03$. The bridge size distribution is not strongly dependent on the volume fraction of the packings.

For a particular bridge size the number of base particles, which complete the stabilization, is variable with an upper bound, $n + 2$, corresponding to a string-like bridge. The mean number of base particles, n_{base}, is plotted as a function of the bridge size in Fig. 5. There is a crossover in behaviour at $n \sim 8$; small bridges are predominantly string-like and larger bridges have more complex structures with relatively fewer base particles. Again this property is not strongly dependent on the volume fraction of the packings in the range we have considered. We did not observe any 'domes' or 'canopies' although this could be an artefact of the relatively small sizes of the deposits. For a particular bridge configuration, a triangulation of the base particles can be used to construct a unique bridge axis as the mean of the triangle normals. With respect to this axis, geometrical descriptors, such as the radius of gyration or the aspect ratio, also show a cross-over that indicates the significance of a sub-population of string-like bridges.

A string-like bridge has uniquely defined end particles and, therefore, a well-defined extension. The mean squared separation, $\langle r_n^2 \rangle$, of the end particles for string-like bridges scales with the number of stabilizing bonds according to $\langle r_n^2 \rangle \sim (n-1)^\gamma$ with $\gamma = 1.33$. This 'superdiffusive' behaviour is illustrated in Fig. 6. The population of string-like bridges we observe, in reorganized three-dimensional deposits, is thus

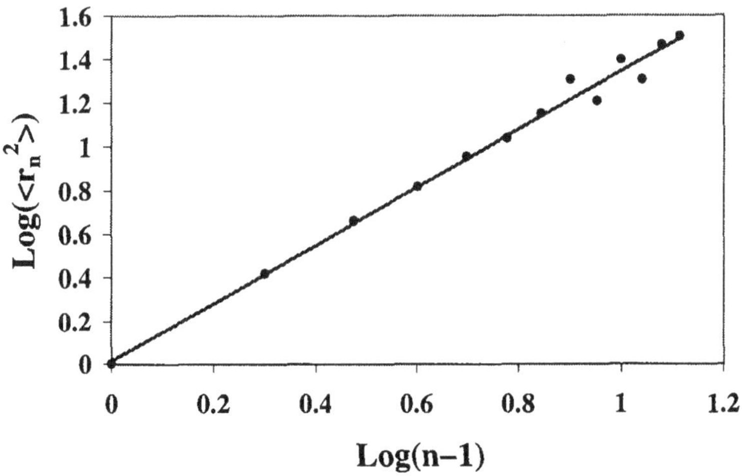

Fig. 6. The mean squared displacement, $\langle r_n^2 \rangle$, for string-like bridges as a function of the number of mutually stabilizing bonds $n-1$. Bridges are part of restructured deposits with steady state volume fraction $\phi = 0.58$.

distinct from the random walk structures that have been identified as the cause of blocking at the outlet of a two-dimensional hopper [15].

Figure 7 shows the distribution function of base extensions for all bridges in packings that are part of the restructuring steady state with $\phi = 0.58$. The extension, b_x, is the projection, in a plane perpendicular to the external field, of the radius of gyration of the base particle configuration (about the bridge axis). Clearly, this measure is related to the ability of a bridge to span an opening and, therefore,

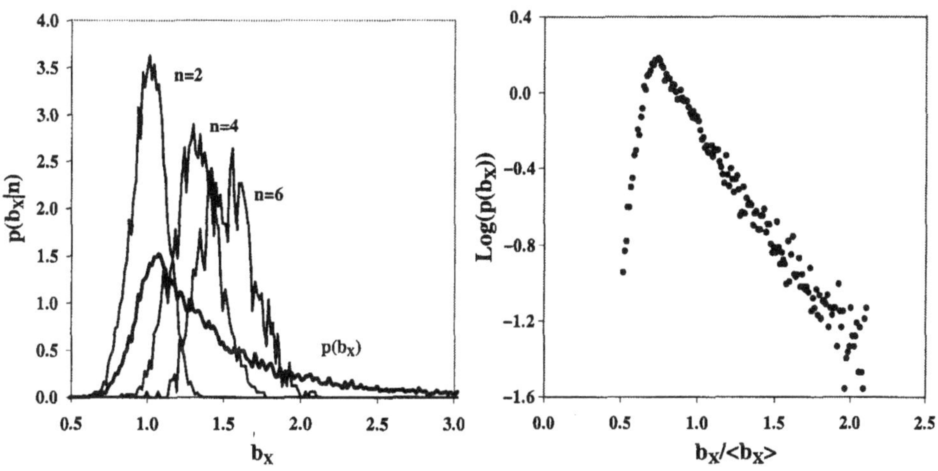

Fig. 7. The distribution of base extensions for bridges that are part of restructured deposits with steady state volume fraction $\phi = 0.58$. The left-hand figure also shows the distributions conditional on the bridge size, n, for $n = 2, 4, 6$. The right-hand figure shows the logarithm of the density as a function of the normalized variable $b_x/\langle b_x \rangle$.

is an indicator of the jamming potential for a bridge. We have also shown, in Fig. 7, distribution functions that are conditional on the bridge size n (for $n = 2, 4, 6$). The conditional distributions are sharply peaked, and are bounded at finite b_x, but the total distribution has a long tail at large extensions reflecting the existence of large bridges. In the second part of Fig. 7 we have plotted the logarithm of the probability density against a normalized variable, $b_x/\langle b_x \rangle$, where $\langle b_x \rangle$ is the mean extension of bridge bases. This figure emphasizes the exponential tail of the distribution function and also shows that bridges with small base extensions are unfavoured. The absence of bridges with base extensions that are considerably smaller than the mean extension is a reflection of the angular constraints that exist in hard particle structures. Small base extensions reduce the number of possible stable configurations for bridges with fixed size n. The form of the distribution in Fig. 7 can be interpretted, clearly, in terms of a partition, $p(b_x) = \sum_n p(b_x \mid n)p(n)$, since the conditional probabilities have restricted ranges, reflecting hard particle volume and angular constraints, and the size distribution has a well-defined scaling that reflects the particular bridge creation and anhiliation processes that are included in the restructuring. In this form, it is clear that the tail of the distribution of b_x arises from the summation and not from bridges with a particular size. It is interesting to note that the form of the normalized distribution in Fig. 7 is similar to the distribution of the normal forces in dense packings of hard particles, e.g. Ref. 12.

4. Discussion

Bridges and arches are significant elements of the mesostructure in many granular solids processing scenarios, e.g. Refs. 5 and 6. These structures, which extend beyond the scale of single particles, are strongly associated with important macroscopic properties of materials and with flow instabilities. We have shown that bridge structures are included throughout the non-sequentially reorganized deposits we have constructed. Bridges have well-defined statistics and, to a first approximation, they are distributed homogeneously within the deposits. We have identified a subpopulation of bridges, which have string-like configurations, that dominate for low bridge sizes. At present, it is unclear whether these structures are a property of the particular reorganization scheme considered here or whether they are a fundamental feature of non-sequential reorganization in hard sphere deposition. The bridge size statistics we have presented do not depend strongly on the volume fraction of the deposits but other measures, such as the bridge orientations (which we will report elsewhere), do vary with packing density, i.e. with the expansion amplitude of the reorganization process. Additionally, it is clear that non-sequential structures like bridges, that become trapped in the close-packed systems, frustrate local ordering in packings of monosized spheres. Thus, the onset of ordering must coincide with changes in the distribution of bridges; for driving amplitudes that are smaller than those used to construct the deposits considered above, we have observed the sudden onset of ordering [9]. We have not examined correlations of the bridges in the series of packings generated by the reorganization process.

Clearly, based on an assumption that the bridges in bulk are the same as those close to an opening, the statistics of extended structures in hard particle deposits is sufficient to estimate the probability that a bridge will form a span at an outlet of a fixed size. In three dimensions, this probability is not the same as the probability that a bridge will form a blockage or a 'jam.' However, initial investigations [13] indicate that data, analogous to the complement of the cumulative form of the distribution in Fig. 7, are in qualitative agreement with observations of the jamming probability. We hope to present details of these analyses in a future report.

Acknowledgments

G. C. Barker acknowledges support from the Fundación Antorchas, and the hospitality of Prof. J. Raul Grigera, during a visit to IFLYSIB, La Plata, Argentina where some of this work was completed. L. A. Pugnaloni acknowledges support from the International Union of Pure and Applied Biophysics during a visit to IFR, UK.

References

[1] Barker, G. C. and Mehta, A., *Phys. Rev.* **A45**, 3435 (1992).
[2] Barker, G. C. and Mehta, A., *Phys. Rev.* **E47**, 184 (1993).
[3] Bernal, J. D., *Nature* **183**, 141 (1959).
[4] Berryman, J. G., *Phys. Rev.* **B27**, 1053 (1983).
[5] Brown, R. L. and Richards, J. C., *Trans. Inst. Chem. Eng.* **38**, 243 (1960).
[6] Cooper, P., *Bulk Solids Handling* **8**, 162 (1988).
[7] Cumberland, D. J. and Crawford, R. J., *The Packing of Particles* (Elsevier, Amsterdam, 1987).
[8] Mehta, A. and Barker, G. C., *Phys. Rev. Lett.* **67**, 394 (1991).
[9] Mehta, A. and Barker, G. C., *J. Phys. Condensed Matter* **12**, 6619 (2000).
[10] Nedderman, R. M., Tuzun, U., Savage, S. B. and Houlsby, G. T., *Chem. Eng. Sci.* **37**, 1597 (1982).
[11] Nowak, E. R., Knight, J. B., Ben-Naim, E., Jaeger, H. M. and Nagel, S. R., *Phys. Rev.* **E57**, 1971 (1998).
[12] O'Hern, C. S., Langer, S. A., Liu, A. J. and Nagel, S. R., *Phys. Rev. Lett.* **86**, 111 (2001).
[13] Pugnaloni, L. A., unpublished.
[14] Seidler, G. T., Martinez, G., Seeley, L. H., Kim, K. H., Behne, E. A., Zaranek, S., Chapman, B. D., Heald, S. M. and Brewe, D. L., *Phys. Rev.* **E62**, 8175 (2000).
[15] To, K., Lai, P.-Y. and Pak, K. K., *Phys. Rev. Lett.* **86**, 71 (2001).
[16] Torquato, S., Truskett, T. M. and Debenedetti, P. G., *Phys. Rev. Lett.* **84**, 2064 (2000).

CHAPTER 2

Inelastic Hard Spheres with Random Restitution Coefficient: A New Model for Heated Granular Fluids

ALAIN BARRAT and EMMANUEL TRIZAC

Laboratoire de Physique Théorique, Bâtiment 210, Université de Paris-Sud, 91405 Orsay Cedex, France*

We consider a vertically shaken granular system interacting elastically with the vibrating boundary, so that the energy injected vertically is transferred to the horizontal degrees of freedom through inter-particle collisions only. This leads to collisions which, once projected onto the horizontal plane, become essentially stochastic and may have an effective restitution coefficient larger than unity. We therefore introduce the model of inelastic hard spheres with random restitution coefficient α (larger or smaller than unity) to describe granular systems heated by vibrations. In the non-equilibrium steady state, we focus in particular on the single particle velocity distribution $f(v)$ in the horizontal plane, and on its deviation from a Maxwellian. We use Molecular Dynamics simulations and Direct Simulation Monte Carlo (DSMC) to show that, depending on the distribution of α, different shapes of $f(v)$ can be obtained, with very different high energy tails. Moreover, the fourth cumulant of the velocity distribution (which quantifies the deviations from Gaussian statistics) is obtained analytically from the Boltzmann equation and successfully tested against the simulations.

Keywords: Granular gases; non-equilibrium steady-states; kinetic theory; velocity distributions.

1. Introduction

Granular matter can exist in many very different states, all of which are currently the subject of much interest [9]. On the one hand, dense granular matter can be studied at rest, and in particular many open problems concern the transmission of forces through a sandpile. On the other hand, since thermal energy is negligible with respect to gravitational or kinetic energy, any dynamical behaviour has to be a response to a certain external energy input; for example, tapping leads to compaction [10], while a strong, continuous energy input by vibrations produces granular gases in continuous motion, for which kinetic energy is much larger than the gravitational one [6, 11, 13, 16, 19, 22]. These vibrated systems are out of equilibrium but the energy input can compensate the dissipation due to inelastic

*Unité Mixte de Recherche UMR 8627.

First published in *Advances in Complex Systems*, Vol. 4, No. 4 (2001), pp. 299–307.

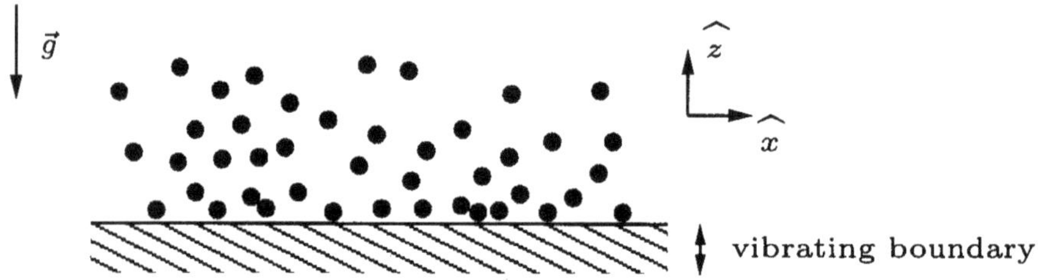

Fig. 1. Schematic view of the system under consideration. The grains are subject to gravity, and submitted to the vibration of an horizontal plate.

collisions between grains and therefore lead to stationarity. While many experiments study the appearance of patterns or inhomogeneities, others, on which we will concentrate here, focus on the velocity distributions and its deviations from the Maxwell–Boltzmann distribution (which would correspond to a system with neither dissipation nor energy injection, i.e. with elastic collisions).

2. System Studied and Modelisation

We want to study a three-dimensional system of grains on a plate, which is shaken vertically (i.e. along the z direction); the energy is therefore injected by a vibrating elastic boundary only in the z direction (Fig. 1). It is partly transferred to the other degrees of freedom, and also dissipated, through the inelastic collisions between grains. The velocities and their distribution are then studied in the horizontal (xy) plane.

2.1. *Usual theoretical approach*

The grains are modeled as smooth inelastic hard spheres (IHS) undergoing binary momentum-conserving collisions with a constant normal restitution coefficient $\alpha < 1$; a collision between two spheres 1 and 2, with velocities $\mathbf{v}_1$ and $\mathbf{v}_2$, dissipates a fraction $(1 - \alpha)$ of the component of the relative velocity $\mathbf{v}_{12} = \mathbf{v}_1 - \mathbf{v}_2$ along the center-to-center direction $\hat{\sigma}$. Once the dissipation has been described in this way, the problem is how to represent the energy injection. A possibility, used and studied by various authors [5, 18, 20, 21, 23], consists of submitting the spheres to a random force, i.e. to random "kicks" at a given frequency between collisions. Energy input then acts in all space directions.

2.2. *A new model*

However, as previously noted, the real energy input occurs only in the vertical direction, and is not transferred *between* but *through* collisions to the horizontal plane. Indeed, a three-dimensional inelastic collision between two spheres globally dissipates energy, but its projection onto the xy plane can in fact gain energy. Such

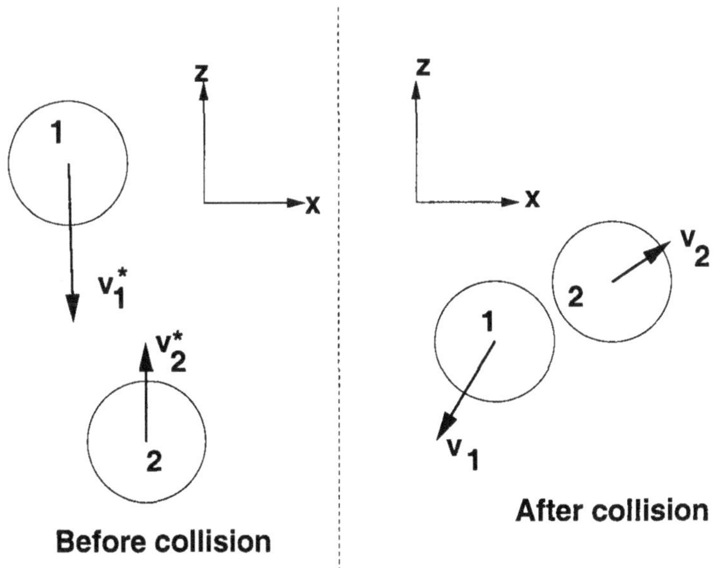

Fig. 2. Example of a globally dissipative collision leading to an energy increase in the horizontal plane. Before collision, the relative velocity in the horizontal plane is almost zero; after collision, it is finite.

an example is schematized in Fig. 2. Before collision, the kinetic energy in the plane is almost zero; it is, however, finite after the collision, so that an effective restitution coefficient defined by the ratio of the relative velocities in the horizontal plane before and after collision would be much larger than one.

In general, therefore, the *effective* restitution coefficient of the *projected* collision can be either smaller or larger than unity. This observation (which is also supported by experimental data [15]) leads to the following effective (projected) simple model [3]:

- two-dimensional hard spheres (of diameter σ) in the xy-plane
- binary momentum-conserving collisions
- *random* normal restitution coefficient α (< 1 or > 1) with distribution $\rho(\alpha)$ (the means over $\rho(\alpha)$ will be denoted by an overline), uncorrelated with the velocities of the particles. At each collision, the coefficient of restitution is randomly drawn from $\rho(\alpha)$.

Since, in a binary collision with restitution coefficient α, the energy change is proportional to $(\alpha^2 - 1)$, we shall consider distributions with $\overline{\alpha^2} = 1$ in order to ensure a stationary, constant temperature regime (at each collision, energy changes, but it is conserved on average). Since the average energy is constant, the granular temperature is also a constant determined by the initial velocity distribution. We will therefore study the distribution of *rescaled* velocities, using analytical and numerical tools.

3. Kinetic Theory

The Molecular chaos approximation factorizes the two-point distribution function:

$$f^{(2)}(\mathbf{v}_1, \mathbf{v}_2, \underbrace{|\mathbf{r}_{12}| = \sigma}_{\text{contact}}, t) = \chi f(\mathbf{v}_1, t) f(\mathbf{v}_2, t)\,, \tag{3.1}$$

where χ accounts for excluded volume effects (for elastic hard spheres, χ coincides with the density dependent pair correlation function at contact). We are then able to write the (Enskog–)Boltzmann equation in the steady state, averaged over the distribution of restitution coefficients:

$$\int' d\mathbf{v}_2 d\hat{\sigma} d\alpha (\mathbf{v}_{12} \cdot \hat{\sigma}) \rho(\alpha) \{\alpha^{-2} f(\mathbf{v}_1^*) f(\mathbf{v}_2^*) - f(\mathbf{v}_1) f(\mathbf{v}_2)\} = 0\,. \tag{3.2}$$

The prime on the integration symbol is a shortcut for $\int d\hat{\sigma} \Theta(\mathbf{v}_{12} \cdot \hat{\sigma})$ ($\hat{\sigma}$ is the center-to-center direction and Θ is the Heavyside function), and we consider collisions which yield $(\mathbf{v}_1, \mathbf{v}_2)$ as post-collisional velocities, for pre-collisional velocities $(\mathbf{v}_1^*, \mathbf{v}_2^*)$:

$$\mathbf{v}_1^* = \mathbf{v}_1 - \frac{1}{2}\left(1 + \frac{1}{\alpha}\right)(\mathbf{v}_{12} \cdot \hat{\sigma})\hat{\sigma}\,, \tag{3.3}$$

$$\mathbf{v}_2^* = \mathbf{v}_2 + \frac{1}{2}\left(1 + \frac{1}{\alpha}\right)(\mathbf{v}_{12} \cdot \hat{\sigma})\hat{\sigma}\,. \tag{3.4}$$

From now on, we will be concerned with the study of the rescaled velocities $\mathbf{c} = \mathbf{v}/v_0$ and of their distribution:

$$f(\mathbf{v}, t) = \frac{n}{v_0^d(t)} P(c)\,, \tag{3.5}$$

where n is the density and the thermal velocity v_0 is by definition related to the temperature $T(t)$ through $\frac{m}{2} v_0^2(t) = T(t) \equiv \frac{2}{nd} \int d\mathbf{v} \frac{m}{2} v^2 f(\mathbf{v}, t)$ (d is the space dimension).

It is usual to look for solutions in the form of a Sonine expansion [12] around the Maxwell–Boltzmann distribution $\Phi(c) = \pi^{-d/2} \exp(-c^2)$:

$$P(c) = \Phi(c) \left[1 + \sum_{p=1}^{\infty} a_p S_p(c^2)\right]\,, \tag{3.6}$$

where the polynomials $\{S_p\}$ are orthogonal for the Gaussian weight Φ.

Using the methods exposed in Ref. 20, we obtain the leading non-Gaussian correction ($a_1 = 0$ from the definition of temperature) a_2, which is related to the fourth cumulant:

$$a_2 \equiv \frac{\langle c^4 \rangle}{\langle c^4 \rangle_\Phi} - 1 = \frac{16\left(1 - 3\overline{\alpha^2} + 2\overline{\alpha^4}\right)}{9 + 24d + 32(d-1)\bar{\alpha} + (8d - 11)\overline{\alpha^2} - 30\overline{\alpha^4}}\,. \tag{3.7}$$

We will compare this result to numerical simulations in the next section, for $d = 2$. For the high energy tail, no analytical results have been obtained and we will investigate this issue numerically.

4. Numerical Simulations

4.1. *Methods*

We use two complementary approaches:

- The Direct Monte Carlo Simulation method (DSMC) [4] generates a Markov chain with the same probabilities of transition as the Boltzmann equation; it produces therefore an "exact" numerical solution of the Boltzmann equation.
- Molecular Dynamics (MD) integrate the *exact* equations of motion of the hard spheres, with no reference to the Boltzmann equation; we consider N spheres of diameter σ, in a box of linear size L in dimension d (here $d = 2$), with periodic boundary conditions [1, 8]; the comparison with DSMC allows us to test the molecular chaos approximation.

4.2. *Results*

The first results show the validity of the Sonine expansion and of the theoretical values for a_2. Figure 3 shows a comparison between DSMC results and the theoretical expansion; the agreement is perfect at small a_2, and satisfying at low velocities (note that the truncated Sonine expansion is a low-velocities expansion) for larger a_2.

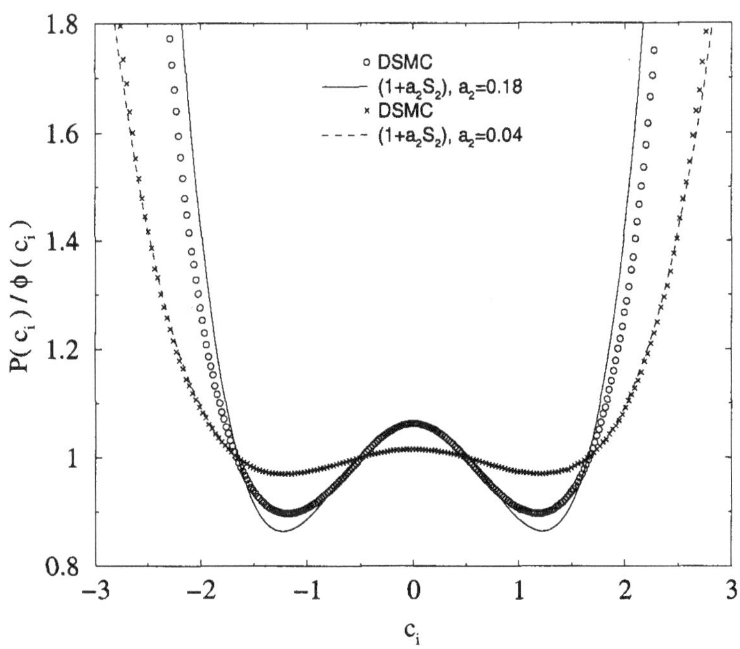

Fig. 3. Comparison of the $P(c)/\Phi(c)$ measured in DSMC (symbols) with the Sonine expansion with the calculated a_2 (lines), for two different $\rho(\alpha)$: flat distribution of $\alpha^2 \in [0.5, 1.5]$ ($a_2 = 0.04$), and flat distribution of $\alpha^2 \in [0, 2]$ ($a_2 = 0.18$).

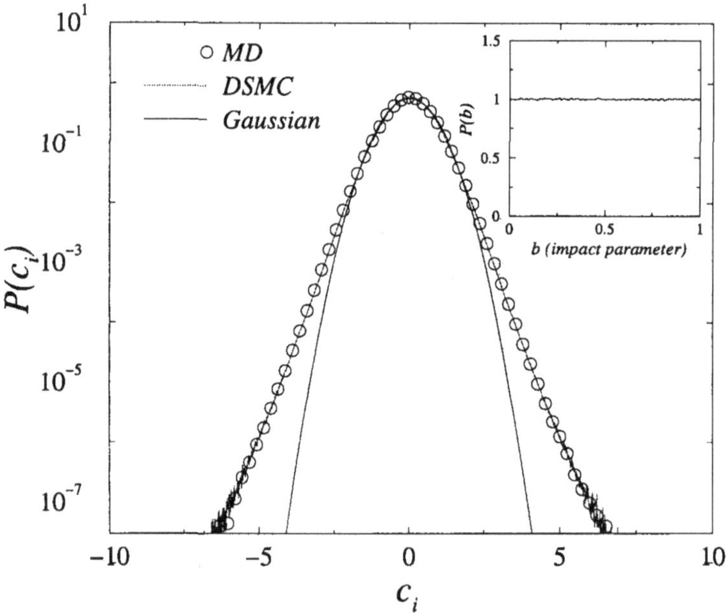

Fig. 4. Velocity distribution $P(c)$, for a flat distribution of $\alpha^2 \in [0,2]$, for MD and DSMC. MD: $50\,10^3$ particles (30% packing fraction); DSMC: $500\,10^3$ particles. A Gaussian is also shown for comparison. Inset: distribution of normalized impact parameters in MD.

Moreover, Fig. 4 shows that MD and DSMC simulations are in perfect agreement (shown only for a particular choice of $\rho(\alpha)$, but checked for other choices of $\rho(\alpha)$). Moreover, the curves obtained in MD simulations with small or large packing fractions (up to 40%) are indistinguishable (not shown). Important deviations from the Maxwell–Boltzmann distribution are obtained, but the inset shows that the distribution of impact parameters in Molecular Dynamics simulations is flat; this is a hint that no violation of molecular chaos is observed and that the factorization of the 2-particle correlation function Eq. (3.1) holds. In MD simulations, inhomogeneities and/or violations of molecular chaos could *a priori* appear, contrarily to DSMC. The fact that no such phenomenon is observed [24] is in contrast with the phenomenology at constant α [14] or with randomly driven IHS [17]; for a constant dissipative restitution parameter, colliding particles emerge with more parallel velocities than in the elastic case $\alpha = 1$ and, when they recollide, their velocities are still more parallel. The possibility of having $\alpha > 1$ seems to have removed this mechanism for the creation of velocity correlations violating molecular chaos, and to produce an efficient randomization of the velocities. This validates the theoretical approach based on the Boltzmann equation.

Let us now turn to the study of the large velocity tails. Figure 5 shows fits to stretched exponentials (*over 6 orders of magnitude*):

$$P(c) \propto \exp(-c^B)$$

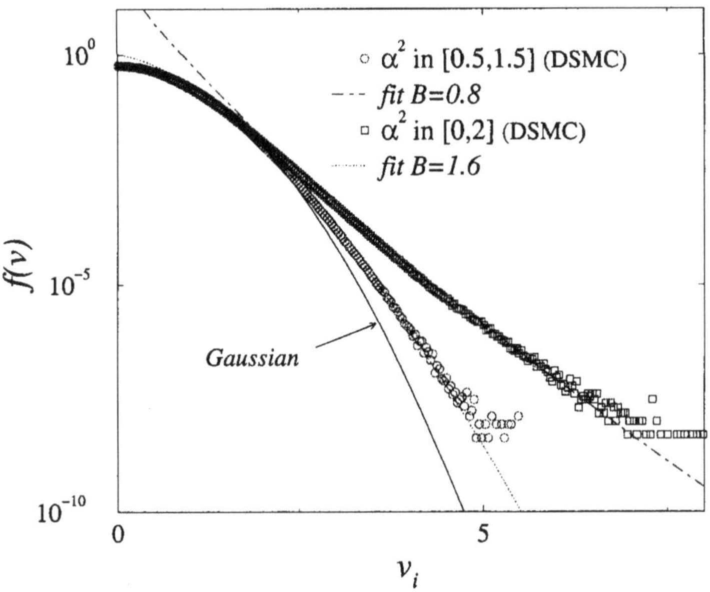

Fig. 5. Fits to stretched exponential forms $\exp(-c^B)$ of the velocity distributions, for a flat distribution of $\alpha^2 \in [0, 2]$ and $\in [0.5, 1.5]$.

with a wide range of possible values for B. In particular, a convenient choice of $\rho(\alpha)$ is compatible with $B = 1.6$, which has been found in some experiments [13, 19] (close to $B = 3/2$ obtained in Ref. 20 for randomly driven IHS fluids). The possibility of obtaining such different values of B may question the relevance of this exponent as an intrinsic quantity for granular gases in steady states.

5. Conclusions and Perspectives

We have introduced the idea of a random restitution coefficient in the IHS model, in order to account for the fact that, for a vertically vibrated layer of granular material, the energy is injected only along the vertical axis, and transferred through collisions in the horizontal directions; the projection in two dimensions of a three-dimensional collision can correspond to a gain in the two-dimensional energy, and therefore to an effective restitution coefficient larger than unity, even if the genuine α is necessarily smaller than unity, i.e. corresponds to a dissipative collision.

We have subsequently studied this model in two dimensions, with a probability distribution $\rho(\alpha)$ for the restitution coefficient. We have focused on the velocity distributions, and in particular on the deviation from the Maxwellian:

- At low velocities, the Sonine expansion technique is used. We obtained analytically the expression of the fourth cumulant a_2 and tested it against Molecular Dynamics (MD) and Monte Carlo Direct Simulations (DSMC). The theoretical predictions for a_2 are quite accurate, with a slight overestimation for a_2 that probably corresponds to the approximations made during the calculation (non-linear terms $\mathcal{O}(a_2^2)$ and higher order Sonine polynomials neglected). Moreover, the comparison between numerical data and the second order Sonine expansion shows a remarkable agreement for small values of a_2.

- The high energy tails, studied with DSMC simulations, can be fitted by functions of the form $\exp(-Ac^B)$, with $B < 2$ depending on $\rho(\alpha)$. It would certainly be interesting to have theoretical predictions concerning B. Note that once a functional form has been chosen for $\rho(\alpha)$, very different tails can be observed depending on the range of variation for α. This feature might question the relevance of the exponent B as an intrinsic quantity for granular gases in steady states.

The comparison of MD and DSMC results shows a remarkable agreement (even with a packing fraction as high as 40% in MD), and the study of the impact parameter in MD shows no violation of molecular chaos [24]. This is to be compared with the situation of free cooling [14] but also with MD results on heated inelastic hard spheres with constant restitution coefficient [17], in which microscopic pre-collisional velocity correlations develop and molecular chaos is violated. A thorough investigation of short scale velocity correlations would require the computation of various pre-collisional averages involving moments of the relative velocities, and has not been performed. Our results, however, suggest that the dynamical correlations inducing recollisions [7] and responsible for the violation of molecular chaos may not be a generic feature of driven granular gases exhibiting a non-equilibrium stationary state, at least at low densities.

In the model introduced here, the random restitution coefficient is uncorrelated with the relative velocities of the particles; this somehow unrealistic feature could be improved in more refined models. Such correlations, which seem difficult to quantify from first principles, might affect the high energy tail or induce pre-collisional velocity correlations. It would be very interesting to be able to link a realistic energy injection mechanism with a precise distribution of restitution coefficients. Two-dimensional collisions, projected in one-dimension, are currently being investigated with analytical, numerical and experimental [15] tools in order to quantify these correlations and study the random restitution coefficient model with correlations [2], and apply it to the experimental observations of Ref. 19.

Finally, a hydrodynamic study of the present random α model, in which the conservation of the energy is valid on average only, while density and momentum are conserved locally, is left for future investigations.

References

[1] Allen, M. P. and Tildesley, D. J., *Computer Simulations of Liquids* (Clarendon Press, Oxford, 1987).
[2] Barrat, A. and Trizac, E., in preparation.
[3] Barrat, A. Trizac, E. and Fuchs, J. N., *Eur. Phys. J.* **E5**, 161 (2001).

[4] Bird, G., *Molecular Gas Dynamics* (Oxford University Press, New York, 1976); *Molecular Gas Dynamics and the Direct Simulation of Gas Flows* (Clarendon Press, Oxford, 1994).
[5] Cafiero, R. Luding, S. and Herrmann, H. J., *Phys. Rev. Lett.* **84**, 6014 (2000).
[6] Clement, E. and Rajchenbach, J., *Europhys. Lett.* **16**, 133 (1991).
[7] Dorfman, J. R. and van Beijeren, H., The kinetic theory of gases, in *Statistical Mechanics, Part B: Time-Dependent Processes*, Berne, B. J. ed. (Plenum Press, New York, 1977), Chap. 3.
[8] See e.g. Herrmann, H.J., in *Disorder and Granular Media*, Bideau, D. and Hansen, A., eds. (Elsevier Science Publisher, 1993).
[9] See e.g. Jaeger, H. M. and Nagel, S. R., *Science* **255**, 1523 (1992); *Granular Matter: An Interdisciplinary Approach*, Mehta, A., ed. (Springer-Verlag, New York, 1994); Jaeger, H. M., Nagel, S. R. and Behringer, R. P., *Rev. Mod. Phys.* **68**, 1259 (1996); *Proceedings of the NATO Advanced Study Institute on Physics of Dry Granular Media*, Herrmann, H. J. *et al.* eds. (Kluwer Academic Publishers, Netherlands, 1998).
[10] Knight, J. B., Fandrich, C. G., Lau, C. N., Jaeger, H. M. and Nagel, S. R., *Phys. Rev.* **E51**, 3957 (1995); Nowak, E. R., Knight, J. B., Povinelli, M., Jaeger, H. M. and Nagel, S. R., *Powder Technol.* **94**, 79–83 (1997); Nowak, E. R., Knight, J. B., Ben-Naim, E., Jaeger, H. M. and Nagel, S. R., *Phys. Rev.* **E57**, 1971–1982 (1998); Jaeger, H. M., in *Physics of Dry Granular Media*, Herrmann, H. J., Hovi, J.-P. and Luding, S. eds. (Kluwer Academic, Dordrecht, The Netherlands, 1998), pp. 553–583.
[11] Kudrolli, A. and Henry, J., *Phys. Rev.* **E62**, R1489 (2000).
[12] Landau, L. and Lifshitz, E., *Physical Kinetics* (Pergamon Press, 1981).
[13] Losert, W., Cooper, D. G. W., Delour, J., Kudrolli, A. and Gollub, J. P., *Chaos* **9**, 682 (1999) and cond-mat/9901203.
[14] Luding, S., *GAMM 99 Proceedings*, June 1999.
[15] Menon, N., private communication.
[16] Olafsen, J. S. and Urbach, J. S., *Phys. Rev. Lett.* **81**, 4369 (1998); *Phys. Rev.* **E60**, R2468 (1999).
[17] Pagonabarraga, I., Trizac, E., van Noije T. P. C. and Ernst, M. H., to appear in *Phys. Rev. E*, cond-mat/0107570.
[18] Puglisi, A., Loreto, V., Marini Bettolo Marconi, U. and Vulpiani, A., *Phys. Rev.* **E59**, 5582 (1999).
[19] Rouyer, F. and Menon, N., *Phys. Rev. Lett.* **85**, 3676 (2000).
[20] van Noije, T. P. C. and Ernst, M. H., *Gran. Matter* **1**, 57 (1998).
[21] van Noije, T. P. C., Ernst, M. H., Trizac, E. and Pagonabarraga, I., *Phys. Rev.* **E59**, 4326 (1999).
[22] Warr, S., Jacques, G. T. H. and Huntley, J. M., *Powder Tech.* **81**, 41 (1994); Warr, S., Huntley, J. M. and Jacques, G. T. H., *Phys. Rev.* **E52**, 5583 (1995).
[23] Williams, D. R. and MacKintosh, F. C., *Phys. Rev.* **E54**, R9 (1996).
[24] For more extreme distributions of α and higher densities, violations of molecular chaos seem to appear, however, without modification of the velocity distribution: Soto, R., private communication.

CHAPTER 3

Spin-Models of Granular Compaction: From One-Dimensional Models to Random Graphs

JOHANNES BERG
Abdus Salam International Centre for Theoretical Physics, 34100 Trieste, Italy

ANITA MEHTA
S N Bose National Centre for Basic Sciences, Block JD Sector III, Salt Lake, Calcutta 700 098, India

We discuss two athermal types of dynamics suitable for spin-models designed to model repeated tapping of a granular assembly. These dynamics are applied to a range of models characterized by a 3-spin Hamiltonian aiming to capture the geometric frustration in packings of granular matter.

Keywords: Granular matter; spin models; glassy systems; disordered systems.

1. Introduction

Theory and experiment have both contributed extensively to the study of compaction in granular media in recent years. The experiments of the Chicago group [18, 19] have in particular inspired a large body of theorists to model the tapping of various systems. In many of these formulations, including the one we present here, an analogy is made between the volume of a granular system and the Hamiltonian of a spin system, along the lines first proposed by Edwards [11]; minimising the energy of the spin system then corresponds to minimising the volume of the granular system.

Grains can interlock only in specific ways if they are undeformable. It is thus often the case that locally compatible grain orientations could result in globally unfavourable ones, since different (and individually well-packed) clusters of grains could be unfavourably oriented with respect to each other. This frustration is therefore an essential ingredient in the modelling of granular media; it can be modelled in terms of orientational disorder of individual grains [15, 22] in lattice-based models, or, as in the present work, the orientational disorder of plaquettes in either lattice or off-lattice models of granular media.

First published in *Advances in Complex Systems*, Vol. 4, No. 4 (2001), pp. 309–319.

The second ingredient of models of granular compaction is the dynamics designed to model a series of successive taps to the system. A tap applied to a granular assembly for a brief moment feeds kinetic energy into the system and causes the particles to move with respect to each other. Then, the particles fall into a mechanically stable configuration.

In the following sections, we investigate the effect of two model tapping dynamics, called thermal and random tapping, on related models of granular media; the first two of these are on a lattice, while the second is embedded in a structure of random graphs.

2. Random Tapping and Thermal Tapping

A single tap applied briefly to a granular assembly feeds kinetic energy into the system and gives particles the freedom to move with respect to each other, thus momentarily decreasing the density. After this phase, the particles move (subject to gravity) to a new mechanically stable configuration and remain there until perturbed by further taps. Therefore a recurring theme in modelling taps is the alternation of periods of random perturbation of the system and periods in which the system is allowed to settle into a mechanically stable state.

Models including versions of this principle have included nonsequential Monte Carlo reorganisation schemes [14], lattice-based models of shaken sandboxes [21], the ratio of upward to downward mobility of particles on a lattice [8], or variable rates of absorption and desorption [7].

We now discuss two different ways of transferring this mechanism to spin models, which we will call thermal tapping and random tapping respectively. Both consist of a 'dilation' phase where the system is perturbed at random, and of a quench phase, where the system relaxes after this. The two dynamics differ only in the dilation phase; in both cases the *quench* phase is modelled by a quench of the system at $T = 0$, which lasts until the system has reached a blocked configuration, i.e. each site i has $s_i = \text{sgn}(h_i)$ or $h_i = 0$. Thus, at the end of each tap the system will be in a blocked configuration.

In thermal tapping [5] the dilation phase is modelled by a single sequential Monte-Carlo-sweep of the system at a dimensionless temperature T: A site i is chosen at random and flipped with probability 1 if its spin s_i is antiparallel to its local field h_i or $h_i = 0$ and with probability $\exp(-h_i/T)$ if it is not. This procedure is repeated N times.

In random tapping [4, 9, 10], however, the dilation phase consists of flipping a certain fraction p of randomly chosen spins, regardless of the value of their local field.

The two dynamics differ only in one point: In the case of thermal tapping, the dynamics during the dilation phase is correlated with the energy landscape. Sites with a large absolute value of the local field h_i have a low probability of flipping into the direction against the field. Such spins may be thought of as being highly constrained by their neighbours, while sites with a low absolute local field correspond to 'loosely constrained' particles.

In the case of random tapping, the configuration reached at the end of the dilation phase is only one of the many configurations having an overlap $1-2p$ with the configuration at the beginning of the tap. The sampling of these configurations, however, is uniform, and not correlated with the energy landscape of the model.

In both cases we use as an initial condition a configuration obtained by quenching the system from a configuration where the spins are chosen independently to be ± 1 with equal probabilities.

3. The Ferromagnetic 3-Spin Hamiltonian

In this section we discuss a simple Hamiltonian, which, we argue, captures some of the salient features of granular compaction. We consider a 3-spin Hamiltonian where N binary spins $S_i = \pm 1$ interact in triplets

$$H = -\rho N = -\sum_{i<j<k} C_{ijk} S_i S_j S_k \,, \tag{3.1}$$

where the variable $C_{ijk} = 1$ with $i < j < k$ denotes the presence of a plaquette connecting sites i, j, k and $C_{ijk} = 0$ denotes its absence. It has a trivial ground state where all spins point up and all plaquettes are in the configuration $+++$, giving a contribution of -1 to the energy. Yet, *locally*, plaquettes of the type $--+$, $-+-$, $+--$ (satisfied plaquettes) also give the same contribution. This results in a competition between local and global satisfaction of the plaquettes. Locally, any of the satisfied plaquettes are equivalent (thus favouring a paramagnetic state), yet globally a ferromagnetic state may be favoured, since there are few configurations satisfying all plaquettes where four configurations $+++$, $--+$, $-+-$, $+--$ occur in equal proportions. In this case, most ground states will be ferromagnetic, corresponding to a state with long-range order and a possibly crystalline state of the granular medium [2, 3, 16].

However, *two* spin flips are required to take a given plaquette from one satisfied configuration to another. Thus, an energy barrier has to be crossed in any intermediate step between two satisfied configurations. In the context of granular matter, this mechanism aims to model the situation where compaction follows a temporary dilation; for example, a grain could form an unstable ('loose') bridge with other grains before it collapses into an available void beneath the latter. This mechanism, by which an energy barrier has to be crossed in going from one metastable state to another, has recently been argued to be an important ingredient in models of granular compaction [16].

This feature is also shared by models with a two-spin interaction, with transitions e.g. from a state $++$ to $--$. However, in such models domains of a given magnetisation are formed, and the dynamics may be described as the evolution of the walls between these domains. We emphasise that in granular matter, the slow dynamics which is observed experimentally in the regime of high densities, is not necessarily due to the formation of domains; it is in fact due to the extensive number of particle rearrangements which are needed to fill any available voids.

We will see below an instance of this in the 3-spin model we present, where the system remains in a disordered state, with an ongoing slow dynamics; the ordered state is never reached, and domain coarsening to this end is also not observed.

The crucial feature of the model responsible for the slow dynamics is the degeneracy of the four configurations of plaquettes with $s_i s_j s_k = 1$ resulting in a competition between satisfying plaquettes *locally* and *globally*. In the former case, all states with even parity may be used, resulting in a large entropy, while in the latter, only the $+++$ state may be used. A dynamics based on local quantities will thus *fail* to find the magnetised configurations of low energy.

This mechanism has a suggestive analogy with the concept of geometrical frustration in granular matter, if we think of plaquettes as granular clusters. When grains are shaken, they rearrange locally, but locally dense configurations can be mutually incompatible. Voids may appear between densely packed clusters as a consequence of these mutually incompatible cluster orientations, leading to a decrease in the global packing fraction of the assembly. The process of compaction in granular media can in this sense be viewed as an optimisation process involving the competition between the compaction of *local* clusters and the simultaneous minimization of voids *globally*.

4. One-Dimensional Models

We first introduce two *one-dimensional* variants of the ferromagnetic 3-spin Hamiltonian as toy models; these illustrate some properties of the 3-spin Hamiltonian as well as the difference between random and thermal tapping.

The first one (model A) simply bunches three successive spins to a plaquette, so

$$H = -\sum_i s_i s_{i+1} s_{i+2} \tag{4.1}$$

with cyclic boundary conditions. The statistical mechanics of this model is trivial; it has a transfer matrix

$$T = \begin{pmatrix} e^{\beta} & 0 & e^{-\beta} & 0 \\ e^{-\beta} & 0 & e^{\beta} & 0 \\ 0 & e^{-\beta} & 0 & e^{\beta} \\ 0 & e^{\beta} & 0 & e^{-\beta} \end{pmatrix} \tag{4.2}$$

with the largest eigenvalue equal to $2\cosh(\beta)$. A thermal dynamics at finite temperature will thus reveal a paramagnet, with a transition to one of the four groundstates $++++++\cdots$, $+--+--+--\cdots$, $--+--+--+\cdots$, $-+--+--+-\cdots$ (the latter three being related by translation) at $T = 0$. The obvious problem with this Hamiltonian for modelling shaken granular media is that one-flip stable states, which arise where there is a single frustrated plaquette, are also domain walls between neighbouring segments of one of the four (ordered) ground states, which we

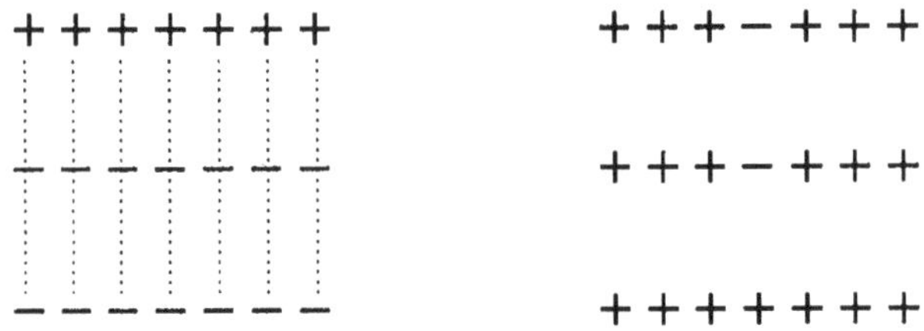

Fig. 1. Left: A ground state of model B. The 3-spin interactions are indicated by the dotted lines. Right: An excited state, which for $J > 2$ is stable against single spin-flips.

have argued are not the dominant defects impeding the compaction of granular media.

This is remedied to a certain extent in the second model (model B), which consists of three threads of ferromagnetically interacting spins, which are in turn linked by a three-spin interaction

$$H = -\sum_{i,a} s_i^a s_{i+1}^a - J \sum_i s_i^1 s_i^2 s_i^3 \,, \tag{4.3}$$

where a, which labels the thread, runs from 1 to 3. Again there are four ground states consisting of three lines with all spins up, or two lines with all spins down and one with all spins up. For $J > 2$ there are now excitations involving a single position only, such as the one shown in Fig. 1, which are stable against single spin flips.

The entries of the transfer-matrix of this model are

$$\langle s^1 s^2 s^3 | T | q^1 q^2 q^3 \rangle = \exp\{\beta J s^1 s^2 s^3 + \beta(s^1 q^1 + s^2 q^2 + s^3 q^3\}$$

with the largest eigenvalue equal to

$$1/2 e^{-\beta(6+J)} \Big(3e^{5\beta} + e^{9\beta} + 3e^{\beta(5+2J)} + e^{\beta(9+2J)} + \sqrt{e^{10\beta}(3+e^{4\beta})(1+e^{2\beta})^2 + 4e^{2\beta(3+J)}(1-e^{4\beta})} \Big) \,.$$

Under a thermal dynamics, and given a sufficiently slow rate of decreasing the temperature, both models reach equilibrium. After a single quench, models A and B reach a density (termed the single particle relaxation threshold (SPRT) in Ref. 4) found to be ~0.63 (this coincides with the result for the corresponding 2-spin model, see Refs. 9 and 10). Both systems under further thermal tapping show a very slow increase of the density (decrease of the energy) towards the ground state, as shown in Figs. 2 and 3, respectively. Under random tapping, however, *no* increase of the density is observed beyond that reached by a single quench in either of models A and B.

For model A, the reason for this behaviour is straightforward. In the case of the 2-spin (Ising) model, a domain wall may be moved with a single spin flip of zero energy. In the case of model A, however, a shift of the domain wall by one lattice site will result in a new frustrated plaquette, and only a further shift will restore

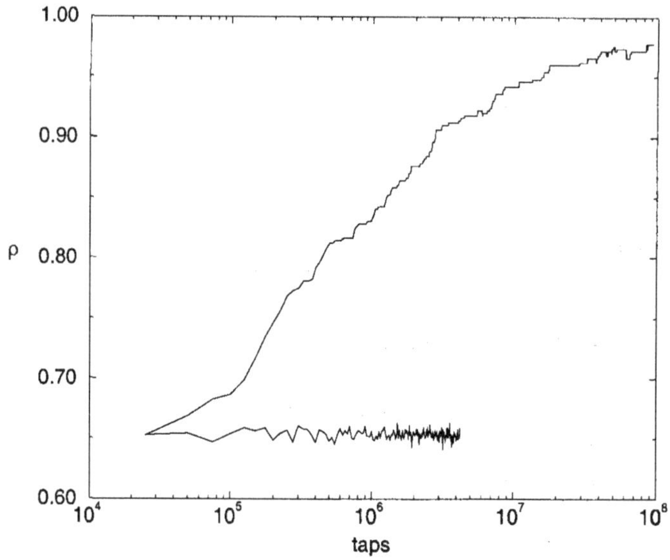

Fig. 2. We compare thermal and random tapping for the 3-spin model A. Whereas thermal tapping reaches the highest density (ordered state), random tapping does not take the system beyond the density reached by a single quench. We use $\rho = -H$ and the data stem from a system of size $N = 1002$ with $p = 0.01$ in the case of random tapping and $T = 1/3$, in the case of thermal tapping. Different values of p give qualitatively the same result.

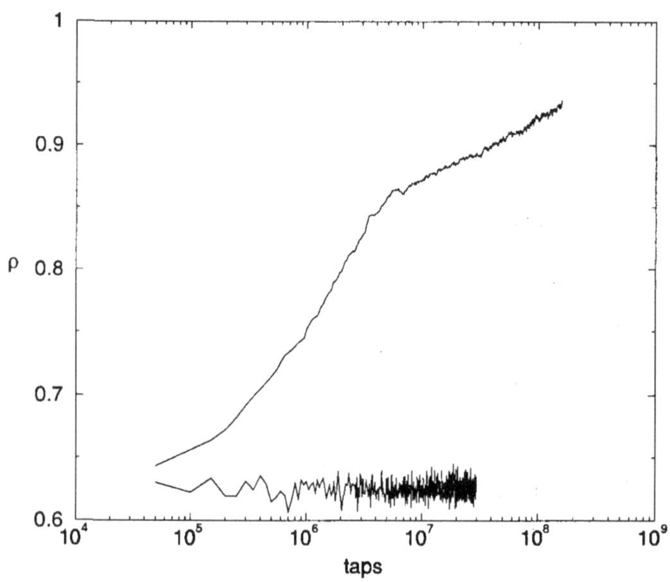

Fig. 3. Model B also shows different behaviour for random and for thermal tapping; only thermal tapping takes the system to densities larger than that reached by a single quench (single particle relaxation threshold). We use $\rho = -H/8$ and $J = 5$ and a system of size $N = 1002$ with $p = 0.01$ in the case of random tapping, with $T = 1$ in the case of thermal tapping.

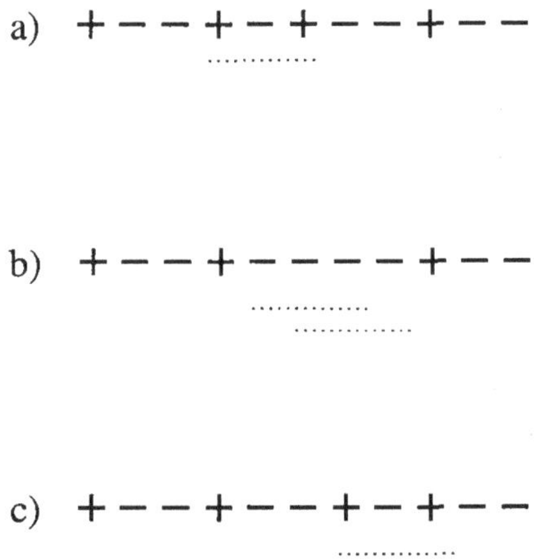

Fig. 4. (a) A domain wall in model A. The frustrated plaquette is marked by the dotted line. (b) Shifting the domain wall by one step results in the creation of a second frustrated plaquette. (c) Only a further shift restores the energy.

the energy to its previous value. This mechanism is illustrated in Fig. 4. It is clear that a random tapping dynamics thus cannot efficiently move domains, which is a necessary step in domain growth as well as the annihilation of smaller domains. The dilation phase of a thermal tap, on the other hand, is a mechanism by which domain walls may be moved.

The same line of argument holds for model B, where, for example, it takes four flips and a temporary expense of energy, to move the defect shown in Fig. 1. As discussed in Sec. 3, the mechanism of the system having to expend energy (i.e. lower the density) before being able to move to a new state of lower or equivalent energy, is one of the main motivations for the use of the 3-spin Hamiltonian.

5. The Random Graph Model

We now turn to a different manifestation of the ferromagnetic 3-spin Hamiltonian and consider (3.1) on a random graph.

A random graph [6] consists of a set of nodes and bonds, with the bonds connecting each node at random to a finite number of others, thus from the point of view of connectivity appearing like a finite-dimensional structure. Each bond may link up two sites (a graph) or more (a so-called hypergraph).

In a similar fashion, graphs — strictly speaking hypergraphs — with plaquettes connecting three or more nodes each, may be constructed. Choosing the connectivity matrix in the Hamiltonian (3.1) $C_{ijk} = 1(0)$ randomly with probability $2c/N^2$ $(1 - 2c/N^2)$, results in a random 3-hypergraph, where the number of plaquettes connected to a site is distributed with a Poisson distribution of average c.

In the context of modelling the compaction of granular matter, random graphs are the simplest structures with a finite number of neighbours. This finite connectivity is a key property, which goes beyond the simple fact that the grains in a deposit are in contact with a finite number of neighbouring grains. For example, cascades found experimentally during the compaction process may be explained by interactions between a finite number of neighbouring sites, where one local rearrangement sets off another one in its neighbourhood, and so on [4].

Another reason for the use of random graphs lies in the disordered structure of granular matter even at high densities. A random graph is the simplest object where a neighbourhood of each site may be defined *without* the consequent appearance of global symmetries such as would appear in the case of a regular lattice. Additionally, the locally fluctuating connectivity may be thought of as modelling the range of coordination numbers of the grains [1].

The absence of domains and domain walls in this case stems of course from the lack of spatial structure. Nevertheless, in the case of the Hamiltonian (3.1), there is an ordered ground state corresponding to all spins being up.

The behaviour of this model under both random and thermal tapping has been described in Refs. 4 and 5, respectively. We briefly recapitulate the results and then discuss the difference between the two dynamics in this case.

The dynamical behaviour may be divided into three regimes. The first one only lasts for the duration of a single tap, and consists of the alignment of all spins with their local field. The density reached by this process has been termed the single particle relaxation threshold (SPRT) [4].

In the second regime, which we term the compaction phase, the system seeks to eliminate the remaining frustrated plaquettes. This is a slow process, since at the end of each tap, all spins are aligned with their local fields. The analogy with geometric frustration is that grains are now locally stable and configurations are well packed; in order for any remaining voids to be filled after this, more than one particle around it would have to reorganise. This regime is characterised by a density which increases logarithmically as $\rho(t) \sim \rho(\infty) - a/\log(t)$, with the number of taps. A more detailed expression of this law [18] is

$$\rho(t) = \rho_\infty - \frac{\rho_\infty - \rho_0}{1 + 1/D \ln(1 + t/\tau)}, \tag{5.1}$$

which may also be written in the simple form $1 + t(\rho)/\tau = \exp\left\{D \frac{\rho - \rho_0}{\rho_\infty - \rho}\right\}$, implying that the dynamics becomes slow (logarithmic) as soon as the density reaches ρ_0.

The asymptotic density is reached when typical states at this density lie within "valleys" separated by extensive free-energy barriers. Once this density is reached, an extensive number of spins have to be flipped (grains to be moved) to go from one valley to the next, the relaxation time diverges and apart from fluctuations no further compaction occurs. These fluctuations about the asymptotic density mark the third phase of the dynamical behaviour. In spin glasses and spin-models of structural glasses this asymptotic density marks a dynamical phase transition [12, 17]. Configurations with higher densities exist of course (notably the ferromagnetic ground state corresponding to crystalline order), but a dynamics based on local information will not reach them. In the context of this model, we thus identify

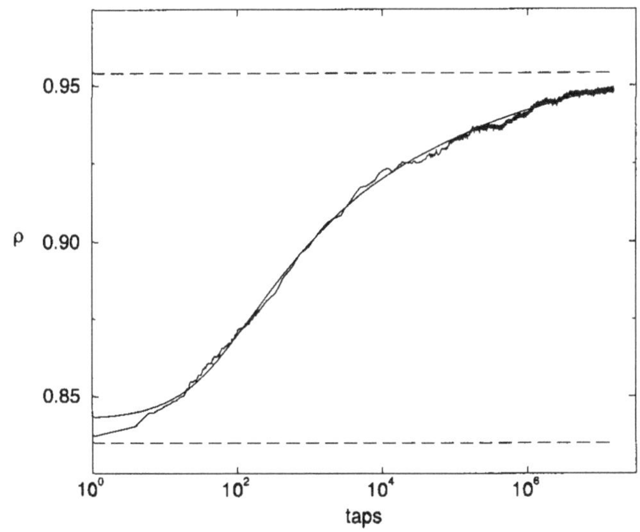

Fig. 5. Compaction curve for thermal tapping at connectivity $c = 3$ for a system of 10^4 spins with $T = 0.4$. The data stem from a single run and the fit (smooth solid line line) follows (5.1) with parameters $\rho_\infty = 0.989$, $\rho_0 = 0.843$, $D = 4.716$ and $\tau = 52.46$. The long-dashed line (top) indicates the approximate density 0.954 at which the dynamical transition occurs, the long-dashed line (bottom) indicates the approximate density 0.835 at which the fast dynamics stops, the *single-particle relaxation threshold.*

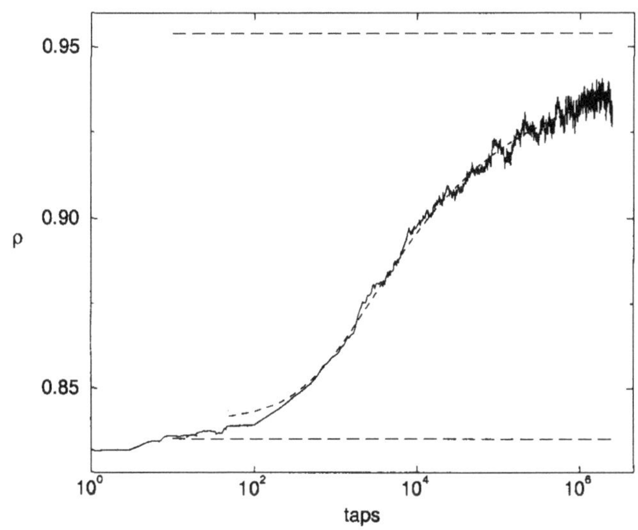

Fig. 6. Compaction curve for random tapping at connectivity $c = 3$ for a system of 10^4 spins (one spin chosen at random is flipped per tap). The data stem from a single run with random initial conditions and the fit (dashed line) follows (5.1) with parameters $\rho_\infty = 0.971$, $\rho_0 = 0.840$, $D = 2.76$ and $\tau = 1510$. The long-dashed line (top) indicates the approximate density of the dynamical transition, the long-dashed line (bottom) indicates the approximate density of the single-particle relaxation threshold.

the random close packing density with a dynamical transition. Here, the phase space turns from a single, paramagnetic state, into a large number of 'pockets' of configurations separated by free-energy barriers, causing a slow dynamics and — at the transition point itself — a breaking of the ergodicity. A simple approximation for the point of the dynamical tranisition has been given in Refs. 4, 5 and 20. The following figures illustrate the fact that the scenario of a rapid attainment of the SPRT, followed by the logarithmically slow approach to the dynamical transition is borne out both by thermal (Fig. 5) and by random tapping (Fig. 6). The two dynamics give similar results in this case, since the irrelevance of geometrical distance on the random graph, does not allow for the presence of domains such as those seen in the previous section. Note that random tapping is, however, much slower in reaching the dynamical threshold. It is important to note also that in both cases, if we increase the tapping intensity, the asymptotic density obtained is *below* that of the random close packing density corresponding to the dynamic transition [18].

6. Conclusion

Spin models of granular compaction consist of two ingredients: a Hamiltonian, which schematically gives the 'density' of the system as a function of the spin configuration, and a dynamics, which aims to model the tapping.

In this paper we discuss the use of 3-spin Hamiltonians, designed to capture the geometrical frustration of grains; locally densely packed configurations may not be compatible with each other at larger length scales. Also, we discuss two different mechanisms designed to mimic the tapping dynamics of granular matter in the context of spin models. Both consist of alternating periods of increasing and decreasing the energy of the spin system in order to model the dilation and quench phase of individual taps. The two mechanisms differ only in the form of the dilation phase: in thermal tapping this consists of a single Monte-Carlo sweep at a temperature T, whereas in random tapping a fraction p if spins are chosen at random and flipped. These two dynamics were investigated for two different classes of 3-spin Hamiltonians, one-dimensional models and random-graph models. In the latter case, the asymptotic state at low tapping amplitudes (random close packing) corresponds to a dynamical phase transition.

Acknowledgments

We thank S. Franz, B. Jones and M. Sellitto for illuminating discussions.

References

[1] Barker, G. C. and Mehta, A., Vibrated powders: Structure, correlations, and dynamics, *Phys. Rev.* **A45**, 3435–3446 (1992).
[2] Barker, G. C. and Mehta, A., Transient phenomena, self-diffusion, and orientational effects in vibrated powders, *Phys. Rev.* **E47**, 184–188 (1993).

[3] Barker, G. C. and Mehta, A., Inhomogeneous relaxation in vibrated granular media: consolidation waves, cond-mat/0010268.

[4] Berg, J. and Mehta, A., *Europhys. Lett.* **56**, 784–791 (2001), cond-mat/0012416.

[5] Berg, J. and Mehta, A., *Phys. Rev.* **E65**, 031305 (2002), cond-mat/0108225.

[6] Bollobas, B., *Random Graphs* (Academic Press, London, 1985).

[7] Brey, J. J., Prados, A. and Sanchez-Rey, B., Simple model with facilitated dynamics for granular compaction, *Phys. Rev.* **E60**, 5685–5692 (1999).

[8] Caglioti, E., Loreto, V., Herrmann, H. J. and Nicodemi, M., A "Tetris-like" model for the compaction of dry granular media, *Phys. Rev. Lett.* **79**, 1575–1578 (1997).

[9] Dean, D. S. and Lefèvre, A., Tapping spin glasses and ferromagnets on random graphs, *Phys. Rev. Lett.* **86**, 5639–5642 (2001).

[10] Lefèvre, A. and Dean, D. S., Tapping thermodynamics of the one dimensional Ising model, *J. Phys.* **A34** (14), L213–L220 (2001).

[11] Edwards, S. F., The role of entropy in the specification of a powder, in *Granular Matter: An Interdisciplinary Approach*, Mehta, A. ed. (Springer-Verlag, New York, 1994).

[12] Franz, S. and Parisi, G., Recipes for metastable states in spin-glasses, *J. Physique* **I5**, 1401–1502 (1995).

[13] Kob, W. and Andersen, H. C., Kinetic lattice-gas model of cage effects in high-density liquids and a test of mode-coupling theory of the ideal-glass transition, *Phys. Rev.* **E48**, 4364–4377 (1993).

[14] Mehta, A. and Barker, G. C., Vibrated powders: A microscopic approach, *Phys. Rev. Lett.* **67**, 394-397 (1991).

[15] Mehta, A. and Barker, G. C., Disorder, memory and avalanches in sandpiles, *Europhys. Lett.* **27**, 501-506 (1994).

[16] Mehta, A. and Barker, G. C., Glassy dynamics in granular compaction, *J. Phys.* **C12**, 6619–6628 (2000).

[17] Monasson, R., Structural glass transition and the entropy of the metastable states, *Phys. Rev. Lett.* **75**, 2847–2850 (1995).

[18] Nowak, E. R., Knight, J. B., Ben-Naim, E., Jaeger, H. M. and Nagel, S. R., Density fluctuations in vibrated granular materials, *Phys. Rev.* **E57** (2), 1971–1982 (1998).

[19] Nowak, E. R., Knight, J. B., Povinelli, M., Jaeger, H. M. and Nagel, S. R., Reversibility and irreversibility in the packing of vibrated granular material, *Powder Technology* **94**, 79–83 (1997).

[20] Ricci-Tersenghi, F., Weigt, M. and Zecchina, R., Simplest random K-satisfiability problem, *Phys. Rev.* **E63**, 026702–026713 (2001).

[21] Stadler, P. F., Luck, J.-M. and Mehta, A., *Europhys. Lett.* **57**, 46–53 (2002), cond-mat/0103076.

[22] Stadler, P. F., Mehta, A. and Luck, J. M., Glassy States in a Shaken Sandbox, in this volume.

CHAPTER 4

Models of Free Cooling Granular Gases

UMBERTO MARINI BETTOLO MARCONI and ANDREA BALDASSARRI
Dip. di Matematica e Fisica and INFM, Univ. di Camerino,
Via Madonna delle Carceri, Camerino, 62032, Italy

ANDREA PUGLISI
Dip. di Fisica, Univ. di Roma "La Sapienza",
P.le Aldo Moro 2, Roma, 00185, Italy

We consider the free evolution of systems of granular particles whose dynamics is characterized by a collision rule which preserves the total momentum, but dissipates the kinetic energy. Starting from an inelastic version of a minimal model proposed by Ulam for a gas of Maxwell molecules, we introduce a new lattice model aimed at investigating the role of dynamical correlations and the onset of spatial order induced by the inelasticity of the interactions. We study, in one- and two-dimensional cases, the velocity distribution, the decay of the energy, the formation of spatial structures and topological defects. Finally, we relate our findings to other models known in other fields.

Keywords: Rapid granular flows; kinetic theory for granular gases; ordering kinetics; inelastic Maxwell gas; shear instability; lattice models.

1. Introduction

Granular systems show rather peculiar and intriguing features both with respect to their static and their dynamical properties. A dilute granular system, subject to tapping, shaking or some other kind of external driving, which supplies the energy dissipated by the inelastic collisions, may behave similarly to a fluid. On the contrary in the absence of external forces it gradually loses its kinetic energy and comes to rest. In addition, it may become spontaneously inhomogeneous and form patterns. Such a behavior during the free cooling process displays interesting analogies and connections with other areas of non-equilibrium statistical mechanics such as ordering kinetics [6], decaying turbulence [11] and diffusion [6].

In the present paper we shall be concerned with the dynamics of assemblies of inelastic grains whose interactions are represented by instantaneous binary collisions. The inelasticity is accounted for through the so-called normal restitution coefficient, α, taken to be velocity independent. A collision between two grains i and j with precollisional velocities v_i and v_j reverses and reduces the component

First published in *Advances in Complex Systems*, Vol. 4, No. 4 (2001), pp. 321–331.

of the relative velocity along the center to center direction $\hat{\sigma}$ by a factor $(1-\alpha)$. The corresponding post-collisional velocities are

$$\begin{aligned} \mathbf{v}_j' &= \mathbf{v}_j + \Theta(-(\mathbf{v}_i - \mathbf{v}_j)\cdot\hat{\sigma})\frac{1+\alpha}{2}((\mathbf{v}_i - \mathbf{v}_j)\cdot\hat{\sigma})\hat{\sigma}\,, \\ \mathbf{v}_i' &= \mathbf{v}_i - \Theta(-(\mathbf{v}_i - \mathbf{v}_j)\cdot\hat{\sigma})\frac{1+\alpha}{2}((\mathbf{v}_i - \mathbf{v}_j)\cdot\hat{\sigma})\hat{\sigma}\,. \end{aligned} \tag{1.1}$$

The Heaviside function Θ here stresses the condition that colliding particles move against one another, a "kinematic constraint" to be satisfied at each collision.

The existing theoretical approaches are based either on realistic descriptions of the grains or on idealized models, which in virtue of their major simplicity lend themselves to analytic solutions or to efficient computations. Furthermore, such an idealized modeling can deliberately sacrifice likeness to reality, in order to identify the physical mechanism responsible for the salient features of the system.

In the following we shall discuss the physics of cooling granular fluids by illustrating its rich phenomenology by means of a series of minimal models based on the simplest rule (1.1) which ensures momentum conservation during inelastic collisions. We start from a mean field version with no positional degrees of freedom for the grains and successively we release such a constraint to consider one-dimensional and two-dimensional systems.

2. Mean Field Model

To appreciate the difference between an ordinary and a granular gas one can start with the following minimal model constituted by N particles without positional degrees of freedom and characterized only by d-dimensional velocities. The evolution of the system is driven by random selection of pairs of velocities $(\mathbf{v}_i, \mathbf{v}_j)$ which are updated according to Eq. (1.1). Since there is no true movement of the grains, the center-to-center direction $\hat{\sigma}$ is randomly chosen with a uniform distribution in the d-dimensional sphere (the "kinematic constraint" can be equivalently disregarded, as it would just randomly avoid half of the collisions happening). A unit time corresponds to N collisions. This model, for two-dimensional velocities, was put forward by S. Ulam [18] for elastic gases ($\alpha = 1$). He showed how the velocity distribution asymptotically converges to the Maxwell distribution, independently from the starting distribution.

An interesting inelastic version of the Ulam model has recently been proposed by Ben Naim and Krapivsky (BK), for one-dimensional (scalar) velocities.

The inelasticity has a major consequence; in the absence of energy injection the total kinetic energy of the fluid decreases, i.e. the velocity distribution becomes narrower with time, and the granular temperature $\epsilon(t) = \langle v^2 \rangle$ decreases. On the other hand it may approach a stationary shape under a suitable rescaling of the velocities.

Due to the infinite connectivity of the model, a mean field treatment which disregards correlations represents the right level of description. In fact, setting $\beta = 2/(1+\alpha)$, the Boltzmann equation turns out to be exact for the model:

$$\partial_t P(v,t) = \beta \int du\, P(u,t) P(\beta v + (1-\beta)u, t) - P(v,t)\,.$$

The first term represents the gain and the second the loss. Notice that in the elastic limit any initial distribution is left unchanged by the dynamics, since the particles upon colliding merely exchange their states. This model is also known as the pseudo-Maxwell model for inelastic gases (the scalar version), as it reproduces the kernel of the collision integral in the Boltzmann equation for the so-called Maxwell molecules, which does not depend upon the relative velocity (see Ref. 10 for a review).

For $0 < \alpha < 1$ the Boltzmann equation (2) possesses a self similar asymptotic solution. By inspection one can see that a solution of Eq. (2) is [2]:

$$P(v,t) = \frac{2}{\pi v_0 \left[1 + \left(\frac{v}{v_0(t)}\right)^2\right]^2}\,, \tag{2.1}$$

where $v_0(t)$ is the mean square velocity, which decreases as $v_0(t) = v_0(0)e^{-t/\tau}$, with $\tau^{-1} = (1-\alpha^2)/2$. Note that according to Eq. (2.1), moments higher than the second diverge: $\langle v^{2n} \rangle = \infty$ for $n > 1$.

In fact we believe that Eq. (2.1) represents the asymptotic solution for a large class of starting initial distributions, for the following arguments:

(1) as shown by Ben Naim and Krapivsky, the dynamics of the moments for a generic starting distribution can be computed, giving $\lim_{t\to\infty}\langle v^{2n}(t)\rangle / \langle v^2(t)\rangle^n = \infty$ for $n > 1$;
(2) we performed numerical simulations of the BK model, collecting evidence of the convergence to the solution (2.1) for several starting velocity distributions, namely uniform, exponential (see Fig. 1, left frame) or Gaussian.

Interestingly, the asymptotic probability density function (PDF) (2.1) does not depend on the restitution parameter α. A similar asymptotic universality is expected for a real one-dimensional granular gas, as shown by recent extensive numerical simulations [3]. However, when the BK model is generalized to vectorial velocities, the tails of the PDF depend on α. Results of our numerical simulations for the two-dimensional case are shown in the right frame of Fig. 1. For $\alpha = 1$ we recover the asymptotic Maxwell distribution predicted by Ulam, whereas for $\alpha = 0$ our data suggest the formation of algebraic tails.

In spite of these interesting features BK models do not bear a strong resemblance to physical reality. Obviously, their mean field character prevents the onset of any kind of inhomogeneities. On the other hand, theoretical approaches based on linear stability analysis predict three distinct and consecutive dynamical regimes [12, 19]: homogeneous, inhomogeneous in the velocity field, inhomogeneous in the velocity

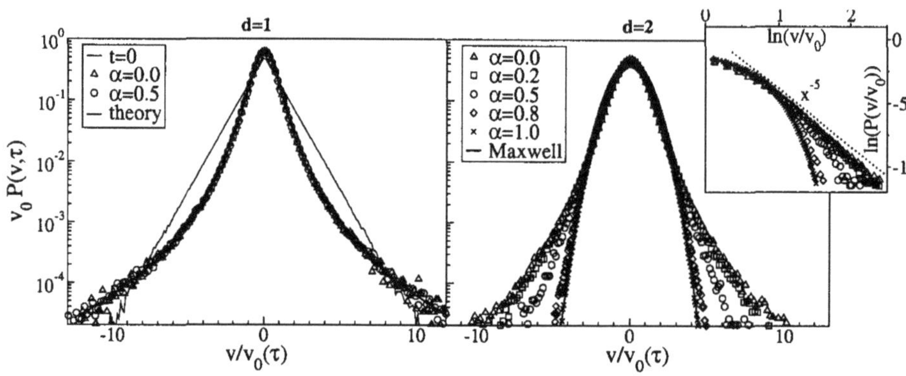

Fig. 1. Asymptotic velocity distributions $P(v,t)$ versus $v/v_0(t)$ for different values of α from the simulation of the inelastic pseudo-Maxwell (Ulam's) model in 1D (left) and 2D (right), where the PDF of a single component is displayed.

and density fields. Realistic tests of this state of affairs are provided by Molecular Dynamics or Event Driven simulations (see, for example, Refs. 4 and 8 in 1D and [9, 12, 15, 17, 19] in 2D or 3D). These simulations agree well with the theory in the homogeneous regime, while in the correlated stage do not provide a clear answer, since they become exceedingly demanding; in fact, the instabilities appear only in large systems and at late times. However, even the homogeneous cooling regime of a one-dimensional granular gas seems very different from that predicted by the BK model. In Fig. 2 (left frame), we show the velocity distribution before the onset of extensive correlations in the velocity fields for a quasi-elastic system. In this case, and more generally for larger inelasticity, we observe a suppression of the tails, in contrast with the algebraic behavior of Eq. (2.1).

3. Lattice Models

In the following sections, we shall show how a simple extension of the BK model, obtained by placing immobile particles, endowed with a velocity, on the nodes of a regular lattice (with periodic boundary conditions) is able to capture the formation of domains in the velocity field. The evolution of the system is obtained by choosing a random nearest neighbor pair and updating their velocities according to the transformation Eq. (1.1), where now $\hat{\sigma}$ represents the unit vector pointing from site $\mathbf{j}$ to site $\mathbf{i}$.

As we said, in the lattice model the particles are fixed to their lattice positions, so that there is no relation between their velocities and their displacements. Nevertheless, the introduction of the lattice avoids "unphysical" collisions between velocities which do not satisfy the "*kinematic constraint.*"

The velocity field initially prepared in a state characterized by random uncorrelated velocities, remains homogeneously random during a first dynamical stage, in agreement with the so-called Homogeneous Cooling Regime (or Haff regime) observed in simulations of inelastic hard bodies. Afterwards, one observes the formation of spatial gradients in the velocity field which becomes macroscopic. This is analogous to the formation of magnetic domains in standard quench processes.

In fact, the free cooling process bears a strong resemblance to a quench from an initially stable disordered phase to a low temperature phase in a magnetic system [6]. In the granular case the relaxation occurs after the removal of the external driving force, by which the initial kinetic energy was fed to keep the system stationary. Since many possible configurations are compatible with the linear and angular momentum conservation and compete in order to minimize the energy dissipation, the system does not relax immediately towards a motionless state, but displays a behavior similar to that observed in a coarsening process.

We have studied the one- [2] and two-dimensional [1] version of this lattice model and some results will be briefly exposed below.

4. One-Dimensional Models

One-dimensional models represent a favorite playground for theoretical physicists and in fact systems of inelastic hard rods on a ring have intensively been studied [3, 8]. Here we want to just briefly show the results for the velocity distributions obtained with our lattice model. At odds with the BK (scalar) model, the one-dimensional lattice model seems to recover quantitatively the distributions measured in the inelastic hard rod system, in both the homogeneous and the inhomogeneous phase.

In Fig. 2 we show the fair collapse obtained with comparing velocity PDF of the lattice model and of the inelastic hard rod gas:

- In the left frame the velocity PDF in the early homogeneous regime of a nearly elastic system ($\alpha = 0.99$) is shown; for larger inelasticities the formation of the peaks is less evident, but a suppression of the tails is generically observed.

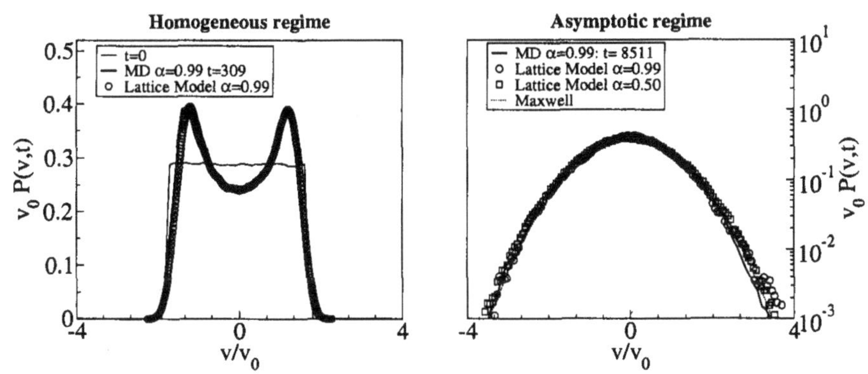

Fig. 2. Rescaled velocity distributions for the 1D MD and in the 1D lattice gas, during the homogeneous (left) and the inhomogeneous phase (right). In the left frame is also shown the initial distribution (both models). The distributions refer to systems having the same energy. Data refer to $N = 10^6$ (both models) particles with $\alpha = 0.99$ and $\alpha = 0.5$ (for the lattice model in the inhomogeneous regime).

- For larger times the asymptotic PDF turns Gaussian in both models, contrary to the expectation, based on the Burgers equation, that the tails of the velocity PDF should be of the form $\exp(-v^3)$.

Other features, such as shock fronts, observed in recent extensive simulations of hard rods [3], have their counterparts in our one-dimensional lattice model and will be discussed in detail in a forthcoming publication [2].

5. Two-Dimensional Lattice Model

Encouraged by the fair agreement between the lattice scalar model and the one-dimensional granular gas, we studied a two-dimensional (triangular) lattice model [1].

The cross-over between homogeneous and inhomogeneous regimes is manifest in the change of the energy decay law. As shown in Fig. 3, at first the average energy per particle $\epsilon(t) = \sum v_i(t)^2/N$ decreases exponentially at a rate $\tau^{-1} = (1-\alpha^2)/4$. This behavior reflects the fact that each velocity evolves independently of the others [13]. On the other hand, for times larger than $t \sim t_c$, of the order of τ, a second regime arises, where correlations play a major role and the average energy per particle decays as $\epsilon(t) \sim t^{-1}$.

Correspondingly, the velocity PDF changes as already explained in the case of the one-dimensional lattice model. In two dimensions, during the early regime the velocity distribution deviates sensibly from a Maxwell distribution, but shows more pronounced high velocity tails. These tails seem to be determined by the lack of spatial correlations up to t_c. In fact, as soon as the energy begins to decay as t^{-1} the velocity distribution changes to a Gaussian.

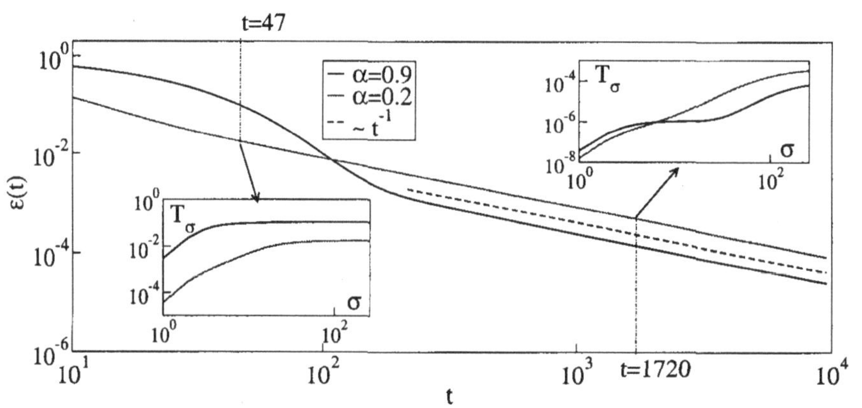

Fig. 3. Energy decay for $\alpha = 0.9$ and $\alpha = 0.2$ (1024^2 sites). The dashed line $\sim 1/t$ is a guide to the eye for the asymptotic energy decay. In the insets we reported the scale dependent temperature, T_σ, defined in the text, as a function of the coarse graining size σ for $t = 47$ and $t = 1720$. The total energy per particle and T_σ remain nearly indistinguishable in the early incoherent regime, but for $\sigma < L(t)$ the thermal energy becomes much smaller than the kinetic energy, a clear indication of the onset of macroscopic spatial order.

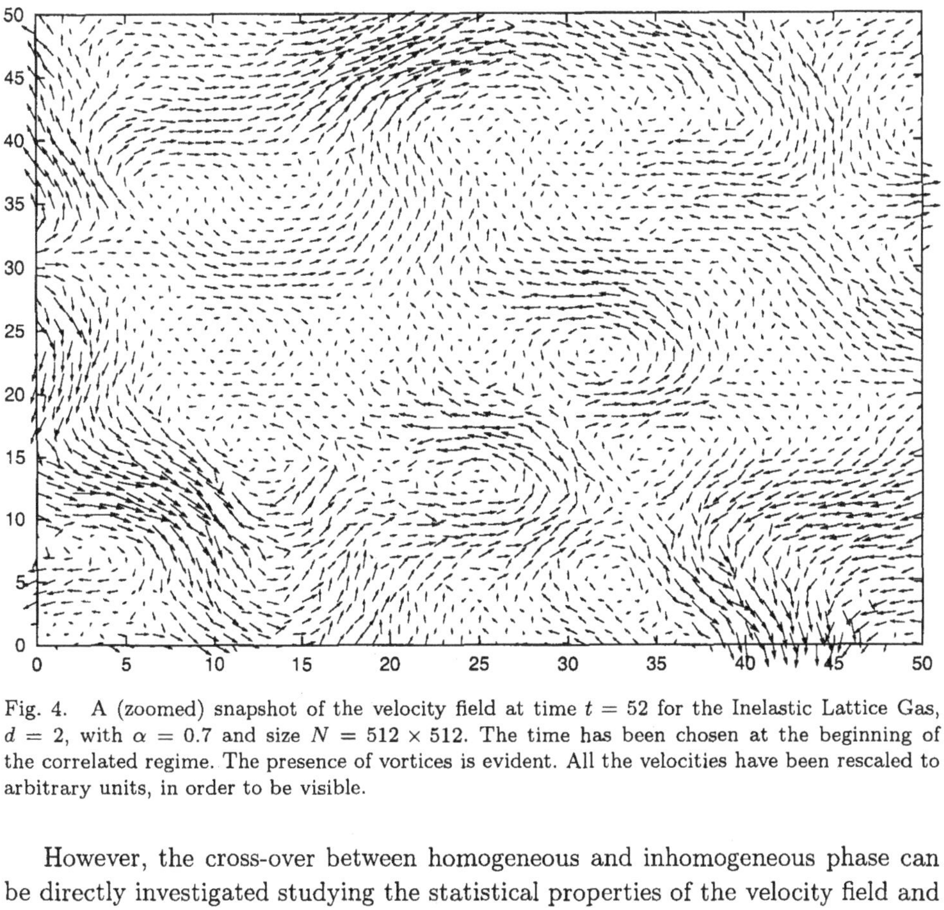

Fig. 4. A (zoomed) snapshot of the velocity field at time $t = 52$ for the Inelastic Lattice Gas, $d = 2$, with $\alpha = 0.7$ and size $N = 512 \times 512$. The time has been chosen at the beginning of the correlated regime. The presence of vortices is evident. All the velocities have been rescaled to arbitrary units, in order to be visible.

However, the cross-over between homogeneous and inhomogeneous phase can be directly investigated studying the statistical properties of the velocity field and its correlations.

The inelasticity of the collisions and the presence of the Θ have the effect of reducing the quantity $(-(\mathbf{v}_i - \mathbf{v}_j) \cdot \hat{\sigma})$, i.e. inducing an alignment of the velocities. The most dramatic consequence, in two dimensions, is the spontaneous formation of vortices as shown in Fig. 4. The characterization of such structures is achieved by means the equal-time structure functions (shown in Fig. 5), i.e. the Fourier transforms of the velocity correlation functions:

$$S^{t,l}(k,t) = \sum_{\hat{k}} v^{t,l}(\mathbf{k},t)\mathbf{v}^{t,l}(-\mathbf{k},t)\,,$$

where the superscripts t, l indicate the transverse and longitudinal components of the field with respect to the wave-vector $\mathbf{k}$ and the sum $\sum_{\hat{k}}$ is over a circular shell of radius k. These structure factors display fairly good data collapse, i.e. they can be rewritten in terms of a single scaling variable $S^{t,l}(k,t) = s^{t,l}(kt^{1/2})$, identifying two growing lengths $L^{t,l}(t)$ both proportional to $t^{1/2}$.

The form of the energy decay, the distribution of the velocity field and the growth $L^{t,l}(t)$ seem to suggest that the evolution can be represented by some effective

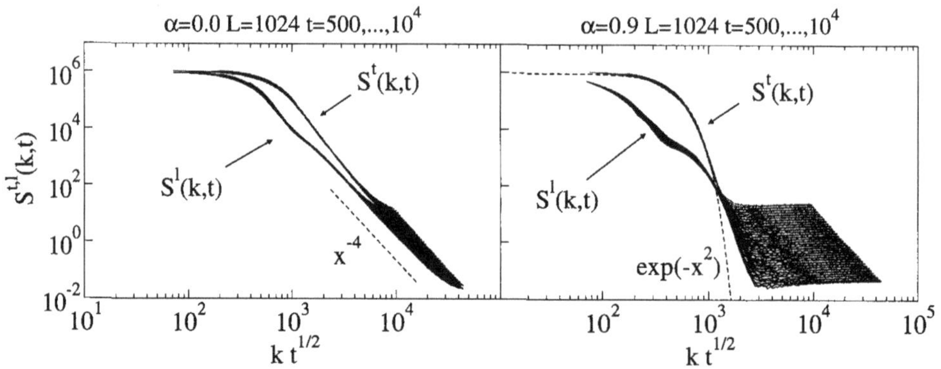

Fig. 5. Data collapse of the transverse (S^t) and longitudinal (S^l) structure functions for $\alpha = 0.$ and $\alpha = 0.9$ (system size 1024^2 sites, times ranging from $t = 500$ to $t = 10^4$). The wavenumber k has been multiplied by $\sqrt{t}$. Notice the presence of the plateaus for the more elastic system. For comparison we have drawn the laws x^{-4} and $\exp(-x^2)$.

diffusive dynamics. In fact, in the absence of the kinematic constraint the evolution of the velocity field would be satisfactorily described by an effective diffusion equation.

The analysis of the structure functions $S^{t,l}(k,t)$, instead, indicates the presence of a long-wavelength region which is diffusive in character, whereas at intermediate wavelength the structure functions decay accordingly as $k^{-\beta}$ with $\beta \sim 4$. Finally, a plateau region is observed (for $\alpha > 0$) where $S^{t,l}(k,t)$ remain nearly constant with respect to k, but decay in time with a power law t^{-2}. The observed intermediate and small wavelength behaviors have no counterparts in the pure diffusive model, where one observes always Gaussian structure functions.

The existence of a $L^{-2}(t)k^{-4}$ region in the structure functions is consistent with generalized Porod's law [6, 7] and is the signature of the presence of topological defects, vortices in this $d = 2$ case, a salient feature of the cooling process. Vortices form spontaneously and represent the boundaries between regions which selected different orientations of the velocities during the quench and are an unavoidable consequence of the conservation laws which forbid the formation of a single domain (conservation of linear and angular momentum).

With the random initial conditions adopted, vortices are born at the smallest scales and subsequently grow in size by pair annihilation, conserving the total charge. By locating the vortex cores, we measured the vortex density $\rho_v(t)$, which represents an independent measure of the domain growth, and in fact, it decays asymptotically ($t \gg t_c$) as an inverse power of time, i.e. $L_v(t) = \rho_v^{-1/2} \propto t^{1/2}$.

The vortex distribution turns out to be not uniform for α not too small. Its inhomogeneity is characterized by the correlation dimension d_2, defined through the definition of the cumulated correlation function:

$$H(R) = \frac{\sum_{i<j} \Theta(R - (\mathbf{r}_i - \mathbf{r}_j))}{N_v^2} \sim R^{d_2} \,,$$

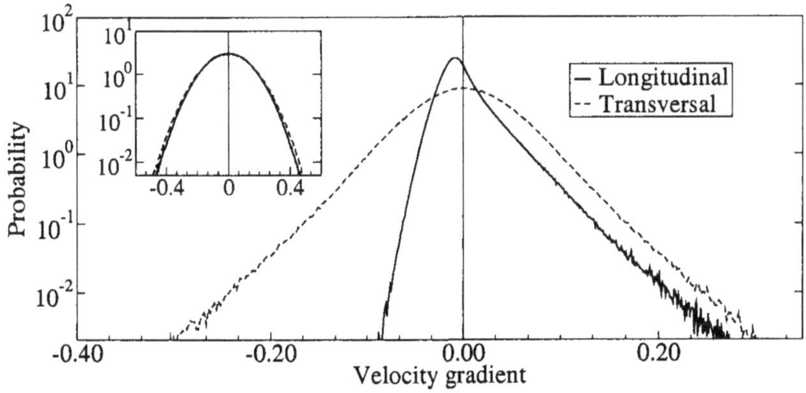

Fig. 6. Probability densities of the longitudinal and transverse velocity increments. The main figure shows the PDF of the velocity gradients ($R = 1$). The inset shows the Gaussian shape measured for $R = 40$ (larger than $L(t)$ for this simulation: $\alpha = 0.2$, $t = 620$, system size 2048^2).

where $\mathbf{r}_i$ are the core locations. For $\alpha \to 1$ the vortices are clusterized ($d_2 < 2$) i.e. do not fill the space homogeneously, whereas at smaller α their distribution turns out to be homogeneous ($d_2 \to 2$).

Vortices are not the only topological defects of the velocity fields. In fact, we observe shocks, similarly to recent experiments in rapid granular flows [16]. Shocks have a major influence on the statistics of the velocity field, i.e. on the probability distributions of the velocity increments. The probability density function (PDF) of the longitudinal increment is shown in Fig. 6 for $R = 1$ (longitudinal velocity gradient) in the main frame, and for $R = 40 > L(t)$ in the inset. For small $R \ll L(t)$ the longitudinal increment PDF is skewed with an important positive tail, whereas for $R \gg L(t)$ it turns Gaussian. The distribution of transverse increments

$$(\mathbf{v}_{i+R} - \mathbf{v}_i) \times \hat{\mathbf{R}} \tag{5.1}$$

instead, is always symmetric, but non-Gaussian distributed for small R. A similar situation exists in fully developed turbulence [5].

Whereas vortices and shocks explain the power law decay of the structure factors, the plateau in the tail of the structure function is related to the so-called internal noise, i.e. to the presence of short range spatial fluctuations induced by the collisions [19]. A small α determines a rapid locking of the velocities of neighboring elements to a common value, while in the case of $\alpha \to 1$, short range small amplitude disorder persists within the domains, breaking simple scaling of $S^{t,l}$ for large k and having the effect of a self-induced noise.

The internal noise observed in the study of the longitudinal and transverse structure factor can be characterized by means of an average local granular temperature T_σ:

$$T_\sigma = \langle |\mathbf{v} - \langle \mathbf{v} \rangle_\sigma|^2 \rangle_\sigma \,, \tag{5.2}$$

where $\langle \cdots \rangle_\sigma$ means an average on a region of linear size σ.

If we call $L(t)$ a characteristic correlation length of the system, since when $\sigma \gg L(t)$ the local average tends to the global (zero) momentum, then $\lim_{\sigma \to \infty} T_\sigma = E$ For $\sigma < L(t)$, instead, $T_\sigma < E$. The behavior of T_σ in the uncorrelated (Haff) regime and in the correlated (asymptotic) regime for two different values of α is presented in the inset of Fig. 3.

A very important observation is the following: for quasi-elastic systems, T_σ exhibits a plateau for $1 \ll \sigma \ll L(t)$ that identifies the strength of the internal noise and individuates the mesoscopic scale necessary to found a hydrodynamic description. The local temperature ceases to be well defined for smaller α; this clearly suggests the absence of scale separation between microscopic and macroscopic fluctuations in the strongly inelastic regime [1, 14].

6. Conclusions

To conclude, the models of granular media we introduced and studied display a rich variety of dynamical behaviors. In particular, we have observed the intriguing feature concerning the velocity PDF: with the exception of the mean field model, the PDF is non-Gaussian in the early stage and turns Gaussian asymptotically. Such a feature is ascribed to the build up of fluctuations of the flow field. We have also proposed a new two-dimensional lattice model very effective in computing its two-time correlations, structure factors, etc. We have stressed the analogy between cooling and phase ordering kinetics. The mesoscopic description of the model remains still to be completed in order to understand why this model is unusual with respect to other phase ordering systems; in fact, the one-point observables display a behavior consistent with a diffusive model, but the two-point quantities show anomalous static properties. Future directions of our work will be represented by the study of driven lattice models, their hydrodynamic description and the connection with Burger's equation.

References

[1] Baldassarri, A., Marini Bettolo Marconi, U. and Puglisi, A., cond-mat/0105299, submitted for publication.

[2] Baldassarri, A., Marini Bettolo Marconi, U. and Puglisi, A., cond-mat/0111066, submitted for publication.

[3] Ben-Naim, E., Chen, S. Y., Doolen, G. D. and Redner, S., *Phys. Rev. Lett.* **83**, 4069 (1999).

[4] Ben-Naim, E. and Krapivsky, P. L., *Phys. Rev.* **E61**, R5 (2000).

[5] Benzi, R., Biferale, L., Paladin, G., Vulpiani, A. and Vergassola, M., *Phys. Rev. Lett.* **67**, 2299 (1991).

[6] Bray, A. J., *Adv. Phys.* **43**, 357 (1994).

[7] Bray, A. J. and Puri, S., *Phys. Rev. Lett.* **67**, 2670 (1991).

[8] Du, Y., Li, H. and Kadanoff, L. P., *Phys. Rev. Lett.* **74**, 1268 (1995).

[9] Chen, S., Deng, Y., Nie, X. and Tu, Y., *Phys. Lett.* **A269**, 218 (2000).

[10] Ernst, M. H., *Phys. Rep.* **78**, 1 (1981).

[11] Frisch, U., *Turbulence* (Cambridge University Press, Cambridge, 1995).

[12] Goldhirsch, I. and Zanetti, G., *Phys. Rev. Lett.* **70**, 1619 (1993).
[13] Haff, P. K., *J. Fluid Mech.* **134**, 401 (1983).
[14] McNamara, S., Hydrodynamic modes of a uniform granular medium, *Phys. Fluids* **A5**, 3056 (1993).
[15] Orza, J. A. G., Brito, R. and Ernst, M. H., cond-mat/002383; Brito, R. and Ernst, M. H., *Eur. Phys. Lett.* **43**, 497 (1998).
[16] Rericha, E., Bizon, C., Shattuck, M. D. and Swinney, H. L., cond-mat/0104474.
[17] Trizac, E. and Barrat, A., *Eur. Phys. J.* **E3**, 291 (2000).
[18] Ulam, S., *Adv. Appl. Math.* **1**, 7 (1980).
[19] van Noije, T. P. C., Ernst, M. H., Brito, R. and Orza, J. A. G., *Phys. Rev. Lett.* **79**, 411 (1997).

CHAPTER 5

The Steady State of the Tapped Ising Model

DAVID S. DEAN and ALEXANDRE LEFEVRE

IRSAMC, Laboratoire de Physique Quantique, Université Paul Sabatier, 118 route de Narbonne, 31062 Toulouse Cedex 04, France

We consider a tapping dynamics, analogous to that in experiments on granular media, on the simple one-dimensional ferromagnetic Ising model. When unperturbed, the system undergoes a single spin flip falling dynamics where only energy lowering moves occur. With this dynamics the system has an exponentially large number of metastable states and gets stuck in blocked or jammed configurations as do granular media. When stuck, the system is tapped, in order to make it evolve, by flipping in parallel each spin with probability p (corresponding to the strength of the tapping). Under this dynamics the system reaches a steady state regime characterized by an asymptotic energy per spin $E(p)$, which is determined analytically. Within the steady state regime we compare certain time averaged quantities with the ensemble average of Edwards based on a canonical measure over metastable states of fixed average energy. The ensemble average yields results in excellent agreement with the dynamical measurements.

Keywords: Dense granular media; Ising model; tapping; thermodynamics.

1. Introduction

The Hamiltonian of the one-dimensional Ising model is given by

$$H = -\sum_i S_i S_{i+1} \,. \tag{1.1}$$

A thermal dynamics which evolves this system into the canonical ensemble is Glauber dynamics [9] and has the merit of being soluble (in the sense one can calculate the time dependence of the energy and certain correlation functions). At zero temperature spin flips which reduce the energy are accepted and those which leave it unchanged are accepted with probability 1/2. The only blocked configurations are the two ground states (all spins up or down). The system coarsens and the average domains length grows as $\sqrt{t}$ [2].

Granular systems have a large number of blocked or jammed configurations and one expects that the entropy of these configurations is extensive [8]. It is well known [16] that spin glasses have an extensive entropy of metastable states, which are responsible for the glassy dynamics of these systems. Indeed spin glass type systems have been proposed as paradigms of granular material [5, 7]. With this

First published in *Advances in Complex Systems*, Vol. 4, No. 4 (2001), pp. 333–343.

in mind the one-dimensional Ising model does not look like an obvious candidate to model the behavior of granular systems. However, consider now the following modification of Glauber dynamics: spin flips are accepted if they lower the energy but not if they leave it unchanged. This is somewhat natural in the context of granular media as frictional forces must be overcome. With this dynamics for a general Ising spin system with Hamiltonian

$$H = -\frac{1}{2}\sum_{ij} J_{ij} S_i S_j \,,$$

the total number of metastable states is given by

$$N_{\mathrm{MS}} = \mathrm{Tr} \prod_{i=1}^{N} \theta\left(\sum_{j \neq i} J_{ij} S_i S_j \right), \tag{1.2}$$

where N is the number of spins, and θ is the Heaviside step function. Note, with our definition of metastability $\theta(0) = 1$. This equation may be understood as follows. The energy change ΔE_i due to the spin flip $S_i \to -S_i$ is given by $\Delta E_i = \sum_{j \neq i} J_{ij} S_i S_j$. Hence, a configuration is single spin flip stable if all ΔE_i for that configuration are non-negative.

It should be pointed out here that the definition of metastable states is of course dependent on the dynamics of the system, in contrast with micro-states in classical statistical mechanics.

The Edwards entropy [8] per spin of blocked or jammed configurations is then given by

$$s = \frac{\ln(N_{\mathrm{MS}})}{N} \,. \tag{1.3}$$

In complex systems such as spin glasses s is $O(1)$ and independent of N in the thermodynamic limit [16]. The Edwards entropy per spin at fixed average energy per spin E is given by

$$s(E) = \frac{\ln(N_{\mathrm{MS}}(E))}{N} \,, \tag{1.4}$$

where

$$N_{\mathrm{MS}}(E) = \mathrm{Tr} \prod_{i=1}^{N} \theta\left(\sum_{j \neq i} J_{ij} S_i S_j \right) \delta(H - NE) \,. \tag{1.5}$$

One may calculate $s(E)$ for the Ising ferromagnet with the definition of metastability used here [6, 10] and one finds

$$s(E) = \frac{(1-E)}{2}\log(1-E) - \frac{(1+E)}{2}\log(1+E) + E\log(2) + E\log(-E) \,. \tag{1.6}$$

This function has its maximum value is at $E_m = -1/\sqrt{5}$, in the large N limit this is where the metastable states are concentrated giving $s = \ln((1+\sqrt{5})/2)$. Hence, this rather simple modification of zero temperature Glauber dynamics gives us a simple one-dimensional spin system with an extensive entropy of blocked configurations — as one would expect in a granular media.

We now proceed by constructing the simplest tapping dynamics imaginable on this system.

A falling or relaxational dynamics is defined as follows. A spin is chosen at random (random sequential update); if flipping this spin strictly lowers the energy of the system then the spin is flipped. If not, another spin is chosen and the same criterion applied until all of the spins are locally stable. The system is now in a blocked or single spin flip metastable state.

Once the system is in a metastable state it is tapped, that is to say each spin is flipped in parallel with a probability p. Generally the system will no longer be in a metastable state and the falling dynamics is once again activated until again the system becomes blocked. Clearly the fact that the system is in a metastable state when tapped models a system where the relaxation process to a metastable state is much shorter than the time between taps. This is the simplest form of tapping dynamics as there is only one time scale (the time between taps) in the problem.

When this tapping dynamics is simulated, one finds that the system, after some *ergodic* time, enters a regime where the energy per spin has becomes constant depending on p:

$$E(t) \to E(p)\,. \tag{1.7}$$

The simulation results are shown in Fig. 1, one sees here that $E(p)$ is a monotonically increasing function of p. This is rather reminiscent of the dependence of

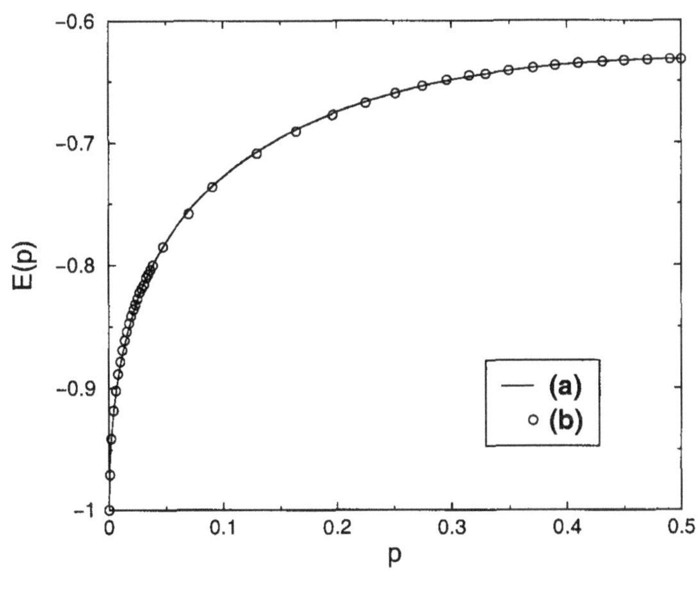

Fig. 1. Comparison between numerical simulations of tapping experiments (b) and the analytical result (a) obtained with (2.7).

the asymptotic compactivity of vertically shaken granular media in the Chicago group's experiments [13] — as the tapping strength (parameterized by $\Gamma = a/g$ the ratio of the maximal vertical acceleration a of the system through a sine cycle to the acceleration due to gravity g) is reduced, the compactivity increases (in the reversible part of the experiment). In numerical simulations of more realistic models of granular media, the enhancement of compactification on reducing the tapping strength has also been observed [12].

2. Determination of the Steady State Energy

In order to determine $E(p)$ we shall consider a simple mean-field theory for the falling dynamics of the ferromagnet on a Bethe lattice of connectivity c. If one concentrates on the bonds between connected sites (i, j), we say that the bond is satisfied if it gives a negative contribution to the energy i.e. $-J_{ij}S_iS_j < 0$. In the case of the ferromagnet this contribution to the energy is clearly either 1 or -1. Hence, here a bond is satisfied/unsatisfied if its contribution to the energy is $-1/1$. For a given site define x to be the difference between the number of unsatisfied and satisfied bonds. Hence x is the local field on the spin at this site and $x \in -c$, $-c+2, \ldots, c-2$, c. If $x > 0$ then the spin can flip bringing about the change $x \to -x$. In addition we denote by $P(x, k)$ the probability that the site of interest has local field x after a total of k attempted random sequential spin flips under the zero temperature falling dynamics (that is to say the dynamics in between taps).

If we denote by k the number of attempted spin flips during the falling dynamics, the possible changes at site i between k and $k+1$ are:

- The spin at site i is chosen and $x > 0$, then the spin at site i will flip and x goes to $-x$.
- The spin at site i is chosen and $x \leq 0$, then the spin at site i cannot flip and x does not change.
- A neighbor of site i with positive local field is chosen and so flips. In this case, x goes to $x+2$ or to $x-2$ depending whether or not the bond with site i was satisfied or not satisfied.
- A neighbor of site i with negative or zero local field is chosen and so does not flip. In this case, x does not change.
- One chooses neither the spin at site i nor any of its neighbors, and so x stays x.

An schematic example of the falling dynamics for a system of connectivity $c = 3$ is shown in Fig. 2. We define f_+ and f_- the probabilities that a given spin can flip conditional on the fact that the bond with a given neighboring site is not satisfied or satisfied respectively, one can show that

$$f_\pm = \frac{\sum_{x>0} P(x)(c \pm x)}{\sum_x P(x)(c \pm x)} . \tag{2.1}$$

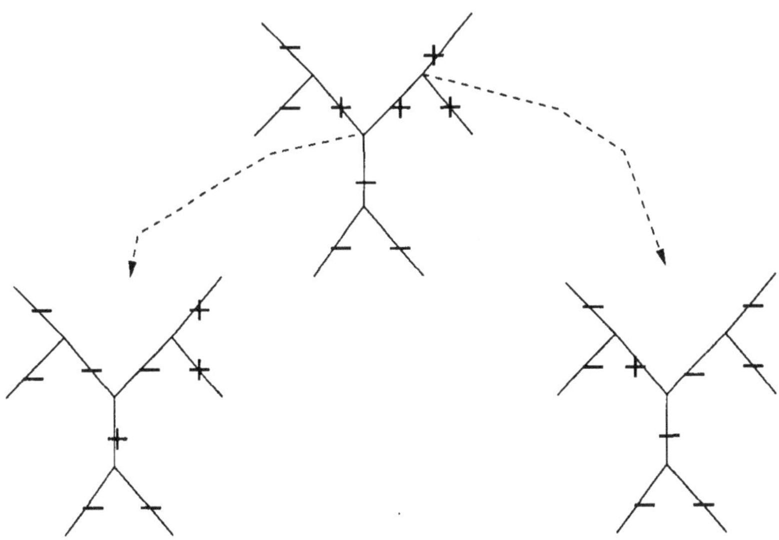

Fig. 2. An example of the falling dynamics for a spin system with $c = 3$. The symbols +/− indicate unsatisfied/satisfied bonds. The broken arrowed lines start at the spin that is flipped and lead to the following configuration.

Assuming that the distribution at every site is given by $P(x, k)$ and assuming independence between the values of x from site to site (the mean-field approximation) we obtain, and on taking the limit $N \to \infty$ and introducing the continuous time $\tau = k/N$,

$$\begin{aligned}\frac{dP(x)}{d\tau} &= \theta(-x)P(-x) + \theta(-x)P(x) - (c+1)P(x) \\ &+ P(x)\left(\frac{c+x}{2}(1-f_+) + \frac{c-x}{2}(1-f_-)\right) \\ &+ P(x+2)\,\frac{c+x+2}{2}f_+ + P(x-2)\,\frac{c-x+2}{2}f_- \,. \end{aligned} \tag{2.2}$$

The case where $c = 2$ (the one-dimensional case) is accessible to analytic solution and we proceed by defining

$$\begin{aligned} u(\tau) &\equiv P(-2,\tau)\,, \\ v(\tau) &\equiv P(0,\tau)\,, \\ w(\tau) &\equiv P(2,\tau)\,. \end{aligned}$$

When $\tau \to \infty$ the system is in a metastable state and hence $w(\infty) = 0$. One finds also that

$$v(\infty) = (v(0) + 2w(0))e^{-\frac{4}{\lambda(0)+2}}\,, \tag{2.3}$$

where $\lambda(0) = v(0)/w(0)$, and clearly $w(\infty) = 1 - v(\infty)$. If we consider an initial configuration where a spin differs from its neighbor with probability a, then the induced initial conditions are: $u(0) = (1-a)^2$, $v(0) = 2a(1-a)$ and $w(0) = a^2$.

The initial energy per spin E_0 is

$$E_0 = -1 + 2a \tag{2.4}$$

and one finds that the energy obtained after the first fall is

$$E_f = -1 + v(\infty) = -1 + 2ae^{-2a}\,. \tag{2.5}$$

This mean-field result for the one-dimensional system can be shown to be exact by combinatorial methods [4].

We see that final value of the energy E_f depends strongly on the initial configuration. Also the system does not fall into a state of energy corresponding to the maximum of $s(E)$ at $E_m = -1/\sqrt{5} \approx 0.44721$. Hence, even if the total number of metastable states is dominated (in the statistical sense) by those of energy E_m, generic initial conditions always seem to lead to an energy lower than this [1]. In [14] the value of E_f for a variety of zero temperature dynamics (sequential, greedy and reluctant) in the fully connected Sherrington–Kirkpatrick (SK) spin glass model [15] was studied, similar behavior was found.

Tapping the system with tapping probability p, starting from the values $\{u(\infty), v(\infty), w(\infty)\}$, we obtain the new *tapped* values $\{u'(0), v'(0), w'(0)\}$. Defining $q \equiv (1-p)$, the relations between the old and *tapped* probabilities are

$$\begin{aligned} u'(0) &= (1-3pq)u(\infty) + pqv(\infty)\,, \\ v'(0) &= 2pqu(\infty) + (1-2pq)v(\infty)\,, \\ w'(0) &= pq\,. \end{aligned} \tag{2.6}$$

In the steady state regime under tapping, the probability $v_s(p)$ (the subscript s indicating steady state) for sites to have zero local field is solution of the fixed-point equation

$$\begin{aligned} v_s(p) &= (4pq + v_s(p)(1-4pq)) \\ &\quad\times \exp\left(-\frac{4pq}{4pq + v_s(p)(1-4pq)}\right)\,. \end{aligned} \tag{2.7}$$

This equation can be solved numerically to calculate $E(p) = -1 + v_s(p)$, the result is shown in Fig. 1 in comparison with the numerical simulations, which we see is excellent.

3. The Steady State Measure

We now come to the question of what is the nature of the steady state regime under tapping. Edwards has proposed [8] that in the regime of slight tapping or shearing, the measure is a flat one over all blocked configurations satisfying the macroscopic constraints. Here the system is characterized by the average value of its energy and hence, in its canonical form [3, 8] the Edwards hypothesis suggests the measure

$$\langle O\rangle_{\mathrm{Edw}} = \frac{\sum_{\mathcal{C}} O(\mathcal{C})\exp(-\beta(p)H[\mathcal{C}])}{Z}\,, \tag{3.1}$$

where $\{\mathcal{C}\}$ is the ensemble of metastable configurations and

$$Z = \int dE N_{\rm MS}(E) \exp(-\beta(p)NE)\,. \tag{3.2}$$

There has been much recent study of the validity of this measure in non-thermal systems and on balance has been shown to be quite successful (although there are examples where it does not hold). Clearly the Edwards measure has the advantage of being the simplest and in a sense the most natural starting point. Unfortunately there is no general theory concerning its applicability and for the moment it has to be checked case by case.

In the context of the model here one may calculate a variety of observables with the Edwards measure analytically and also the numerical simulation of the tapping process is relatively straight forward [11].

The partition function for the system is then given by

$$Z = \operatorname{Tr} \prod_{i=1}^{N} \exp(\beta S_i S_{i+1}) \theta(S_{i-1}S_i + S_i S_{i+1})\,, \tag{3.3}$$

where the function $\theta(x) = 0$ for $x < 0$ and $\theta(x) = 1$ for $x \geq 0$ enforces the metastability (each spin is stable or marginally stable in its local field). Performing a local change of variables to new Ising spins $\sigma_i = S_i S_{i+1}$ we obtain

$$Z = \operatorname{Tr} \prod_{i=1}^{N} \exp(\beta \sigma_i) \theta(\sigma_i + \sigma_{i+1})\,. \tag{3.4}$$

Hence, we find that $Z = \operatorname{Tr} T^N$ where T is the transfer matrix

$$T = \begin{pmatrix} a & 1 \\ 1 & 0 \end{pmatrix} \tag{3.5}$$

with $a = \exp(\beta)$. From this we find that

$$E = \frac{a}{\sqrt{a^2 + 4}} \tag{3.6}$$

which simply determines β.

By a straight forward transfer matrix calculation one may calculate:

- The internal energy fluctuation per spin $c = (\langle \mathcal{E}^2 \rangle - \langle \mathcal{E} \rangle^2)/N$ where $\mathcal{E}$ is the total internal energy, and consequently $\langle \mathcal{E} \rangle = NE$. One obtains

$$c = -E(1 - E^2)\,. \tag{3.7}$$

- The correlation functions $C(r) = \langle S_i S_{i+r} \rangle$ and $D(r) = \langle S_i S_{i+1} S_{i+r} S_{i+r+1} \rangle$. We find that the connected part of $D(r)$ is given by

$$D_c(r) = (1 - E^2) \left(\frac{E+1}{E-1} \right)^r \tag{3.8}$$

and

$$C(r) = A\left(\frac{u}{t_0}\right)^r + (1-A)\left(\frac{1}{ut_0}\right)^r \tag{3.9}$$

with

$$A = \frac{u^2t_0^2 - 1}{(u^2-1)(t_0^2+1)}, \tag{3.10}$$

where $u = (a + \sqrt{a^2-4})/2$ and $t_0 = (a + \sqrt{a^2+4})/2$.

- The distribution of domain sizes. If $P(r)$ the probability that a given domain has length r we find

$$\begin{aligned} P(r) &= \left(\frac{1+E}{1-E}\right)\left(\frac{-2E}{1-E}\right)^{r-2} \qquad r \geq 2 \\ &= 0 \qquad\qquad\qquad\qquad\qquad r < 2\,. \end{aligned} \tag{3.11}$$

  Hence, the distribution of domain sizes is geometric for $n > 2$, the fact that $P(1) = 0$ is a consequence of metastability as a domain of length 1 would be a single spin surrounded by two antiparallel neighbors which is unstable.

One can compare the results of numerical simulations of tapping with the above theoretical ones. For a given value of the energy, let us say E, we have tapped the system with a strength p such that in the steady state $E = E(p)$. The system is tapped for a sufficiently large number of times, say t_s, to ensure that the average of the internal energy $E(t)$ measured becomes stationary. Once in this steady state regime, the quantities of interest are measured over a measurement time (number of taps) $t_m = 10^5$. The systems were of size of order 10^5 spins and the results were also averaged over $N_s = 5000$ realizations. Hence, mathematically, the average value of a quantity A is calculated, as one would in a Monte Carlo simulation of a thermal system, as

$$\langle A \rangle = \frac{1}{N_s}\sum_{i=1}^{N_s}\frac{1}{t_m}\sum_{t=t_s+1}^{t_s+t_{mes}} A_i(t)\,. \tag{3.12}$$

In Fig. 3, we compare the fluctuation of the energy c calculated using Edwards' measure, as a function of E, against those obtained from the simulations, the agreement is very good. For small values of p the error bars in our measurements are very small and the agreement with Eq. (3.7) is excellent. For larger values of p the error bars are large as the statistical fluctuations are larger, however from Fig. 3 we see that the value given by Eq. (3.7) is within the error bars.

The correlation functions $C(r)$ and $D(r)$ have also been computed numerically. In Fig. 4 we have plotted the results in comparison with those expected from Eqs. (3.8) and (3.9). Here again, the comparison is excellent (remark that the agreement is better for lower energies, as again the statistical fluctuations due to the tapping are much smaller for low p than for high p).

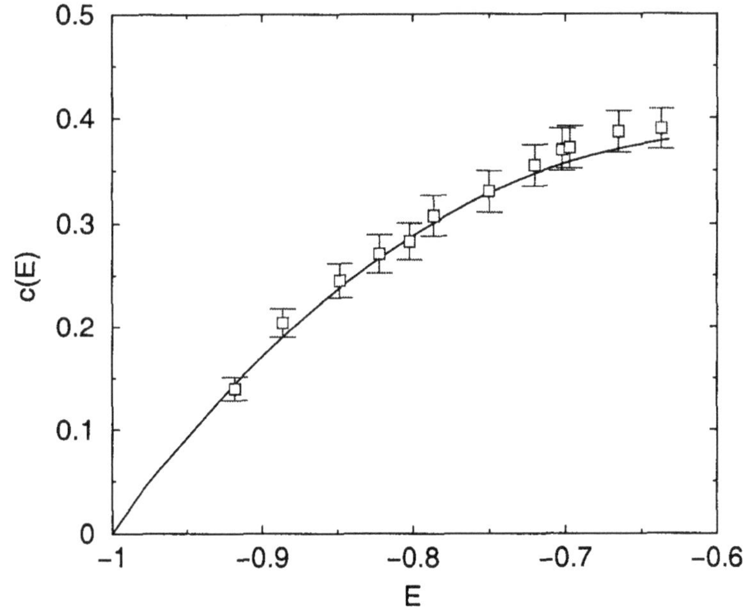

Fig. 3. The internal energy fluctuation $c(E)$ versus E. The solid line corresponds to the value obtained from Eq. (3.7) and the symbols are the results obtained from tapping simulations made on 5000 systems of 20000 spins.

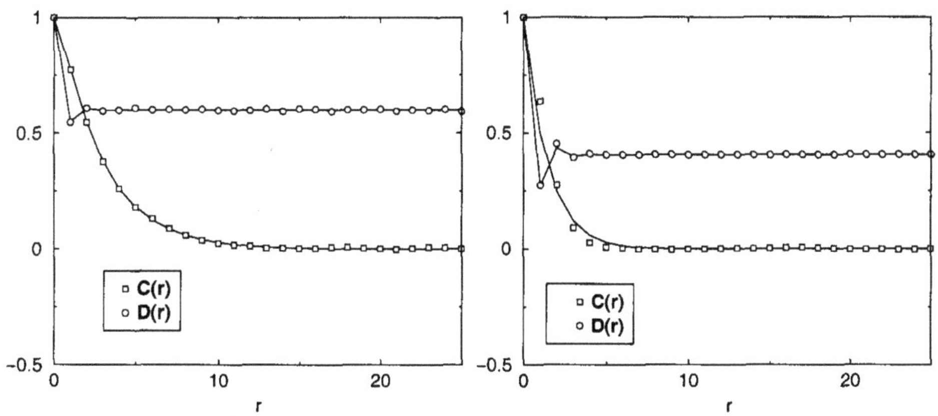

Fig. 4. Comparison between the expected $C(r)$ and $D(r)$ from the theoretical calculation with the results from numerical simulations for $E = -0.78$ (left) and $E = -0.63$ (right). The symbols are the results of the tapping experiments and the solid lines correspond to that predicted by Eqs. (3.9) and (3.8).

The distribution $P(r)$ of domain sizes, is shown in Fig. 5 and has a perfect exponential decay for $r \geq 2$. This guarantees that the result Eq. (3.11) is in perfect agreement with the simulations (as the energy E is fixed).

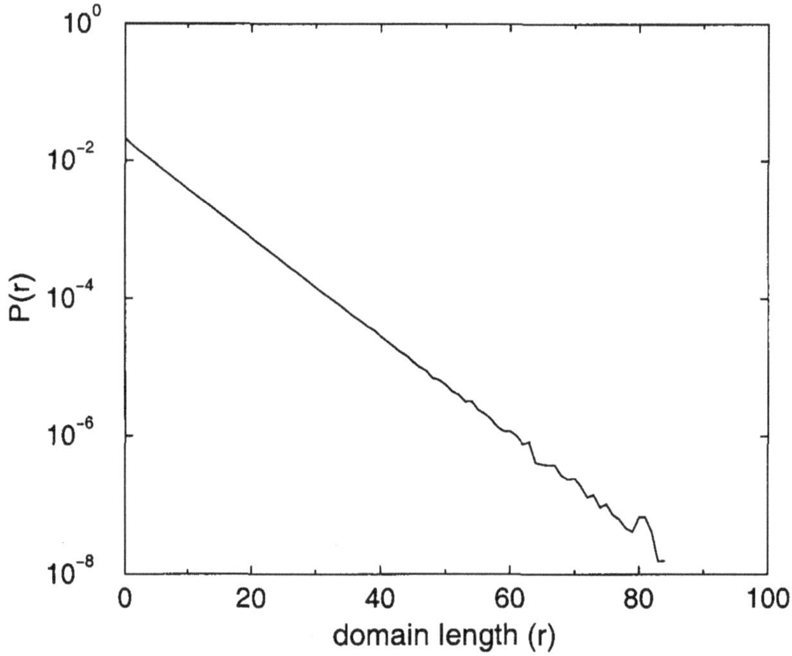

Fig. 5. Distribution of domain lengths from tapping simulations for $p = 0.1$. The vertical scale is logarithmic. The slope is $b = 0.165 \pm 0.001$, in excellent agreement with that obtained from Eq. (3.11).

4. Conclusion

We have seen that a slight and natural modification of the zero temperature dynamics of the one-dimensional Ising model gives one a system with an extensive entropy of metastable states as one would expect in a granular system. A natural tapping dynamics yields steady state behavior similar to that of vertical tapping on granular media [13], hence this system is a simple testing ground to explore the behavior of mechanically perturbed complex systems. In the steady state regime we have seen that the flat measure proposed by Edwards [8] predicts certain macroscopic quantities of the steady state with a high degree of precision, even though no obvious form of detailed balance or ergodicity is present in the system. However it should be noted that the fact that the mean-field calculation (2.7) appears to be exact also supports the validity of the geometric distribution of domains sizes Eq. (3.11) predicted by the Edwards measure.

References

[1] Barrat, A., Kurchan, J., Loreto, V. and Sellito, M., *Phys. Rev. Lett.* **85**, 5034 (2000); Barrat, A., Kurchan, J., Loreto, V. and Sellito, M., *Phys. Rev.* **E63**, 051301 (2001); Javier Brey, J., Prados, A. and Sanchez-Rey, B., *Physica* **A275**, 310 (2000); Coniglio, A. and Nicodemi, M., cond-mat 0107134.

[2] Bray, A. J., *Adv. Phys.* **43**, 357 (1994).
[3] Coniglio, A. and Nicodemi, M., cond-mat 0010191.
[4] Dean, D. S. and Lefèvre, A., *Phys. Rev.* **E64**, 046110 (2001); Prados, A. and Brey, J. J., cond-mat 0106236.
[5] Dean, D. S. and Lefèvre, A., *Phys. Rev. Lett.* **86**, 5639 (2001).
[6] Dean, D. S., *Eur. Phys. J.* **B15**, 493 (2000).
[7] Edwards, S. F. and Mehta, A., *J. Phys.* **50**, 2489 (1989); Berg, J. and Mehta, A., cond-mat 0012416; Berthier, L., Cugliandolo, L. F. and Iguian, J. L., *Phys. Rev.* **E63**, 051302 (2001).
[8] Edwards, S. F., *Granular Media: An Interdisciplinary Approach*, Mehta, A. ed., (Springer-Verlag, New York, 1994).
[9] Glauber, R. J., *J. Math. Phys.* **4**, 294 (1963)
[10] Lefèvre, A. and Dean, D. S., *Eur. Phys. J.* **B21**, 121 (2001).
[11] Lefèvre, A. and Dean, D. S., *J. Phys.* **A34**, L213–L220 (2001).
[12] Mehta, A. and Barker, G. C., *Phys. Rev. Lett.* **67**, 394 (1991); Philippe, P. and Bideau, D., *Phys. Rev.* **E63**, 051304 (2001); Caglioti, A., Loreto, V., Herrmann, H. and Nicodemi, M., *Phys. Rev. Lett.* **79**, 1575 (1997).
[13] Nowak, E. R., Knight, J. B., Ben-Naim, E., Jaeger, H. M. and Nagel, S. R., *Phys. Rev.* **E57**, 1971 (1998).
[14] Parisi, G., cond-mat 9501045.
[15] Sherrington, D. and Kirkpatrick, S., *Phys. Rev. Lett.* **35**, 1792 (1975).
[16] Tanaka, F. and Edwards, S. F., *J. Phys.* **F13**, 2769 (1980); Bray, A. J. and Moore, M. A., *Phys. Rev. Lett.* **58**, 57 (1987); Derrida, B. and Gardner, E., *J. Phys. France* **47**, 959 (1986).

CHAPTER 6

The Effect of Avalanching in a Two-Species Ripple Model

REBECCA B. HOYLE
Department of Mathematics and Statistics, University of Surrey, Guildford, Surrey GU2 7XH, UK

ANITA MEHTA
S.N. Bose National Centre for Basic Sciences, Block JD, Sector III Salt Lake, Calcutta 700 098, India

This paper discusses a simple two-species ripple model with avalanching. The effect of the avalanching term is investigated numerically, and is found to be crucial in producing realistic ripple profiles.

Keywords: Aeolian; sand ripples; avalanching; continuum model; two-species.

1. Introduction

Aeolian sand ripples are formed by the action of the wind on the sand bed in the desert or at the seashore. They are a few centimetres in wavelength with crests perpendicular to the prevailing wind direction. Early theoretical work on ripple formation [1, 2, 3] has been followed more recently by models that treat these ripples as being composed of two layers of sand grains: the 'bare surface' made up of relatively immobile grain clusters, and a layer of mobile grains moving on top [10, 16, 21, 23]. There are important differences in these approaches; in those of Terzidis, Claudin and Bouchaud [21] and Valance and Rioual [23] for example, the nature of the interaction between the flowing and stuck layers differs significantly from that described in our approach [10], particularly in the area of *nonlocality*, which forms an important ingredient of our model. Recently, these ideas have been used to investigate the formation of sand dunes via continuum approaches very similar to our own [17], while discrete methods have been used to look at formation of vortex ripples in water [18]. Finally, laboratory experiments on ripple formation have recently been devised [8, 20], which enable the testing of theoretical hypotheses on systems more manageable than those provided by nature.

First published in *Advances in Complex Systems*, Vol. 4, No. 4 (2001), pp. 345–352.

One of the important ingredients of our earlier model was its inclusion of bistability at the *angle of repose.* As is well known, a sandpile can be either static or flowing if its surface slope is within a given range above the angle of repose; the upper bound of this range is the *maximum angle of stability*, after which the sandpile avalanches — that is, spontaneous flow sets in. This phenomenon has also been represented in a discrete model of sandpile avalanches [14], where it has been used for the interpretation of avalanche shapes found in experiment [7]. Here we consider the effect of 'avalanching' in a two-species continuum model of sand ripples based on our earlier model [10]. As explained above we use the term 'avalanching' to describe spontaneous flow, when sand grains are shed very rapidly from the immobile layer into the flowing layer, as the surface slope approaches the *maximum angle of stability* γ. Though this process is less dramatic in ripples than in sandpiles or dunes, and we do not expect to see discrete avalanche events with large sections of the sand surface falling away, this rapid grain shedding nonetheless turns out to have important consequences for the development and shaping of ripples.

2. Ripple Equations

We consider two-dimensional sand ripples comprising a surface defined by the local height of clusters, $h(x,t)$, covered by a thin layer of flowing mobile grains whose local density is $\rho(x,t)$, where x is a horizontal space coordinate and t is time. The ripples evolve under the influence of a constant flux of saltating grains, which impact the sand bed at an angle β to the horizontal, knocking grains out of the bare surface, causing them to hop along the ripple surface and land in the layer of flowing grains. Granular relaxation mechanisms then smooth the ripple surface. We aim for a minimal model capturing the essential physics of ripple formation; thus the model equations used here are a simplification of those studied in our earlier work [10]. They take the following form:

$$\begin{aligned} h_t &= D_h h_{xx} - f(x,t) \\ &\quad - \begin{cases} \lambda\rho(|h_x| - \tan\delta) & 0 \le |h_x| \le \tan\delta\,, \\ \nu(|h_x| - \tan\delta)(\tan^2\gamma - h_x^2)^{-\frac{1}{2}} & \tan\delta \le |h_x| < \tan\gamma\,, \end{cases} \\ \rho_t &= D_\rho \rho_{xx} + \chi(\rho h_x)_x + \int_{-\infty}^{+\infty} p(a) f(x-a,t)\,da \\ &\quad + \begin{cases} \lambda\rho(|h_x| - \tan\delta) & 0 \le |h_x| \le \tan\delta\,, \\ \nu(|h_x| - \tan\delta)(\tan^2\gamma - h_x^2)^{-\frac{1}{2}} & \tan\delta \le |h_x| < \tan\gamma\,, \end{cases} \end{aligned} \tag{2.1}$$

where D_h, D_ρ, λ, ν and χ are positive constants, δ is the *angle of repose*, and $p(a)$ is the distribution of hop lengths a for grains knocked out of the ripple surface by the saltation flux, and where $f(x,t) = a_p J(\sin\beta + h_x \cos\beta)$ with a_p the average cross-sectional area of a sand grain and $J > 0$ a measure of the saltation flux intensity.

Naturally the flowing grain density can never be less than zero, so we also impose $\rho_t \ge 0$ anywhere that we have $\rho = 0$.

The rate of knocking out of grains by the saltation flux is assumed proportional to the component of the saltation flux perpendicular to the ripple surface [9]. The hopping out of the layer of clusters is modelled by the term $-f(x,t)$ in the equation for h_t, and grains landing in the flowing layer are modelled by the term $\int_{-\infty}^{+\infty} p(a)f(x-a,t)da$ in the equation for ρ_t [10]. The hop length distribution $p(a)$ can be measured experimentally [15, 22]. Here we assume a normal distribution with mean $\bar{a}$ and variance σ^2.

Where the sand bed is shielded from the saltation flux by upwind ripple peaks it is said to be in *shadow*. No grains are knocked out of the surface in these regions and the equation for h_t is modified by neglecting the term $-f(x,t)$.

The remaining terms [10] describe the granular relaxation mechanisms that smooth the ripple surface [4]. These terms, especially those modelling tilting, are based on the formulation of coupled continuum equations first presented in Ref. 13, although alternative versions, with different assumptions made, exist in Refs. 5 and 6. The term $D_h h_{xx}$ represents the diffusive rearrangement of clusters while the term $D_\rho \rho_{xx}$ represents the diffusion of the flowing grains. The flux-divergence term $\chi(\rho h_x)_x$ models the flow of surface grains under gravity. The current of grains is assumed proportional to the density of flowing grains and to their velocity, which in turn is proportional to the local slope to leading order [9]. The term $\lambda\rho(|h_x| - \tan\delta)$ represents the tendency of flowing grains to stick onto the ripple surface at slopes less than the angle of repose, δ. The tilt and avalanching term $\nu(|h_x| - \tan\delta)(\tan^2\gamma - h_x^2)^{-\frac{1}{2}}$ models the tendency of clusters in the bare surface to shed grains into the flowing layer when tilted beyond the angle of repose, δ. The rate of shedding of grains becomes very large as slopes approach the maximum angle of stability γ — this is our representation of the phenomenon of *avalanching.*

The model is nondimensionalised by setting $x \to x_0\tilde{x}$, $t \to t_0\tilde{t}$, $a \to x_0\tilde{a}$, $\rho \to \rho_0\tilde{\rho}$, $h \to h_0\tilde{h}$, where $x_0 = D_h/a_pJ\cos\beta$, $t_0 = D_h/(a_pJ\cos\beta)^2$, $h_0 = D_h\tan\gamma/a_pJ\cos\beta$, $\rho_0 = a_pJ\sin\beta/\lambda\tan\delta$. The renormalised equations are

$$
\begin{aligned}
h_t &= h_{xx} - \hat{f}(x,t) \\
&\quad - \begin{cases} \rho\dfrac{\tan\beta}{\tan\delta}\left(|h_x| - \dfrac{\tan\delta}{\tan\gamma}\right) & 0 \le |h_x| \le \dfrac{\tan\delta}{\tan\gamma}\,, \\ \hat{\nu}\left(|h_x| - \dfrac{\tan\delta}{\tan\gamma}\right)(1-h_x^2)^{-\frac{1}{2}} & \dfrac{\tan\delta}{\tan\gamma} \le |h_x| < 1\,, \end{cases} \\
\rho_t &= \frac{D_\rho}{D_h}\rho_{xx} + \hat{\chi}(\rho h_x)_x + \frac{h_0}{\rho_0}\int_{-\infty}^{+\infty} p(a)f(x-a)da \\
&\quad + \frac{h_0}{\rho_0}\begin{cases} \rho\dfrac{\tan\beta}{\tan\delta}\left(|h_x| - \dfrac{\tan\delta}{\tan\gamma}\right) & 0 \le |h_x| \le \dfrac{\tan\delta}{\tan\gamma}\,, \\ \hat{\nu}\left(|h_x| - \dfrac{\tan\delta}{\tan\gamma}\right)(1-h_x^2)^{-\frac{1}{2}} & \dfrac{\tan\delta}{\tan\gamma} \le |h_x| < 1\,, \end{cases}
\end{aligned}
\tag{2.2}
$$

where the tildes have been dropped and where $\hat{f}(x,t) = h_x + \tan\beta/\tan\gamma$, $\hat{\nu} = \nu t_0/h_0$ and $\hat{\chi} = \chi h_0/D_h$.

A steady solution $h_t = \rho_t = 0$ to these equations is given by $h = h_c$, where h_c is any constant, and $\rho = 1$. The stability of this solution can be investigated by setting $h = h_c + \hat{h}e^{\eta t+ikx}$ and $\rho = 1 + \hat{\rho}e^{\eta t+ikx}$, where $\hat{h}, \hat{\rho} \ll 1$ are constants, and linearising in $\hat{h}$ and $\hat{\rho}$. To leading order in the wavenumber k, assumed to be small as ripples are large scale features in comparison with surface roughness, the growth rate eigenvalues are $\eta = -h_0 \tan\beta/\rho_0 \tan\gamma$, which is associated with the relaxation of the flowing grain density ρ to its equilibrium value, and $\eta = Ak^2$, where

$$A = -1 + \bar{a} - \frac{\rho_0}{h_0}\hat{\chi}, \tag{2.3}$$

which is associated with ripple growth. For ripples to emerge, we must have $A > 0$.

3. Numerical Results

The effect of avalanching on ripple profile was studied numerically. The nondimensionalized equations (2.2) were integrated numerically using periodic boundary conditions and compact finite differences [11, 12]. Both *shadowing* and the requirement for ρ not to be less than zero were taken account of in the code. For the run with avalanching behaviour we chose nondimensional quantities as follows: $\bar{a} = 3.0$, $\sigma = 0.1$, $D_\rho/D_h = 1.0$, $h_0/\rho_0 = 80.0$, $\hat{\chi} = 0.2$, $\hat{\nu} = 5.0$, $\beta = 10°$, $\delta = 30°$, $\gamma = 35°$. The nondimensional length of the integration domain was $x_{\max} = 150.0$, and the saltation flux was coming from the left. The angles were chosen to agree with observational evidence [3, 19, 24], the ratio h_0/ρ_0 to ensure a thin layer of flowing grains in comparison to the ripple height, and the remaining parameters to allow ripple growth. The simulation was started with a flat surface, h and ρ constant, with surface roughness modelled by the addition of small amplitude noise to both variables. The output was rescaled back into physical coordinates using $D_h = 1.0$ and $\lambda = 10.0$. We also performed a run with avalanching switched off, $\hat{\nu} = 0$, but with all other parameters the same, and a further run where only tilting was allowed.

In the case where avalanching was included ripples developed from the initial surface roughness, and a process of ripple merger ensued (similar to that described in previous ripple models [10, 16]), leading to a final state with one large, slowly-growing, ripple in the periodic box. The final large rightward-moving ripple, shown in Fig. 1 at time $t = 68.7$, has a fairly pointed crest with slope 24.1° on the windward side and −30.5° on the lee side. The stoss (windward) slope is long and comparatively shallow, whereas the lee slope is shorter and relatively straight. Note that the ripple length is short compared to the ripple spacing, which here is given by the length of the periodic domain of integration. This is because there is a long shadow zone on the lee side of the ripple, where no saltation flux hits the sand bed, and hence where no ripple can be built up. At earlier stages of evolution the

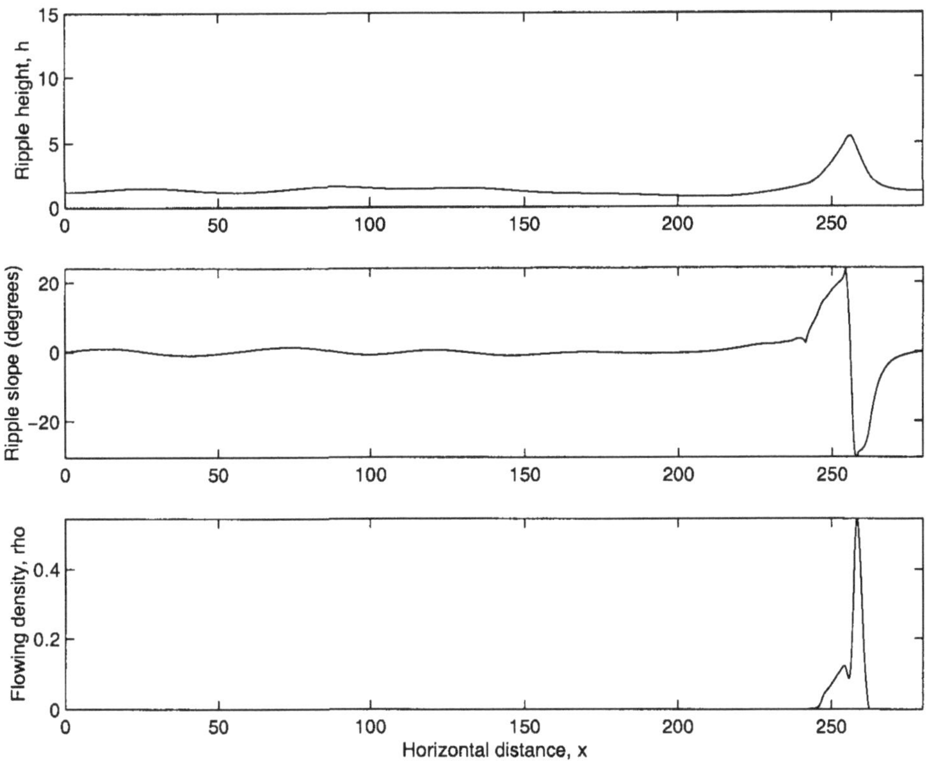

Fig. 1. Profiles of the ripple surface height h, ripple surface slope $\tan^{-1}(h_x)$ and flowing grain density ρ against horizontal distance x at time $t = 68.7$ for a ripple subject to avalanching. Parameter values are given in the text.

ripple length and spacing are comparable, because the ripple crest is lower, and the shadow zone shorter. We show the fully-evolved ripple in Fig. 1 because the ripple shape and slopes are more clearly visible than in the trains of smaller ripples. The same is true for the ripples illustrated in Figs. 2 and 3.

The density of flowing grains, also shown in Fig. 1, has a maximum just to the right of the ripple crest where grains have avalanched out of the ripple surface as they fall down the short straight lee slope. There is also a much smaller secondary peak in flowing grain density just to the left of the ripple crest associated with the comparatively high surface slope there.

In the case where avalanching was suppressed (Fig. 2) the large ripple develops much higher surface slopes on the stoss side (39.3°), and looks more symmetrical as the stoss slope is shorter. Also, in contrast to the case of the avalanching ripple, the ripple is steeper on the stoss side than the lee (−28.2°), which is contrary to observations [19]. The density of flowing grains shows a double peak, one on each side of the crest, with the left-hand peak being the higher as the surface is steeper there and so grains tilt out more rapidly.

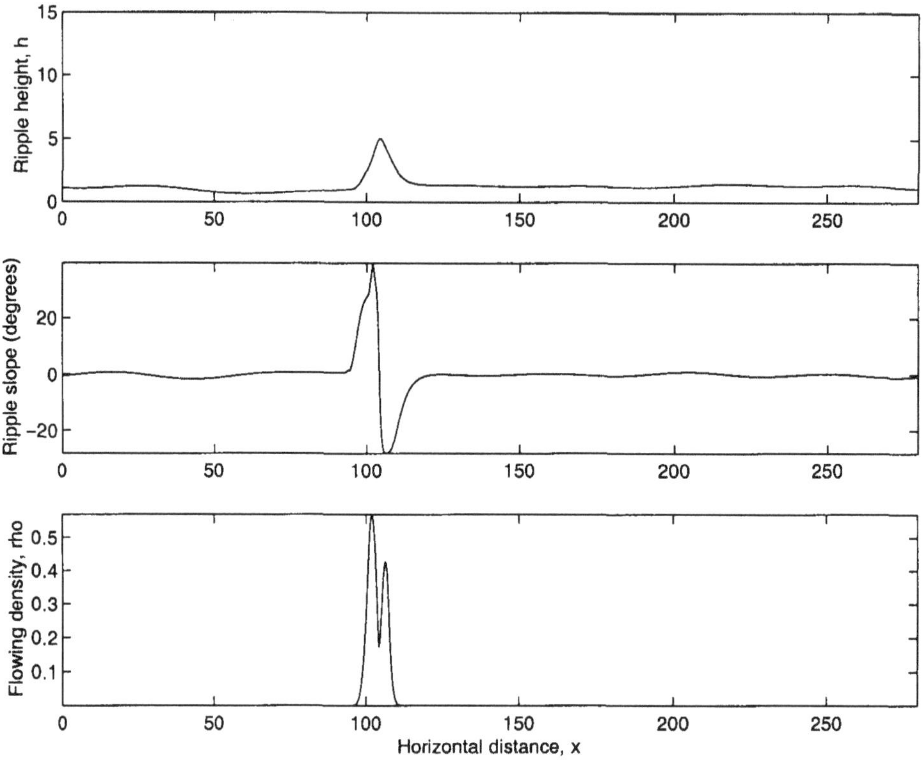

Fig. 2. Profiles of the ripple surface height h, ripple surface slope $\tan^{-1}(h_x)$ and flowing grain density ρ against horizontal distance x at time $t = 66.7$ for a ripple *not* subject to avalanching or tilting. Parameter values are the same as those for the ripple with avalanching except that the coefficient of avalanching ν is set to zero.

In order to compare the effect of simple tilting with that of avalanching, we also performed a run where the denominator of the avalanching term was set to 1, while all other parameters remained the same. In this model, grains are still tilted out of the surface at slopes greater than the angle of repose, but the rate of shedding does not become large as the slope approaches the maximum angle of stability, as indeed there is no effective maximum angle in this case. The results are shown in Fig. 3, and are similar to those obtained in the case where avalanching is entirely suppressed, except that the stoss slope is slightly less steep (32.4°). The lee slope is again shallower (−27.8°) than the stoss, and the ripple looks relatively symmetric. The flowing grain density again shows a double peak around the ripple crest, with the left-hand peak being the higher.

Our numerical results indicate that grain avalanching, as opposed to simple tilting, maintains slopes below the maximum angle of stability, and leads to an asymmetrical ripple in accord with observations [19], with a longer shallower stoss slope and a short, steep lee slope.

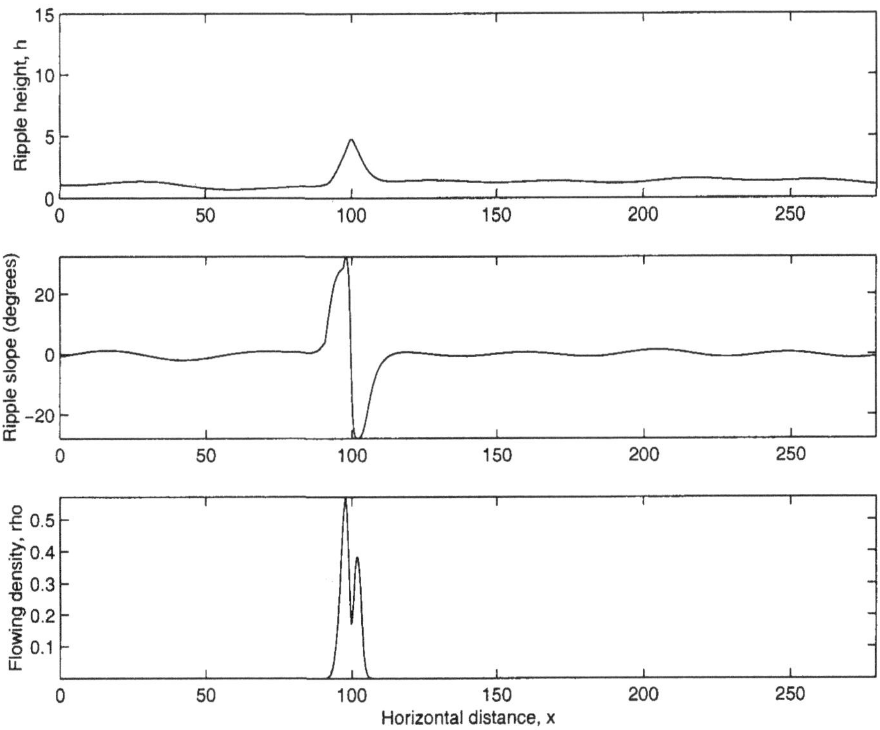

Fig. 3. Profiles of the ripple surface height h, ripple surface slope $\tan^{-1}(h_x)$ and flowing grain density ρ against horizontal distance x at time $t = 58.9$ for a ripple subject to tilting, but not avalanching. Parameter values are the same as those for the ripple with avalanching.

4. Conclusion

It is observed in nature [19] that ripple slopes are generally quite shallow. Our results show that even the subtle avalanching that occurs in ripples is enough to decrease surface slopes, and that the avalanching term in our model can account for shallow ripples by maintaining surface gradients away from the maximum angle of stability γ. We have also shown that avalanching produces asymmetrical ripples where the stoss slope is longer and shallower than the short steep lee slope, again in agreement with observations [19].

In contrast, ripples where avalanching behaviour is suppressed or replaced with simple tilting, tend to develop high surface slopes, with stoss and lee slopes of similar lengths, and higher gradients on the stoss side than the lee, all of which are contrary to observation. Avalanching thus appears to be a crucial ingredient in sculpting ripple profiles.

Acknowledgments

We thank Neal Hurlburt and Alastair Rucklidge for the use of their compact finite differences code.

References

[1] Anderson, R. S., *Sedimentology* **34**, 943 (1987).
[2] Anderson, R. S., *Earth-Sci. Rev.* **29**, 77 (1990).
[3] Bagnold, R. A., *The Physics of Blown Sand and Desert Dunes* (Methuen and Co., London, 1941).
[4] Biswas, P., Majumdar, A., Mehta, A. and Bhattacharjee, J. K., *Phys. Rev.* **E58**, 1266 (1998).
[5] Bouchaud, J. P., Cates, M., Prakash, J. R. and Edwards, S., *J. Phys. France* **I4**, 1383 (1994).
[6] Boutreux, T., Raphaël, E. and de Gennes, P. G., *Phys. Rev.* **E58**, 4692 (1998).
[7] Daerr, A. and Douady, S., *Nature* **399**, 241 (1999).
[8] Hansen, J. L., van Hecke, M., Haaning, A., Ellegaard, C., Andersen, K. H., Bohr, T. and Sams, T., *Nature* **410**, 324 (2001).
[9] Hoyle, R. B. and Woods, A. W., *Phys. Rev.* **E56**, 6861 (1997).
[10] Hoyle, R. B. and Mehta, A., *Phys. Rev. Lett.* **83**, 5170 (1999).
[11] Hurlburt, N. E. and Rucklidge, A. M., *Mon. Not. R. Astron. Soc.* **314**, 793 (2000).
[12] Lele, S. K., *J. Comput. Phys.* **103**, 16 (1992).
[13] Mehta, A., Luck, J. M. and Needs, R. J., *Phys. Rev.* **E53**, 92 (1996).
[14] Mehta, A. and Barker, G. C., *Europhys. Lett.* **56**, 626–632 (2001), cond-mat/0006162.
[15] Mitha, S., Tran, M. Q., Werner, B. T. and Haff, P. K., *Acta Mechanica* **63**, 267 (1986).
[16] Prigozhin, L., *Phys. Rev.* **E60**, 729 (1999).
[17] Sauermann, G., Kroy, K. and Herrmann, H. J., *Phys. Rev.* **E64**, 1305 (2001).
[18] Krug, J., *Adv. Complex Syst.* **4**, 353 (2001); see also article "Coarsening of Vortex Ripples in Sand" by J. Krug elsewhere in this volume.
[19] Sharp, R. P., *J. Geol.* **71**, 617 (1963).
[20] Stegner, A. and Wesfreid, J. E., *Phys. Rev.* **E60**, R3487 (1999).
[21] Terzidis, O., Claudin, P. and Bouchaud, J. P., *Eur. Phys. J.* **B5**, 245 (1998).
[22] Ungar, J. and Haff, P. K., *Sedimentology* **34**, 289 (1987).
[23] Valance, A. and Rioual, F., *Eur. Phys. J.* **B10**, 534 (1999).
[24] von Burkalow, A., *Bull. Geol. Soc. Am.* **56**, 669 (1945).

CHAPTER 7

Coarsening of Vortex Ripples in Sand

JOACHIM KRUG
Fachbereich Physik, Universität Essen, 45117 Essen, Germany

The coarsening of an array of vortex ripples prepared in an unstable state is discussed within the framework of a simple mass transfer model first introduced by K. H. Andersen *et al.* (see Ref. 1). Two scenarios for the selection of the final pattern are identified. When the initial state is homogeneous with uniform random perturbations, a unique final state is reached which depends only on the shape of the interaction function $f(\lambda)$. A potential formulation of the dynamics suggests that the final wavelength is determined by a Maxwell construction applied to $f(\lambda)$, but comparison with numerical simulations shows that this yields only an upper bound. In contrast, the evolution from a perfectly homogeneous state with a localized perturbation proceeds through the propagation of wavelength doubling fronts. The front speed can be predicted by standard marginal stability theory. In this case the final wavelength depends on the initial wavelength in a complicated manner which involves multiplication by powers of two and rational ratios such as 4/3.

Keywords: Sand ripples; pattern formation; wavelength selection; coarsening; front propagation; marginal stability; metastable states.

Uns überfüllts. Wir ordnens. Es zerfällt.
Wir ordnens wieder und zerfallen selbst.
Rainer Maria Rilke[a]

1. Introduction

Vortex ripples are a familiar occurrence in coastal waters, where the waves expose the sand surface at the sea bottom to an oscillatory flow. The name reflects the important role of the separation vortices that form on the lee side of the ripples in stabilizing the ripple slopes [5, 6]. In laboratory experiments, one observes the formation of a stable periodic pattern with a ripple wavelength proportional to the amplitude a of the water motion [9, 10, 21].

The purpose of the present contribution is to analyze a simple model that was recently introduced by Andersen and co-workers to describe the stability and evolution of vortex ripple patterns [1]. We focus here on the mathematical aspects of

[a] *It fills us. We arrange it. It collapses. We rearrange it and collapse ourselves.* From the eighth Duino elegy.

First published in *Advances in Complex Systems*, Vol. 4, No. 4 (2001), pp. 353–362.

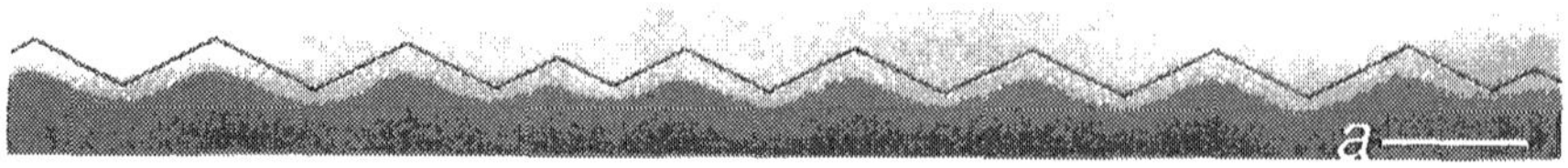

Fig. 1. Experimental image of a ripple pattern obtained in an annular container under oscillatory driving. The line above the sand surface shows that the pattern can be fitted to an array of triangles with constant slope. The amplitude of fluid motion a is indicated. From Ref. 3.

the problem, and refer the reader to the literature for further motivation and a detailed comparison to experiments [1, 3]. The model of interest is introduced in the next section. Sections 3 and 4 discuss the selection of the final ripple pattern for the cases of uniform and localized perturbations of an unstable initial state, while Sec. 5 contains some remarks concerning the description of vortex ripples using continuum equations.

2. The Mass Transfer Model

We consider a fully developed ripple pattern of the kind obtained in a quasi-one-dimensional annular geometry [3, 20, 21] (Fig. 1). Each of the N ripples is described by a single size parameter $\lambda_n(t)$, which can be thought to represent its length, or the amount of sand it contains. The periodic (annular) boundary conditions imply that the total length $L = \sum_{n=1}^{N} \lambda_n$ is conserved.[b] During one period of fluid motion, ripple n exchanges mass (or length) with its neighbors $n-1$ and $n+1$. This leads to the balance equation

$$\frac{d\lambda_n}{dt} = 2f(\lambda_n) - f(\lambda_{n+1}) - f(\lambda_{n-1}). \tag{2.1}$$

The *interaction function* $f(\lambda)$ is the central ingredient in the model. It describes the amount of sand that is transferred *to* a ripple of size λ due to the vortex forming behind this ripple.

The following argument suggests that $f(\lambda)$ should be a nonmonotonic function with a maximum near $\lambda = a$ [1]. Small ripples create a small separation vortex which is unable to erode much of the neighboring ripples, hence $f(\lambda)$ vanishes for small λ. On the other hand, even for a large ripple the size of the vortex cannot be much larger than the amplitude a of the fluid motion. If $\lambda \gg a$, the vortex does not reach beyond the trough to the next ripple, so $f(\lambda)$ vanishes also. The mass transfer is most efficient, and f is maximal, when $\lambda \approx a$. The interaction function can be measured in fluid dynamical simulations [1] and experiments [3], which confirm these qualitative considerations. In the following we shall take $f(\lambda)$ to be an arbitrary, single humped function which vanishes[c] at $\lambda = 0$ and at $\lambda = \lambda_{\max}$, and displays a maximum at $\lambda = \lambda_c$. The physical meaning of $\lambda_{\max}$ is that new

[b] Models in which the ripple lengths and masses are conserved separately have been developed in Ref. 1.

[c] Note that the dynamics (2.1) is invariant under shifts $f \to f + \text{const}$. In general, $\lambda_{\max}$ is therefore determined by the condition $f(\lambda_{\max}) = f(0)$.

ripples are created in the troughs when the ripple spacing exceeds $\lambda_{\max}$. Since we are concerned here with the *coarsening* of ripples, this process plays no role.[d]

Any homogeneous pattern with $\lambda_n \equiv \bar{\lambda}$ is a stationary solution of Eq. (2.1). To investigate its stability, we impose a small perturbation, $\lambda_n = \bar{\lambda} + \epsilon_n$, and linearize Eq. (2.1) in ϵ_n. We find solutions of the form $\epsilon_n \sim \exp[iqn + \omega(q)t]$, where the growth rate of a perturbation of wavenumber q is given by

$$\omega(q) = 2f'(\bar{\lambda})(1 - \cos q)\,. \tag{2.2}$$

This implies that (i) a homogeneous state is stable if $f'(\bar{\lambda}) < 0$, and (ii) an unstable state decays predominantly through perturbations of wavenumber $q = \pi$, in which every second ripple grows and every second ripple shrinks. The model (2.1) thus predicts an entire band of stable homogeneous states with wavelengths $\lambda_c < \bar{\lambda} < \lambda_{\max}$, in accordance with experiments [3, 10, 21]. The existence of a multitude of linearly stable, stationary states is a property that the model shares with many other systems in granular physics.

When investigating the dynamics of unstable states, Eq. (2.1) has to be supplemented by a rule which decides what should happen when the length of a shrinking ripple reaches zero. We impose simply that such a ripple is eliminated, and the remaining ripples are relabeled so that the earlier neighbors of the lost ripple now are next to each other. Since the time derivatives of the lengths of these neighboring ripples jump at the instant of disappearance of the lost ripple, this elimination procedure introduces a distinctly non-smooth element into the dynamics.

3. Wavelength Selection from Generic Unstable States

In this section we are concerned with the evolution out of generic unstable states. In practice this means that the initial state is of the form $\lambda_n = \bar{\lambda} + \delta_n$, where $\bar{\lambda} < \lambda_c$ and δ_n are random numbers smaller than $\lambda_c - \bar{\lambda}$. The evolution is then followed numerically to the point where all surviving ripples are in the stable regime, $\lambda_n > \lambda_c$. Beyond this time no further ripples are eliminated, and hence the mean ripple length $\langle \lambda_n \rangle = L/N$ no longer changes.[e]

Simulations using a wide range of interaction functions indicate that the evolution selects a unique *equilibrium wavelength* λ_{eq} which is essentially independent of the initial conditions,[f] and depends only on the shape of the interaction function. In view of the large number of linearly stable stationary states, and the deterministic character of the dynamics, this is a highly nontrivial property of the model, which requires explanation. While a full understanding is still lacking, we present here a partial solution which appears to yield at least an upper bound on λ_{eq}.

[d] According to (2.1), $d\lambda_n/dt \le 2f(\lambda_n)$, which implies that $\lambda_{\max}$ can never be reached from an initial condition with $\lambda_n < \lambda_{\max}$ for all n.

[e] When all ripples are in the stable regime, Eq. (2.1) describes a diffusive evolution towards a completely homogeneous state.

[f] Recent numerical work suggests that a slight dependence on the initial conditions is in fact present [11].

The key observation is that Eq. (2.1) can be cast into the form of an overdamped mechanical system by introducing the positions of the ripple troughs $x_n(t)$ as basic variables, such that $\lambda_n = x_{n+1} - x_n$. Then the dynamics (2.1) becomes

$$\frac{dx_n}{dt} = -\frac{\partial V}{\partial x_n}, \tag{3.1}$$

where V is a sum of repulsive pair potentials acting between the troughs,

$$V = \sum_n v(x_{n+1} - x_n) = -\sum_n \int_0^{x_{n+1}-x_n} d\lambda\, f(\lambda). \tag{3.2}$$

This is supplemented by the elimination rule, which corresponds in the particle picture to a coalescence process.[g]

Equation (3.1) suggests that the dynamics is driven towards minimizing V. It is then natural to surmise that the final state is determined by the minimum of V *under the constraint of fixed total length* L. Since the final state is clearly homogeneous, the quantity to be minimized is

$$V_{\mathrm{hom}}(\lambda) = N \int_0^{\lambda} d\lambda'\, f(\lambda') = \frac{L}{\lambda} \int_0^{\lambda} d\lambda'\, f(\lambda'). \tag{3.3}$$

This leads to the prediction that $\lambda_{\mathrm{eq}} = \lambda^*$, where λ^* is the solution of

$$\int_0^{\lambda^*} d\lambda\, f(\lambda) = \lambda^* f(\lambda^*). \tag{3.4}$$

This is, of course, the analytic form of the Maxwell construction, as it would be applied to the chemical potential in a coexistence region.

The prediction (3.4) has several desirable properties. First, it is manifestly independent of the initial condition. Second, it guarantees that λ^* is located in the stable region, i.e. $\lambda_c < \lambda^* < \lambda_{\mathrm{max}}$. Third, it is invariant under multiplication of f by an arbitrary factor.[h] Nevertheless, it is wrong; comparison with simulations shows that $\lambda^* > \lambda_{\mathrm{eq}}$ always. Table 1 contains some typical results obtained using a family of piecewise linear interaction functions,

$$f(\lambda) = \begin{cases} \lambda/\lambda_c\,, & \lambda < \lambda_c\,, \\ (\lambda_{\mathrm{max}} - \lambda)/(\lambda_{\mathrm{max}} - \lambda_c)\,, & \lambda \geq \lambda_c\,. \end{cases} \tag{3.5}$$

In this case

$$\lambda^*/\lambda_c = \sqrt{\lambda_{\mathrm{max}}/\lambda_c}\,. \tag{3.6}$$

The two wavelengths λ^* and λ_{eq} appear to become equal, in the sense that $(\lambda^* - \lambda_c)/(\lambda_{\mathrm{eq}} - \lambda_c) \to 1$, when the stable branch of the transfer function becomes very steep, i.e. when $\lambda_{\mathrm{max}} - \lambda_c \ll \lambda_c$. On the other hand, when the stable branch is shallow, the prediction $\lambda_{\mathrm{eq}} = \lambda^*$ fails completely. This can be seen by considering the extreme

[g] Related particle systems have been considered in Ref. 18.

[h] This is required because such a multiplication only affects the time scale of evolution and should not change the final state.

Table 1. Comparison of the wavelength λ^* predicted by Eq. (3.4) with the equilibrium wavelength λ_{eq} obtained in numerical simulations of the model. The simulation results were averaged over 100 runs using 1000 initial ripples with lengths uniformly distributed in $[0.5, 1]$.

$\lambda_{\max}/\lambda_c$	1.1	1.25	1.5	2	4	6	8	∞
λ^*/λ_c	1.0488	1.118	1.225	1.414	2	2.449	2.828	∞
$\lambda_{\text{eq}}/\lambda_c$	1.0477	1.1105	1.196	1.307	1.461	1.516	1.536	1.607

case of an interaction function that remains constant for $\lambda > \lambda_c$, i.e. (3.5) with $\lambda_{\max}/\lambda_c = \infty$. Then, we find numerically that the final wavelength is $\lambda_{\text{eq}} \approx 1.61\ \lambda_c$, while clearly $\lambda^* = \infty$.

An obvious interpretation of finding $\lambda_{\text{eq}} < \lambda^*$ is that the deterministic, overdamped dynamics (3.1) gets stuck in a metastable state before reaching the configuration of minimal "energy" V. This suggests that it should be possible to increase the final wavelength by either making the dynamics less damped, or by introducing noise. The first modification implies that the trough "particles" are supplied with a mass and a momentum variable along the lines of Ref. 18. Preliminary simulations show that this does indeed increase the final wavelength, but not sufficiently to reach λ^*. A noisy version of the model has been described and investigated (for the case of monotonic interaction functions) in Ref. 12. Noise also increases the final wavelength, however in addition it introduces a new coarsening mechanism involving rare fluctuations [12, 23], which in principle drives the wavelength towards $\lambda_{\max}$, as long as ripple creation is not included. A stochastic model with a mechanism for the nucleation of new ripples could display interesting steady states.

4. Front Propagation

We now consider a perfectly ordered, homogeneous, unstable initial condition, $\lambda_n \equiv \lambda^{(i)} < \lambda_c$ for all n, which is destabilized by a local perturbation, e.g. by making a single ripple shorter or longer. Then, two fronts emanate from the perturbed region which propagate into the unstable state and leave in their wake a stable homogeneous configuration at a new wavelength $\lambda^{(f)} > \lambda_c$ (Fig. 2). The elimination of ripples occurs in the vicinity of the fronts.

Since the period-2 mode ($q = \pi$) is the most unstable according to Eq. (2.2), its growth controls the propagation of the fronts. Following the standard theory of front propagation into unstable states [19], we write the propagating perturbation as a traveling wave with an exponential tail:

$$\epsilon_n(t) = (-1)^n \exp[-\alpha(n - ct)]\,, \tag{4.1}$$

where c is the propagation speed and α the decay constant. Inserting this into the linearization of Eq. (2.1) we find the relation

$$c(\alpha) = \frac{2f'(\lambda^{(i)})}{\alpha}(1 + \cosh(\alpha)) \tag{4.2}$$

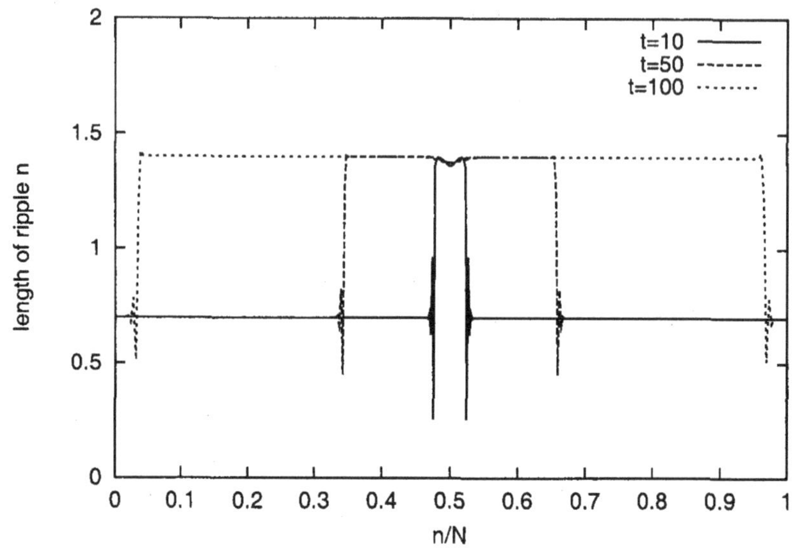

Fig. 2. Front propagation for a parabolic interaction function $f(\lambda) = 2\lambda - \lambda^2$ with initial wavelength $\lambda^{(i)} = 0.7$ and final wavelength $\lambda^{(f)} = 1.4$. The figure shows the ripple wavelength λ_n as a function of the scaled ripple number n/N (note that N decreases with time). Initially the system contains 500 ripples.

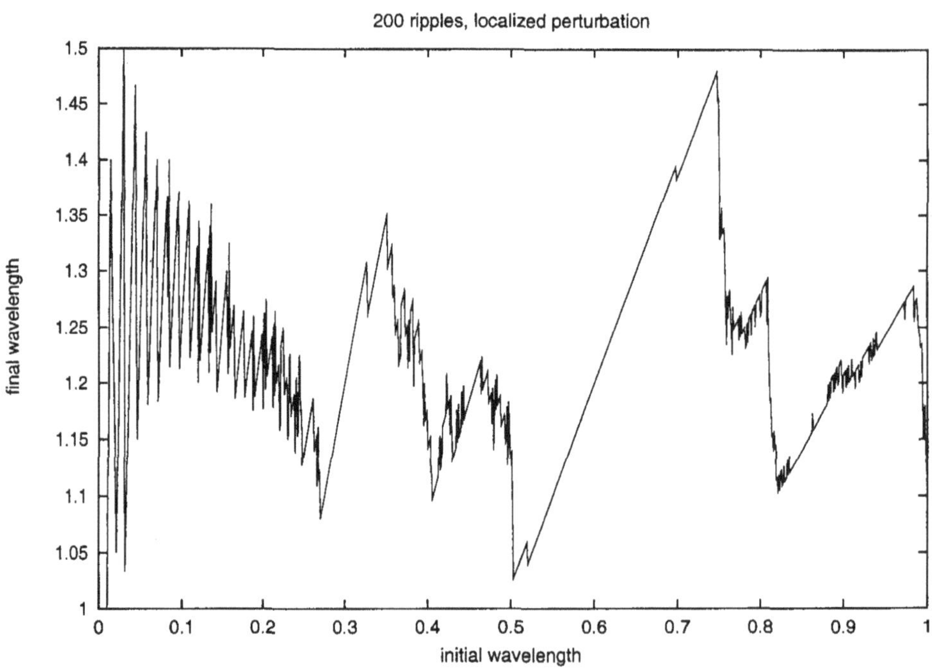

Fig. 3. Final wavelength as a function of initial wavelength for the piecewise linear interaction function (3.5) with $\lambda_{\max}/\lambda_c = 3/2$. For each value of $\lambda^{(i)}$ a system of initially 200 ripples was simulated until all ripples reached the stable regime. All wavelengths are measured in units of λ_c.

between c and α. Localized initial conditions usually propagate at the "marginal stability" speed c^* corresponding to the minimum of Eq. (4.2) [19], and hence we expect that the front velocity is given by $c^* \approx 4.4668\, f'(\lambda^{(i)})$. This prediction is well confirmed by numerical simulations.

We next turn to the relationship between the initial and final wavelengths. In stark contrast to the situation discussed in Sec. 3, here the final selected wavelength depends on the initial state in a surprisingly complex manner (Fig. 3). The most prominent feature in the graph is a straight line of slope 2 which extends from $\lambda_c/2 = 0.5$ to $\lambda_{\max}/2 = 0.75$. In this regime the wavelength selection process is very simple. The growth of the period-2 mode near the front implies that every second ripple is eliminated, hence the wavelength doubles. When $\lambda^{(i)} > \lambda_c/2$, this is sufficient to bring the ripples into the stable band. We therefore conclude that

$$\lambda^{(f)} = 2\lambda^{(i)} \quad \text{for} \quad \lambda_c/2 < \lambda^{(i)} < \lambda_{\max}/2\,. \tag{4.3}$$

When $\lambda^{(i)} < \lambda_c/2$, the ripples are still unstable after the doubling of the wavelength. The simplest scenario for the further evolution is that the new state again becomes unstable with respect to the period-2 mode, so that the wavelength doubles once more. Indeed a segment with slope 4 can be detected in Fig. 3, which starts near $\lambda_c/4$. For smaller initial wavelengths this scenario breaks down because the accumulation of exponentially growing perturbations prevents the intermediate homogeneous states to become established. We have not attempted any further analysis of the complicated behavior seen in Fig. 3 for $\lambda^{(i)} < \lambda_c/4$.

A different kind of complication arises when $\lambda^{(i)} > \lambda_{\max}/2$. In this case the growth of the period-2 mode terminates before the smaller ripples have reached zero length, because the system gets temporarily trapped in a *stationary* period-2 state with alternating ripple lengths $\lambda_A > \lambda_c$ and $\lambda_B < \lambda_c$ (Fig. 4). It is possible to prove that such a state, which has to satisfy the constraints

$$f(\lambda_A) = f(\lambda_B)\,, \qquad (\lambda_A + \lambda_B)/2 = \lambda^{(i)} \tag{4.4}$$

always exists when $\lambda_{\max} < 2\lambda_c$ and $\lambda^{(i)} > \lambda_{\max}/2$. Indeed, consider the function

$$F(\lambda_A) = f(\lambda_A) - f(\lambda_B) = f(\lambda_A) - f(2\lambda^{(i)} - \lambda_A)\,. \tag{4.5}$$

This function vanishes at $\lambda_A = \lambda_B = \lambda^{(i)}$, where its slope $F'(\lambda^{(i)}) = 2f'(\lambda^{(i)}) > 0$. Furthermore, F is odd under reflection around $\lambda_A = \lambda^{(i)}$, and $F(\lambda_{\max}) = -f(2\lambda^{(i)} - \lambda_{\max}) < 0$, because $2\lambda^{(i)} - \lambda_{\max} > 0$. It then follows by continuity that F has to possess two additional zeros, corresponding to a solution of Eq. (4.4) with $\lambda_A > \lambda_B$. It also follows that

$$f'(\lambda_A) + f'(\lambda_B) < 0\,, \tag{4.6}$$

which is the condition for stability *within* the space of period-2 configurations.

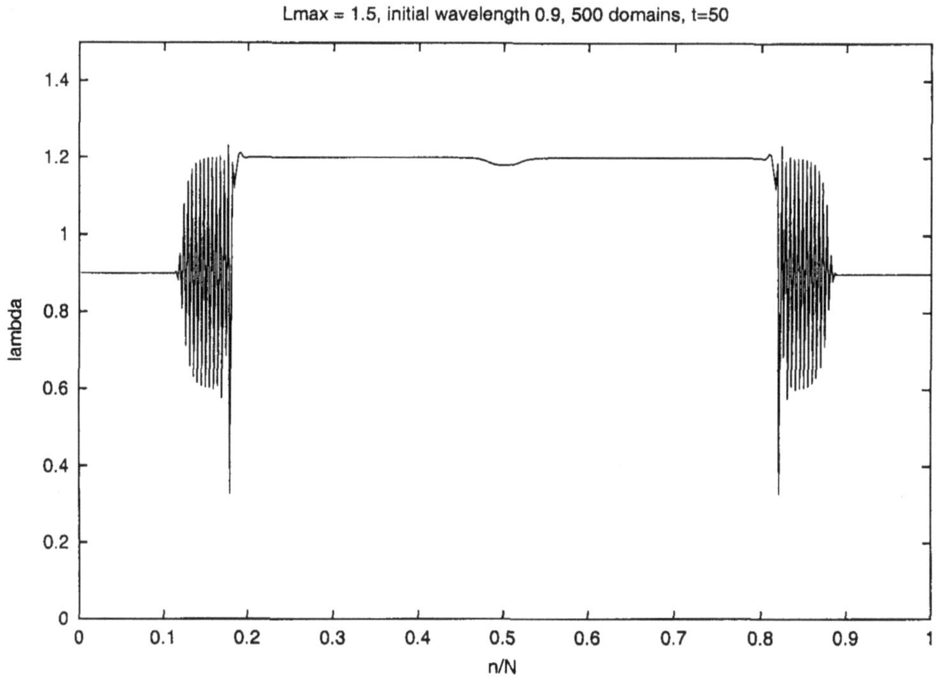

Fig. 4. Front propagation for $\lambda^{(i)} > \lambda_{\max}/2$. The figure shows a system of initially 500 ripples at time $t = 50$. The interaction function was piecewise linear with $\lambda_{\max}/\lambda_c = 3/2$. Note the period-2 state appearing between the front and the homogeneous final state. Here $\lambda^{(i)} = 0.9\,\lambda_c$ and $\lambda^{(f)} = 1.2\,\lambda_c = 4/3\,\lambda^{(i)}$.

A stability analysis of the stationary period-2 state yields the linear growth rate

$$\omega(q) = f'(\lambda_A) + f'(\lambda_B) + \sqrt{(f'(\lambda_A) - f'(\lambda_B))^2 + 4f'(\lambda_A)f'(\lambda_B)\cos^2 q}\,. \qquad (4.7)$$

Since $f'(\lambda_A)f'(\lambda_B) < 0$, the growth rate is maximal at $q = \pi/2$, and it vanishes at $q = 0$ and $q = \pi$. We conclude that the stationary period-2 solution is most unstable with respect to perturbations of period 4. In effect, this implies that one out of four ripples is eliminated, and hence $\lambda^{(f)}/\lambda^{(i)} = 4/3$. This explains the region of slope 4/3 in Fig. 3 starting around $\lambda^{(i)} \approx 0.8$. Other rational ratios can (and do) appear in a similar manner.

5. Continuum Equations for Vortex Ripples?

The model (2.1) was proposed to describe the stability and evolution of fully developed ripple patterns, but it does not address the question of how these patterns emerge from the flat bed. In part, this reflects the fact that the separation vortices appear only once the pattern has reached a certain amplitude, so a different mechanism must control the initial instability [2, 7]. On the other hand, a theoretical description that encompasses the transient evolution from the flat bed *as well as* the fully developed ripple pattern would be highly desirable, in particular for the analysis of two-dimensional systems [9, 10]. In this section we suggest that such a comprehensive description may be difficult to achieve.

For the related problem of wind-driven (*aeolian*) sand ripples, a description in terms of partial differential equations for the (one-dimensional) continuous profile $h(x,t)$ of the sand surface has been developed, e.g. Refs. 4, 13, 17 and 22; a review is given in Ref. 8. Let us collect the properties that such an equation should have for the case of vortex ripples under water: (i) since the pattern does not depend on the thickness of the water layer, the dynamics should be invariant under constant shifts of the height, $h \to h + \text{const.}$, (ii) the oscillatory driving implies symmetry under $x \to -x$, (iii) the slope of the ripples should saturate around the angle of repose, and (iv) the pattern should *not* be invariant under $h \to -h$ (closer inspection of profiles like that in Fig. 1 shows that the peaks are cusp-like while the troughs are rounded, see Ref. 21). Restricting ourselves to terms which are polynomial in the derivatives of h, the simplest equation satisfying these requirements is

$$h_t = -h_{xx} - h_{xxxx} + (h_x)^3_x - b(h_x)^2_{xx} \,, \tag{5.1}$$

where subscripts refer to partial derivatives and b is a positive constant. It is easy to see that the flat bed solution of Eq. (5.1) is unstable, with the fastest growing mode (corresponding to the initial pattern) occurring at wavelength $2\pi\sqrt{2}$. The third term on the right-hand side leads to a selected slope of ± 1, while the last term sharpens the peaks and rounds off the troughs of the ripples.

A detailed study of Eq. (5.1) has been carried out by Politi [16], who shows that the wavelength of the pattern coarsens indefinitely as $\ln t$. Coarsening appears to be a general feature of height equations with polynomial terms [8]. Patterns which do not coarsen can be obtained only at the expense of introducing unbounded growth of the slope, and hence of the amplitude, of the pattern [15]. A class of equations which contains both types of behavior is

$$h_t = -\left\{ \frac{h_x}{1+h_x^2} + \frac{1}{(1+h_x^2)^\nu} \left[\frac{h_{xx}}{(1+h_x^2)^{3/2}} \right]_x \right\}_x \,, \tag{5.2}$$

which arises in the context of meandering instabilities of stepped crystal surfaces [14]. The exponent ν is characteristic of the relaxation mechanism of the steps, the cases of immediate physical relevance corresponding to $\nu = 1$ and $\nu = 1/2$ [14]. The analysis of this equation shows that unbounded amplitude growth occurs for $-1/2 < \nu < 3/2$, and coarsening for $\nu < -1/2$.

We therefore conjecture that local height equations generally cannot describe the emergence and evolution of patterns of constant wavelength *and* amplitude. A general proof, or the discovery of a counterexample, would be of considerable interest. Meanwhile, we believe that models like (2.1) can play a useful part in the analysis of such patterns.

Acknowledgments

I am much indebted to Ken H. Andersen and Tomas Bohr for many enlightening discussions and interactions. Most of this work was performed during a sabbatical stay at CAMP, Denmark's Technical University, Lyngby, and at the Niels Bohr Institute, Copenhagen. The kind and generous hospitality of these institutions is gratefully acknowledged.

References

[1] Andersen, K. H., Chabanol, M.-L. and van Hecke, M., *Phys. Rev.* **E63**, 066308 (2001).
[2] Andersen, K. H., *Phys. Fluids* **13**, 58 (2001).
[3] Andersen, K. H., Abel, M., Krug, J., Ellegaard, C., Søndergaard, L. R. and Udesen, J., *Pattern Dynamics of Vortex Ripples in Sand: Nonlinear Modeling and Experimental Validation*, preprint (cond-mat/0201529).
[4] Anderson, R. S., *Sedimentology* **34**, 943 (1987).
[5] Ayrton, H., *Proc. Roy. Soc. London* **A84**, 285 (1910).
[6] Bagnold, R. A., *Proc. Roy. Soc. London* **A187**, 1 (1946).
[7] Blondeaux, P., *J. Fluid Mech.* **218**, 1 (1990).
[8] Csahók, Z., Misbah, C., Rioual, F. and Valance, A., *Eur. Phys. J.* **E3**, 71 (2000).
[9] Hansen, J. L., van Hecke, M., Haaning, A., Ellegaard, C., Andersen, K. H., Bohr, T. and Sams, T., *Nature* **410**, 324 (2001).
[10] Hansen, J. L., van Hecke, M., Ellegaard, C., Andersen, K. H., Bohr, T., Haaning, A. and Sams, T., *Phys. Rev. Lett.* **87**, 204301 (2001).
[11] Hellén, E. K. O., (unpublished).
[12] Hellén E. K. O. and Krug, J., *Coarsening of Sand Ripples in Mass Transfer Models with Extinction*, preprint (cond-mat/0203081).
[13] Hoyle R. B. and Mehta, A., *Phys. Rev. Lett.* **83**, 5170 (1999).
[14] Kallunki, J. and Krug, J., *Phys. Rev.* **E62**, 6229 (2000).
[15] Krug, J., *Physica* **A263**, 170 (1999).
[16] Politi, P., *Phys. Rev.* **E58**, 281 (1998).
[17] Prigozhin, L., *Phys. Rev.* **E60**, 729 (1999).
[18] Rost, M. and Krug, J., *Physica* **D88**, 1 (1995).
[19] van Saarloos, W., *Phys. Rev.* **A37**, 211 (1988).
[20] Scherer, M. A., Melo, F. and Marder, M., *Phys. Fluids* **11**, 58 (1999).
[21] Stegner, A. and Wesfreid, J. E., *Phys. Rev.* **E60**, R3487 (1999).
[22] Terzidis, O., Claudin, P. and Bouchaud, J. P., *Eur. Phys. J.* **B5**, 245 (1998).
[23] Werner, B. T. and Gillespie, D. T., *Phys. Rev. Lett.* **71**, 3230 (1993).

CHAPTER 8

Dense Granular Media as Athermal Glasses

JORGE KURCHAN

P.M.M.H. Ecole Supérieure de Physique et Chimie Industrielles,
10, rue Vauquelin, 75231 Paris CEDEX 05, France

We briefly describe how mean-field glass models can be extended to the case where the bath and friction are non-thermal. Solving their dynamics, one discovers a temperature with a thermodynamic meaning associated with the slow rearrangements, even though there is no thermodynamic temperature at the level of fast dynamics. This temperature can be shown to match the one defined on the basis of a flat measure over blocked (jammed) configurations. Numerical checks on realistic systems suggest that these features may be valid in general.

Keywords: Granular matter; glass theory; effective temperatures; compaction, aging.

1. Glasses and Dense Granular Matter

An ensemble of many elastic particles of irregular shapes at low temperatures and high densities forms a glass — that is, an out-of-equilibrium system having a relaxation timescale that grows as the system ages. Granular matter would be just an example of this, albeit a rather special one, in that the thermal kinetic energy $\sim k_B T$ per particle is negligible and that the gravity field plays an unusually important role. What in fact distinguishes granular matter from a glass at zero temperature and very high pressure is the non-thermal manner in which energy is supplied to the grains (vibration, tapping or shearing) and lost by them (inelastic collisions). It is because of this difference that we refer to the granular-matter/glass *analogy*, rather than *identity*.

This analogy was already described at the experimental level by Struik [23], who presented settling powders as aging systems on an equal footing with glasses, and made more explicit by the Chicago group [13, 21]. From the theoretical point of view, there has been a free exchange of ideas and models from one field to the other. (See Refs. 2, 5–7, 10, 16, 18–20, 24 and 25 for some examples.)

We can thus view the conceptual passage from glasses to dense granular matter as divided in two steps. The first consists of studying glass models in contact with a heat bath of very low temperature, under a strong gravity field, and considering them from the point of view of the quantities that are measured in granular matter

First published in *Advances in Complex Systems*, Vol. 4, No. 4 (2001), pp. 363–368.

experiments. The second step consists of focusing on which new features are brought in by the non-thermal agitation and friction mechanisms.

As far as the compaction dynamics is concerned, the second question is usually considered less relevant; thus, in the models, vibration is often substituted by a thermal agitation satisfying detailed balance, for example in lattice models by letting particles move upwards with probability $p_{\rm up}$ and downwards with probability $1 - p_{\rm up}$ (a thermal bath with temperature $\propto \ln^{-1}\left[\frac{p_{\rm up}}{1-p_{\rm up}}\right]$). However, if the recent analytical developments in glass theory [12] are to be applied to granular matter, it is unavoidable to face the question of the non-thermal nature of the energy exchange mechanism, as we shall see below.

2. Cage and Structural Temperatures in Glasses

A dozen years ago, a family of models was identified as being schematic *mean-field* versions of structural glasses, somewhat like the Curie–Weiss model is for ferromagnets. Above a critical temperature, the dynamics of these models is given by mode-coupling equations, or generalisations of them. These equations predict that the relaxation of all quantities proceeds in two steps: a rapid one given the movement of particles in a 'cage' formed by its neighbours, and a slow one generated by the rearrangement of cages — the structural or α-relaxations. As the temperature is lowered, the structural relaxation time becomes larger and larger, and it diverges at the critical temperature. (This transition is in fact smeared in real life, a fact that can be understood within the same framework.)

If the systems are quenched below the transition temperature, they fall out of equilibrium; the structural relaxation time is not constant but grows with the ('waiting') time elapsed after the quench, a phenomenon known as aging. Alternatively, one can submit a system below the critical temperature to forces that, like shear stress, can do work continuously. The surprising result in this case is that aging is interrupted (see Refs. 3 and 16); the structural relaxation time saturates to a driving-force dependent value. This rejuvenation effect is known as *shear-thinning* or *thixotropy*, depending on whether it applies in the liquid or the glass phase.

Below the transition temperature, the system is out of equilibrium, either because it is still aging or because of the external forces in the driven case. An old idea [28] in glass physics is to consider that the structural degrees of freedom remain at a higher temperature (of the order of the glass temperature), while the cage motion thermalises with the bath. In order to make this idea sharp, we can ask what would be the reading of a thermometer coupled to the glass. One can show [8] that this is related to the ratio of fluctuations and dissipation, as we now describe.

Consider an observable A, with zero mean and with fluctuations characterised by their autocorrelation function $C_A(t_w + t, t_w) = \langle A(t + t_w)A(t_w)\rangle$. Let us denote $\chi_A(t_w + t, t_w)$ the response $\delta\langle A(t_w + t)\rangle/\delta h$ to a field h conjugate to A, acting from t_w to $t_w + t$.

If above the glass temperature we plot χ_A versus C_A using t as a parameter, we obtain a straight line with gradient $-1/T$; the fluctuation-dissipation theorem. For a system aging or subjected to nonconservative forces below the glass temperature we can still make the same plot, using t (and t_w, in the aging case) as parameters. It turns out that one obtains a line with *two* straight tracts; for values of C_A, χ_A corresponding to fast relaxations the gradient is $-1/T$, while for values corresponding to the structural relaxation the gradient is a constant $-1/T_{\rm dyn}$. The effective temperature $T_{\rm dyn}$ so defined is in fact the temperature read in a thermometer coupled to A tuned to respond to the slow fluctuations [8]. Most importantly, it is observable-independent within each timescale.

These facts were originally found in the mean-field/mode-coupling approximation for glassy dynamics, and later verified numerically (at least within the times, sizes and precision involved) for a host of realistic glass models [3, 12, 26].

The appearence of a temperature $T_{\rm dyn}$ for the slow degrees of freedom, immediately suggested a comparison with an idea proposed by Edwards originally for granular matter [10, 11]. For a glass at very low temperatures it can be stated as follows: as the glass ages and its energy $E(t)$ slowly decreases, the value of all macroscopic observables at time t can be computed from an ensemble consisting of all blocked configurations (the local energy minima) having energy $E(t)$, taken with *equal statistic weights.* This ensemble immediately leads to the definition of an entropy $S_{\rm Edw}(E)$ as the logarithm of the number of blocked configurations, and a temperature $T_{\rm Edw}^{-1} = dS_{\rm Edw}/dE$ [1].

Now, for the mean-field/mode coupling models, *it turns out that* $T_{\rm dyn}$ *and* $T_{\rm Edw}$ *coincide*, and, furthermore, Edwards' ensemble defined above yields the correct values for the observables out of equilibrium [16]. This has been recently checked for more realistic (nonmean-field) models [2, 22].

3. Structural Temperature in (Dissipative) Granular Matter

In order to see what new features are to be found in granular matter, we start with the mean-field/mode coupling models, modifying them in two ways. Firstly, we allow for frictional forces that are non-linear, complicated functions of the velocities. Secondly, we drive the systems with forces that do not derive from a potential ('shear-like') or are strong and periodic in time (vibration and tapping). We expect that the mean-field glass models thus modified will be minimal mean-field granular matter models.

We measure as before correlations and responses, and, in particular diffusion $\langle |x(t+t_w) - x(t_w)|^2 \rangle$ and mobility $\delta |x(t+t_w)|/\delta f$, where f is a force acting from t_w to $t+t_w$. The vibrated or tapped case has to be measured 'stroboscopically'; in order to avoid seeing oscillations we only consider times that correspond to integer numbers n, n' of cycles:

$$C(t_n, t_{n'}) = \langle x(t_n) x(t_{n'}) \rangle \,, \tag{3.1}$$

$$\chi(t_n, t'_n) = \frac{\delta \langle x(t_n) \rangle}{\delta f}, \tag{3.2}$$

where the force acts during an integer number of cycles from $t_{n'}$ to t_n.

In the thermal case we found that above the glass temperature the comparison of correlations and responses yields the bath's temperature (as it should, in an equilibrium situation), and below the glass phase in addition a temperature T_{dyn} for the slow degrees of freedom. For the granular athermal case, this already poses a problem, as not even in a liquid-like fluidised state do we have a well-defined temperature! (In other words, a parametric χ_A versus C_A plot will not give a straight line independent of the observable A.) This will also be true for the 'cage' motion [29] in the dense regime.

Surprisingly enough, the next step came from the treatment of quantum glasses at zero temperature at the mean-field level. It turns out [9, 14] that these systems obey a quantum fluctuation-dissipation theorem in the cage motion, but a classical one in the slow, structural motion; the nature of the bath is irrelevant (in the sense of the renormalisation group) as far as the slow motion is concerned. In the context of granular matter, a similar reasoning [4, 15] shows that while there is no well-defined dynamic temperature associated to the fast relaxations — and in the fluidised regime these are the only relaxations present — the slow structural relaxations still satisfy a fluctuation-dissipation relation, with an observable-independent temperature T_{dyn} (see Figs. 1 and 2).

Once these questions have been clarified at the level of mean-field/mode-coupling models, one feels encouraged to check them numerically and experimentally in realistic systems [27]. Recently [17], a simulation of granular matter subjected to shear has given evidence for the existence of a structural temperature. This dynamical temperature is calculated from the relation between diffusivity and mobility of different tracers, and its independence of the tracer shape is checked. The interest of this setting is that it can be implemented experimentally.

Within the same model, a direct computation of a thermodynamic temperature defined on the basis of the blocked configurations has yielded very good agreement with the dynamical temperature.

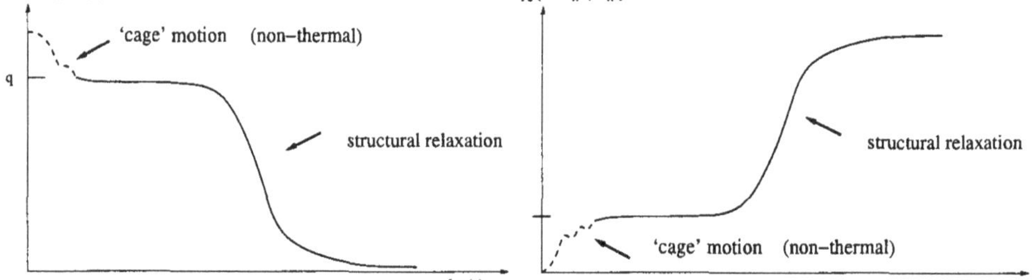

Fig. 1. Sketch of a correlation (left) and a response (right) versus time.

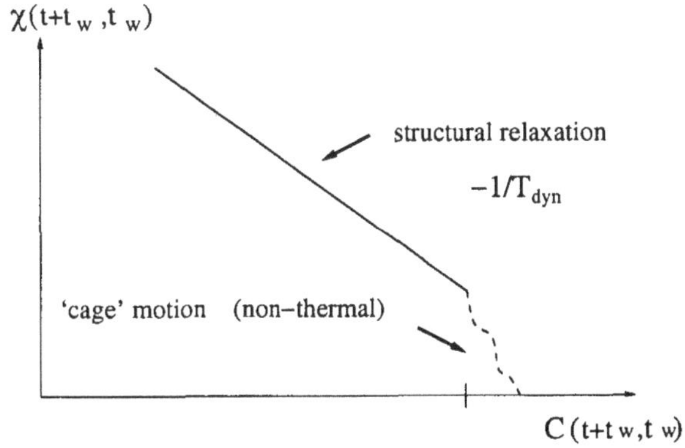

Fig. 2. Effective temperature plot. The dashed tract (fast relaxations) is not straight and is observable-dependent. The full line (structural relaxations) is straight and defines an observable-independent temperature.

4. Conclusion

In conclusion, there has been progress in the theory of statistical ensembles for dense granular matter.

- We have a better idea of how we should understand them, and of their possible domain of validity.
- We have solvable models, and a limit in which we can check if and when these ideas hold strictly.
- We have suggestions for experiments that will test the validity of the approach in each case.

References

[1] A brief historical note: Edwards himself never defined such a temperature, and indeed one of the main aims of his articles cited above is to advocate the use of volume rather than energy as the relevant variable, leading to 'compactivity' (a kind of pressure) rather than to temperature.

However, in those articles he takes for granted a flat average over blocked configurations; this is the aspect of those works which interests us here. Once an 'ergodic' notion is defined, one can choose which ensemble is the most appropiate in each case (fixing volume and/or energy, etc.) and the thermodynamic variables follow.

'Edwards' temperature' here is short for: 'the temperature that is obtained from a flat ensemble *à la* Edwards.'

[2] Barrat, A., Kurchan, J., Loreto, V. and Sellitto, M., *Phys. Rev. Lett.* **85**, 5034 (2000); *Phys. Rev.* **E63**, 051301 (2001).

[3] Barrat, J.-L., Berthier, L. and Kurchan, J., *Phys. Rev.* **E61**, 5464 (2000); Barrat, J.-L. and Berthier, L., *Phys. Rev.* **E63**, 012503 (2001).

[4] Berthier, L., Cugliandolo, L. F. and Iguain, J. L., *Phys. Rev.* **E63**, 051302 (2001).

[5] Caglioti, E., Loreto, V., Herrmann, H. J. and Nicodemi, M., *Phys. Rev. Lett.* **79**, 1575 (1997).

[6] Coniglio, A. and Hermann, H.-J., *Physica* **A225**, 1 (1996).

[7] Coniglio, A. and Nicodemi, M., cond-mat/0010191.
[8] Cugliandolo, L. F., Kurchan, J. and Peliti, L., *Phys. Rev.* **E55**, 3898 (1997).
[9] Cugliandolo, L. F. and Lozano, G., *Phys. Rev. Lett.* **80**, 4979 (1998).
[10] Edwards, S. F., in *Disorder in Condensed Matter Physics*, Blackman and Taguena, eds. (Oxford University Press, 1991).
[11] Edwards, S. F., The role of entropy in the specification of a powder, in *Granular Matter: An Interdisciplinary Approach*, Mehta, A., ed. (Springer, 1994), and references therein.
[12] For a brief review and references on these recent developments in glass theory, see: J. Kurchan, cond-mat/0011110, to appear in the special issue 'Physics of Glasses', C. R. Acad. Sci. Paris, Série IV (2001).
[13] Jaeger, H. M. and Nagel, S. R., *Science* **255**, 1523 (1992); Jaeger, H. M., Nagel, S. R. and Behringer, R. P., *Rev. Mod. Phys.* **68**, 1259 (1996).
[14] Kennett, M. P. and Chamon, C., *Phys. Rev. Lett.* **86**, 1622 (2001).
[15] Kurchan, J., *J. Phys. Condensed Matter* **12**, 6611 (2000).
[16] Kurchan, J., Rheology, and how to stop aging, in *Jamming and Rheology*, Liu, A. J. and Nagel, S. R. eds. (Taylor & Francis, London, 2001), pp. 72, cond-mat 9812347, see also: http://www.itp.ucsb.edu/online/jamming2/.
[17] Makse, H. and Kurchan, J., Thermodynamic approach to dense granular matter: A numerical realization of a decisive experiment, submitted for publication.
[18] Nicodemi, M. and Coniglio, A., *Phys. Rev. Lett.* **82**, 916 (1999).
[19] Nicodemi, M., Coniglio, A. and Hermann, H.-J., *Phys. Rev.* **E55**, 3962 (1977).
[20] Nicodemi, M., *Phys. Rev. Lett.* **82**, 3734 (1999).
[21] Nowak, E., Knight, J. B., Ben-Naim, E., Jaeger, H. M. and Nagel, S. R., *Phys. Rev.* **E57**, 1971 (1998).
[22] Prados, A., Brey, J.J. and Sánchez-Rey, B., *Physica* **A275**, 310 (2000); Lefevre, A. and Dean, D. S., *J. Phys.* **A34**, L213–L220 (2001).
[23] See, for example, Chap. 7 of L. C. E. Struik, *Physical Ageing in Amorphous Polymers and Other Materials* (Elsevier, Houston, 1978).
[24] Sellitto, M. and Arenzon, J. J., Free-volume kinetic models of granular matter, *Phys. Rev. E* (December 2000).
[25] Sellitto, M., *Euro. J. Phys.* **B4**, 135 (1998).
[26] The observable independence of T_{dyn} has only recently been checked in a realistic model: J.-L. Barrat and L. Berthier, to be published.
[27] The question was already experimentally addressed in Ref. 21. The evidence there is not conclusive either for or against a thermodynamic approach [17].
[28] Tool A. Q., *J. Am. Ceram. Soc.* **29**, 240 (1946).
[29] Warr, S. and Hansen, J.-P., *Europhys. Lett.* **36**, 589 (1996).

CHAPTER 9

Transient and Steady-State Dynamics of Granular Shear Flows

W. LOSERT and G. KWON

Department of Physics, IPST and IREAP, University of Maryland, College Park, MD 20742, USA

The initiation and steady-state dynamics of granular shear flow are investigated experimentally in a Couette geometry with independently moveable outer and inner cylinders. The motion of particles on the top surface is analyzed using fast imaging. During steady state rotation of both cylinders at different rates, a shear band develops close to the inner cylinder for all combinations of speeds of each cylinder we investigated. Experiments on flow initiation were carried out with one of the cylinders fixed. When the inner cylinder is stopped and restarted after a lag time of seconds to minutes in the same direction, a shear band develops immediately. When the inner cylinder is restarted in the opposite direction, shear initially spans the whole material, i.e. particles far from the shear surface are moving significantly more than in steady state.

Keywords: Granular matter; shear flow; flow initiation; history dependence; Couette geometry; shear band.

1. Introduction

The properties of ensembles of solid particles such as sand have intrigued researchers and engineers for a long time [6]. While individual particles in such granular materials are solid, the ensemble of particles can flow seemingly like a liquid. Flow of granular materials is distinct from ordinary fluid flow, with velocity gradients that are often confined to thin shear bands, a variable particle density, and the need for dilation in order to allow flow. Even during flow, the shear forces exhibit properties resembling solid friction; the shear force is independent of shear rate and proportional to pressure.

The characteristics of granular shear flow have been studied in detail in recent years. Experimental, theoretical, and numerical work is referenced in Bocquet *et al.* [4] and in a review article by Clement [5]. Many of the experimental studies have looked at steady state shear flow in a Couette geometry with a movable inner cylinder and a stationary outer cylinder in three dimensions [2, 7–9] and in two dimensions [3]. The flow is studied after it is allowed to reach a steady state. The

First published in *Advances in Complex Systems*, Vol. 4, No. 4 (2001), pp. 369–377.

following key points emerge in shear flow experiments for a large range of materials in two and three dimensions:

- The velocity of particles decreases quickly over a few particle diameters away from the shearing wall [2]. See also Refs. 3, 7 and 8.
- The velocity profile, normalized by the shear velocity U, is roughly independent of U *at small shear velocities* (see e.g. Refs. 3 and 8).
- The shear force σ is approximately independent of U, if the granular material is allowed to dilate (see e.g. Ref. 9).
- There are strong inhomogeneities in the force distribution even during flow [1, 3].

In earlier research, Losert *et al.* [4, 7] carefully analyzed the experimentally measured microscopic particle dynamics in a circular Couette geometry and found a power law relation between the *local* RMS velocity fluctuations and the *local* velocity gradient. This relation can be explained as a result of the high density of particles, which leads to caging of particles by their neighbors. This prevents the development of velocity gradients but does not restrict fluctuations. Thus, velocity gradients are confined to a thinner shear band than fluctuations. When this effect is properly taken into account, a hydrodynamic model, which we have introduced in Refs. 4 and 7, quantitatively describes most key properties of granular shear flow discussed above, including both flow properties and shear forces, though it does not address the inhomogeneous force distributions directly. In the model, fluctuations are generated at the shearing wall and decay inside the material. Larger fluctuations lower the density and permit larger velocity gradients.

In this paper we focus on two aspects of the motion of particles in a Couette cell:

- The influence of the Couette cell geometry on the shear flow is studied in a system, in which both inner and outer cylinder are rotated independently. Most other studies were carried out with a rotating inner cylinder and a stationary outer cylinder. The width of the granular layer is $\sim$ 50 particle diameters, a factor of about 5 wider than most other shearcells, and about one order of magnitude larger than the characteristic shear band width. The maximum shear speed is 1.4 m/s, significantly faster than other recent measurements.
- The motion of particles is also investigated at the start of the shear flow, not only after a steady state has been reached. We determine whether the initiation of flow depends on the direction of previously applied shear.

The investigation of the start of a shear flow is aimed to provide insight into the transition between two complex states of granular matter.

The steady granular shear flow state is a far from equilibrium state of a driven, dissipative system. The flow of particles can only be sustained, if energy is constantly pumped into the system and dissipated through friction and inelastic collisions.

In granular material at rest, forces are tranmitted through the granular assembly through an inhomogeneous and often anisotropic network of stress chains. The properties of the contact network, e.g. whether the force at the base of a sandpile is a local maximum or minimum, may depend on the construction history of the material [10].

Here, we investigate the transition from the inhomogeneous stationary state to the flowing state. Two issues will be addressed briefly in this paper, and investigated in more detail in future publications. (1) How quickly is the far from equilibrium steady flowing state selected? (2) Does the history dependent contact network influence particle dynamics at the start of shear?

2. Experimental Setup

The experimental setup is shown in Fig. 1. Both cylinders are connected to computer controlled variable speed DC servo motors (Aerotech Inc.). The shear cell was designed and built for grains with particle diameters $d > 0.75$ mm to limit the effects of charging and humidity. The inner cylinder radius of 102 mm and the outer radius of 146 mm provide the widest practically feasible granular layer of up to 60 particle

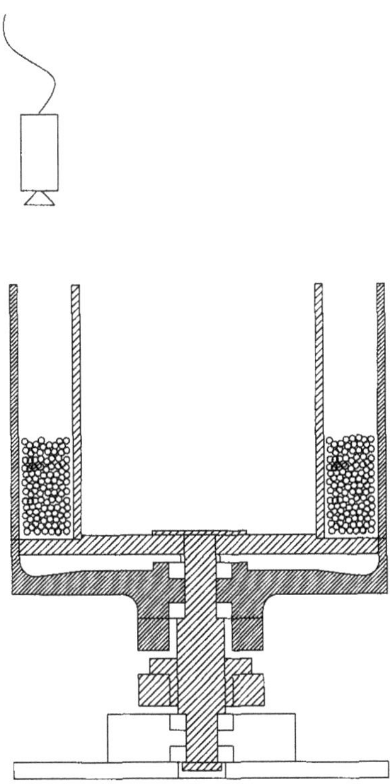

Fig. 1. Schematic of experimental setup. Granular material (1 mm glass beads) are confined in a 44 mm gap between movable inner and outer cylinders. The bottom of the shear cell moves with the inner cylinder.

diameters and the smallest ratio between the cylinder diameters. The bottom of the shear cell is connected to the inner cylinder, unlike in earlier experiments where the inner cylinder was movable and the bottom stationary. Particles are filled between the cylinders up to a layer height of about 80 particle diameters.

In the experiments described below, black spherical glass beads are used (from Jaygo Inc.) with a diameter d between 1.0–1.25 mm ($\rho_m = 2.55$ g/cm^3) yielding a shear zone of $\sim 45d$.

The motion of particles on the top surface is imaged with a fast CCD camera at 30–1000 frames/sec. Particle motion is extracted from image sequences of 1096 images (240 × 512 pixels) using procedures written in IDL (RSI Inc.). A detailed description of the capabilities and limitations of tracking particles on the surface of a granular flow is given in Ref. 4.

The rotation frequency f of the cylinder imposes the shear velocity U at the boundary of the granular material $U = rf$, with $r = 102$ mm the radius of the inner cylinder. The rotation frequency is varied over more than 2 orders of magnitude between $f = 0.01$ Hz and $f = 2.2$ Hz, i.e. $U = 6$ mm/s and $U = 1.4$ m/s for inner cylinder rotation.

3. Experimental Results

3.1. *Shear flows with moving inner and outer cylinder*

Shear flow is investigated for different relative velocities of the inner and outer cylinder. The rotation frequency of the outer cylinder is fixed at −0.19 Hz. The average tangential velocity of particles $V_t(y)$ is shown in Fig. 2 as a function of the distance y from the inner cylinder. The inner cylinder is rotated in the same direction at higher and lower velocity, is kept at rest, or is rotated in the opposite direction.

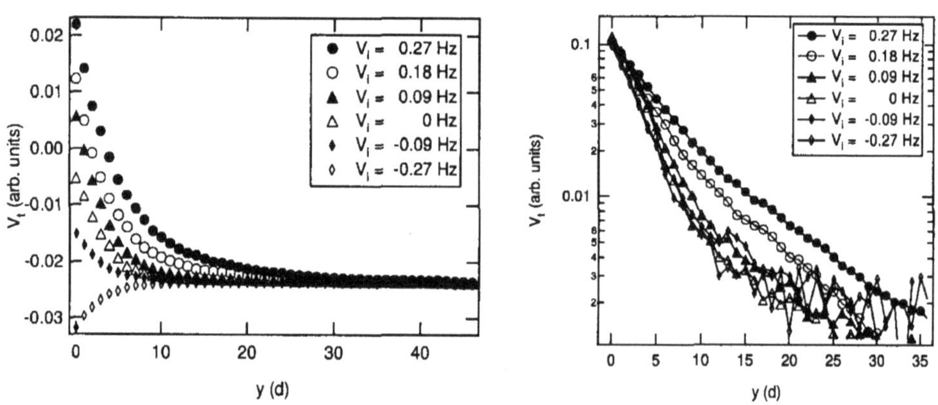

Fig. 2. Average tangential velocity V_t as a function of distance from the inner cylinder y. Linear scale (left) and logarithmic scale (absolute values of velocity) (right). The outer cylinder is rotated at −0.19 Hz. The rotation frequency of the inner cylinder is indicated in the figure.

In all cases a shear band, characterized by large velocity gradients, develops near the inner cylinder. V_t decreases roughly exponentially as a function of y with a characteristic length l_c:

$$V_t = V_0 e^{y/l_c} . \tag{3.1}$$

Note that the bottom moves with the inner cylinder, so the shear band is not an indication of the boundary between inner and outer cylinder in this case. The logarithmic plot (Fig. 2, right) shows that the velocity profile is roughly exponential, with a characteristic length that slowly increases with the relative velocity.

3.2. *High speed steady state shear flow*

The shear band is investigated as a function of shear velocity at shear rates of the inner cylinder of up to $f = 2.2$ Hz, i.e. shear velocities up to $U = 1.4$ m/s, faster than the velocities reached in earlier experiments. Again, V_t decreases roughly exponentially with distance y from the inner cylinder. The characteristic length l_c of the shear band, which was found to be independent of shear rate at low shear rates [3, 4], increases with increasing shear rate at high shear rates as shown in Fig. 3.

We note that this observed increase in l_c at high shear rates may be due to the bottom boundary condition (the bottom is attached to the rotating inner cylinder in this experiment, while it was attached to the stationary outer cylinder in earlier experiments). In future experiments the role of the bottom wall will be investigated by measuring the flow profile as a function of layer height.

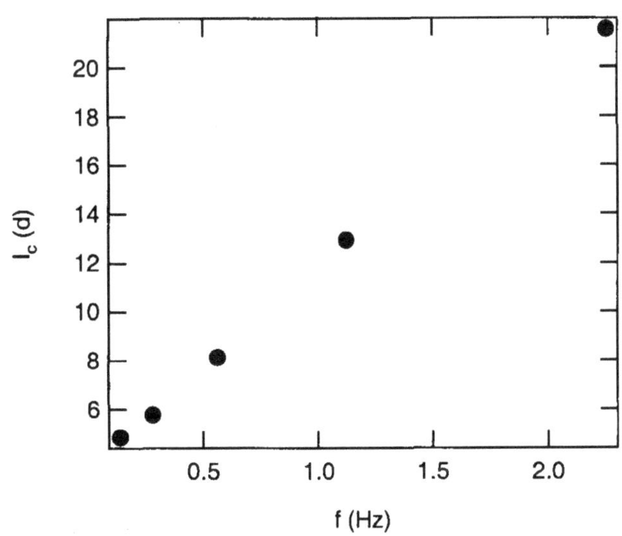

Fig. 3. Characteristic length l_c of the shear band as a function of rotation frequency f at high rotation frequencies of the inner cylinder.

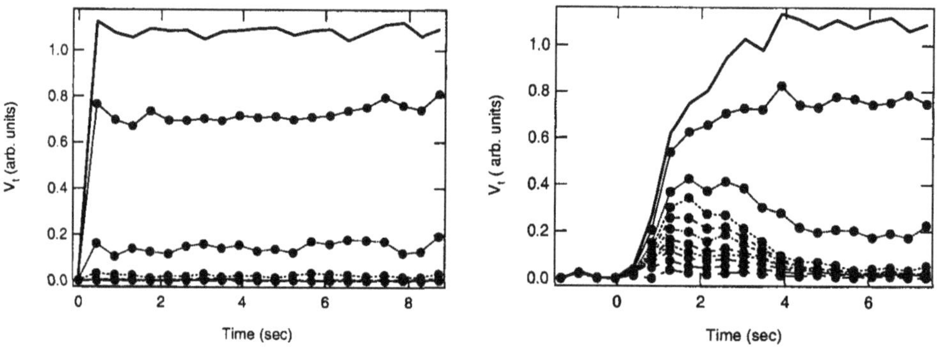

Fig. 4. Start of inner cylinder motion: Average particle velocity as a function of time at different distances from the inner cylinder. Inner cylinder motion is started at $t = 0$ s. Previous motion of the inner cylinder in (left) same direction or (right) opposite direction. (Solid line: inner cylinder; Solid line with symbols: region $0 - 5d$ from inner cylinder; Dashed line with symbols: region $0 - 5d$ from outer cylinder. ($f = 0.02$ Hz).)

3.3. *Sudden start of shear flow*

Here, we analyze the dynamics of particles at the start of motion of the shear cell, i.e. at the transition of granular matter from a stationary state to a flowing state. In the following experiments a steady state granular shear flow is suddenly stopped. The material is then kept at rest for a lag time of several seconds to minutes.

When shear is suddenly restarted at $f = 0.02$ Hz *in the same direction as prior to the stop*, the velocity profile $V_t(y)$ reaches a steady state roughly exponential shape rapidly. Figure 4 (left) shows the instantaneous particle speed averaged over 0.4 s and over 10 radial segments spanning 5 mm each. The roughly exponential steady state is reached within the first 0.4 s. The steady flowing state is therefore reached, while the inner cylinder moves by less than 3 particle diameters.

The 'construction history' of the stationary granular material is modified in a second experiment, by *starting the motion of the inner cylinder in the direction opposite to the prior shear direction*, as shown in Fig. 4 (right). In this case the steady flowing state is reached only after several seconds, during which particles far from the inner cylinder move significantly faster than in steady state.

This extra motion is analyzed in more detail. The total displacement of particles as a function of the displacement of the inner cylinder is shown in Fig. 5 (left). At later times, the total displacement increases linearly with the motion of the inner cylinder, indicating a steady state shear flow. The linear extrapolation of the motion of particles due to a steady state shear flow is indicated as straight lines in Fig. 5 (left). These lines do not extrapolate to zero for zero displacement of the inner cylinder. This indicates particle displacement in addition to the motion due to steady state shear flow. Note that the particle displacement reaches the asymptotic steady shear flow displacement (straight lines) at roughly the same time independent of distance from the inner cylinder.

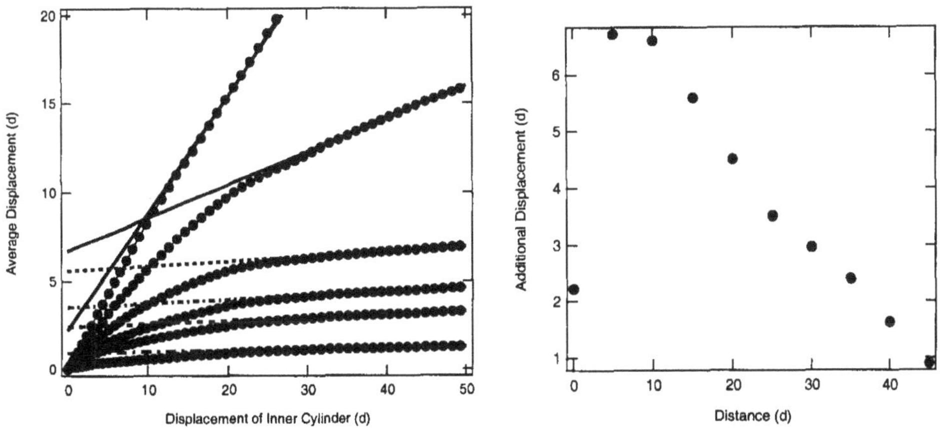

Fig. 5. Shear flow is started in the direction opposite to earlier shear: Average displacement as a function of time (left) for particles at different distances from the inner cylinder. Additional displacement (right), when compared to a steady state shear flow (from linear extrapolation of the total distance travelled to $t = 0$ s).

The additional displacement of particles during the start of shear flow as a function of distance from the inner cylinder is shown in Fig. 5 (right). The additional displacement increases linearly with distance from the outer cylinder until it drops close to the inner cylinder.

Similar behavior is found when the inner cylinder is fixed and the outer cylinder is moved, as shown in Fig. 6. When the outer cylinder motion is started *in the same direction as previous shear* at $f = 0.01$ Hz (Fig. 6, left), the velocity profile rapidly reaches a steady state with a shear band close to the inner wall as discussed in the previous section.

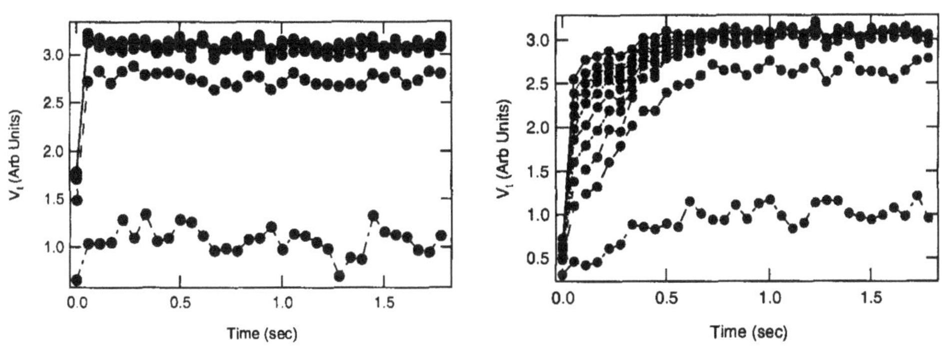

Fig. 6. Start of outer cylinder motion: Average particle velocity as a function of time at different distances from the inner cylinder. Outer cylinder motion is started at $t = 0$ s. Previous motion of the outer cylinder in (left) same direction or (right) opposite direction. (Solid line with symbols: region $0 - 5d$ from outer cylinder; dashed line with symbols: region $0 - 5d$ from inner cylinder ($f = 0.01$ Hz).)

When the outer cylinder is moved *in the direction opposite to previous shear* (Fig. 6, right), the shear band develops more slowly. As in the case of inner cylinder motion, regions that do not have a noticeable velocity *gradient* develop velocity gradients during the start of motion.

The boundary condition at the smooth bottom of the shearcell may also influence the particle dynamics at the start of a shear flow, especially when the inner cylinder (and bottom) reverse direction. Preliminary experiments suggest that an additional displacement of particles far from the inner cylinder is still observed with a fixed bottom, though the quantitative results may differ. This issue will be investigated in further studies, with a a smooth bottom wall connected to the outer cylinder.

4. Summary

This experimental study focuses two aspects of particle motion in granular matter sheared in a Couette geometry: the steady state granular shear flow between *two* moving sidewalls and the dynamics at the start of a granular shear flow.

In a steady state shear flow in a Couette geometry, the shear band is located next to the inner cylinder for all shear conditions investigated, even when only the outer cylinder is rotated. The shear band width increases with increasing shear rate. Further experiments are in progress to distinguish more clearly between three effects that may influence the tangential velocity profile in the shear cell: (1) The effect of the velocity difference between the cylinders. (2) The influence of centrifugal forces. (3) The importance of the bottom boundary condition.

The particle dynamics at the start of a granular shear flow depend on the direction of previously applied shear. If the motion of either the inner or outer cylinder is restarted in the same direction as before, a steady state velocity profile is reached immediately within the resolution of our measurement of < 0.4 s.

If the inner cylinder is restarted in the direction opposite to previously applied shear, particles far from the inner cylinder move significantly more than during steady state shear during a transient time. The additional displacement during the start of shear flow increases with distance from the stationary outer wall and reaches up to 6 particle diameters. Similar behavior is found for a reversal of motion of the outer cylinder.

The principal direction of forces for shear in one direction is roughly perpendicular the direction of forces for shear in the opposite direction. Therefore, this result may reflect a 'softness' of granular matter in the direction perpendicular to the principal force axis.

Acknowledgments

This project benefitted from earlier experiments in collaboration with J. P. Gollub and D. Schalk at Haverford College, which were supported by the National Science Foundation under Grant DMR-0072203.

References

[1] Aharonov, E. and Sparks, D., *Phys. Rev.* **E60**, 6890 (1999).

[2] Anjaneyulu, P. and Khakhar, D. V., *Powder Technology* **83**, 29 (1995).

[3] Behringer, R. P. *et al.*, *Physica* **D133**, 1 (1999).

[4] Bocquet, L., Losert, W., Schalk, D., Lubensky, T. and Gollub, J., to appear in *Phys. Rev. E.*

[5] Clement, E., *Curr. Opinion in Coll. and Interface Sci.* **4**, 294 (1999).

[6] Jaeger, H. M., Naegel, S. R. and Behringer, R. P., *Rev. Mod. Phys.* **68**, 1259 (1996).

[7] Losert, W., Bocquet, L., Lubensky, T. C. and Gollub, J. P., *Phys. Rev. Lett.* **85**, 1428 (2000).

[8] Mueth, D. M. *et al.*, *Nature* **406**, 385 (2000).

[9] Tardos, G. I., Khan, M. I. and Schaeffer, D. G., *Phys. Fluids* **10**, 335 (1998).

[10] Vanel, L. *et al.*, preprint, cond-mat/9906321 (1999).

CHAPTER 10

Liquid-Solid Transition in Bidisperse Granulates

STEFAN LUDING
ICA 1, Pfaffenwaldring 27, 70569 Stuttgart, Germany
and
TU-Delft, Julianalaan 136, 2628 BL Delft, The Netherlands
lui@ica1.uni-stuttgart.de
s.luding@tnw.tudelft.nl
http://www.ica1.uni-stuttgart.de/~lui

Simulation results of dense granulates with particles of different sizes are compared with theoretical predictions concerning the mixture pressure. An effective correlation function is computed which depends only on the total volume fraction and on the dimensionless width of the size-distribution function. From simulation data of elastic and weakly dissipative systems, one can predict how much disorder (size-dispersity) is necessary to avoid ordering effects due to crystallization. Finally, a *global equation of state* is proposed, which unifies both the dilute, disordered gas/fluid and the dense, solid regime.

Keywords: Liquid-solid phase transition; hard sphere gas/fluid/solid; binary mixture; disorder-order.

1. Introduction

The hard-sphere (HS) system is a traditional and simple toy-model for various phenomena, like disorder-order transitions, the glass transition, or simple gases and liquids [4, 8, 23, 30]. A theory that describes the behavior of rigid particles in the gas and disordered fluid regime is the kinetic theory [4, 13], where particles are assumed to be rigid and collisions take place in zero time (they are instantaneous), exactly like in the hard-sphere model. In a more dense system which resembles a solid or a glass, particle-in-cell models or a free volume theory can be applied [12, 16]. In the intermediate transition regime, no satisfactory theoretical description is available at the moment [14, 16, 20, 26].

When dissipation is added to the HS model, one has the inelastic hard sphere (IHS) model, i.e. the simplest version of a granular gas, a member of the more general class of dissipative, non-equilibrium, multi-particle systems [9, 23]. Attempts to describe granular media by means of kinetic theory are usually restricted to certain limits like constant or small density or weak dissipation [2, 7, 11, 25, 27]. In

First published in *Advances in Complex Systems*, Vol. 4, No. 4 (2001), pp. 379–388.

general, granular systems consist of particles with different sizes and properties, a situation which is rarely addressed theoretically [1, 10, 28]. However, the treatment of bi- and polydisperse mixtures is easily performed by means of numerical simulations [5, 20, 21, 22].

In this study, theories and simulations for situations with particles of equal and different sizes are compared. In Sec. 2, the model system is introduced and, in Sec. 3, theoretical results are reviewed and compared with numerical results concerning the pressure. The disorder-order transition and the amount of difference in particle size, necessary to avoid it, is discussed in Sec. 4. The results are summarized and discussed in Sec. 5.

2. Model System

For the numerical modeling of the system, periodic, two-dimensional (2D) systems of volume $V = L_x L_y$ are used, with horizontal and vertical size L_x and L_y, respectively. N particles are located at positions $\mathbf{r}_i$ with velocities $\mathbf{v}_i$ and masses m_i. From any simulation, one can extract the kinetic energy $E(t) = \frac{1}{2}\sum_i m_i \mathbf{v}_i^2$, dependent on time via the particle velocity $\mathbf{v}_i = \mathbf{v}_i(t)$. In 2D, the "granular temperature" is defined as $T = E/N$.

2.1. *Polydispersity*

The particles in the system have the radii a_i randomly drawn from mono-, bi-, and polydisperse size distribution functions

$$w(a) = \begin{cases} \delta(a - a_0) & \text{with } \langle a \rangle = a_0\,, \\ n_1\delta(a - a_1) + n_2\delta(a - a_2) & \text{with } \langle a \rangle = (n_1 + (1 - n_1)/R)a_1\,, \\ \dfrac{1}{2w_0 a_0}\theta_{[(1-w_0)a_0,(1+w_0)a_0]}(a) & \text{with } \langle a \rangle = a_0\,, \end{cases}$$

with the two-sided step-function $\theta_{[x_1,x_2]}(x) = 1$ for $x_1 \le x \le x_2$ and $\theta(x) = 0$ otherwise.

In the bidisperse case, the number fraction is $n_1 = N_1/N$ for particles with size a_1 in a system with $N = N_1 + N_2$ particles in total and N_2 particles with radius a_2. The size ratio $R = a_1/a_2$ is needed to classify a bidisperse size distribution with the volume fraction $\nu = \nu_1 + \nu_2$ as the last relevant system parameter, since the partial volume fractions can be expressed in terms of n_1 and R. The dimensionless kth moment is $A_k = n_1 + (1 - n_1)R^{-k} = \langle a^k \rangle / a_1^k$, scaled by a_1.

Since they are needed later on, the expectation values for the moments of a and their combination, the dimensionless width-correction $\mathcal{A} = \langle a \rangle^2 / \langle a^2 \rangle$, are summarized in Table 1 in terms of a_1, n_1 and R for the bidisperse situations and in terms of a_0 and w_0 in the polydisperse cases. Different values of ν are realized by shrinking or growing either the system or the particles.

Table 1. Moments $\langle a \rangle$, $\langle a^2 \rangle$ and $\mathcal{A} = \langle a \rangle^2 / \langle a^2 \rangle$ of the size distribution functions.

		$\langle a \rangle$	$\langle a^2 \rangle$	$\langle a \rangle^2 / \langle a^2 \rangle$
(i)	monodisperse	a_0	a_0^2	1
(ii)	bidisperse	$A_1 a_1$	$A_2 a_1^2$	A_1^2 / A_2
(iii)	polydisperse	a_0	$(1 + w_0^2/3) a_0^2$	$3/(3 + w_0^2)$

2.2. *Particle interactions*

The particles are assumed to be perfectly rigid and follow an undisturbed straight trajectory until a collision occurs as described below. Due to the rigidity, collisions occur instantaneously, so that an event driven simulation method [15, 19] can be used. Note that no multi-particle contacts can occur in this model.

A change in velocity — and thus a change in energy — can occur only at a collision. The standard interaction model for instantaneous collisions of particles with radii a_i, mass $m_i = (4/3)\pi \rho a_i^3$, and material density ρ is used in the following. (Using the mass of a sphere is an arbitrary choice.) The post-collisional velocities $\mathbf{v}'$ of two collision partners in their center-of-mass reference frame are given, in terms of the pre-collisional velocities $\mathbf{v}$, by $\mathbf{v}'_{1,2} = \mathbf{v}_{1,2} \mp \frac{(1+r)m_{12}}{m_{1,2}} \mathbf{v}_n$, with $\mathbf{v}_n \equiv [(\mathbf{v}_1 - \mathbf{v}_2) \cdot \hat{\mathbf{n}}]\hat{\mathbf{n}}$, the normal component of the relative velocity $\mathbf{v}_1 - \mathbf{v}_2$, parallel to $\hat{\mathbf{n}}$, the unit vector pointing along the line connecting the centers of the colliding particles, and the reduced mass $m_{12} = m_1 m_2/(m_1 + m_2)$. If two particles collide, the change of the translational energy is $\Delta E = -m_{12}(1 - r^2) v_n^2/2$.

3. Simulation and Theory

In the following, we examine situations with different polydispersity. Most of the simulations were performed in the elastic limit, $r = 1$, however, we checked that also simulations with weak dissipation and some weak driving force lead to the same results. Due to the scaling of the pressure with the energy of the system, as introduced below, also homogeneously cooling situations [18] are well fitted by the elastic results, if the factor 2 is replaced by $(1+r)$. A more detailed discussion of the range of applicability of the elastic results with respect to density and dissipation-strength is far beyond the scope of this study.

3.1. *Particle correlations*

In *monodisperse systems*, the particle–particle pair correlation function at contact

$$g_4(\nu) = \frac{1 - 7\nu/16}{(1 - \nu)^2} - \frac{\nu^3/16}{8(1 - \nu)^4}, \tag{3.1}$$

can be derived theoretically from a low density expansion [4, 8, 11, 16, 19], and it depends on the volume fraction only. The first term of g_4 is denoted as $g_2 =$

$(1 - 7\nu/16)/(1 - \nu)^2$, on which the polydisperse equations are based (see below). The particle–particle correlation function as a function of the distance is obtained from the simulations by averaging over several snapshots, normalized to the value $g(r \gg 2a) = 1$ for long distances [20]. At densities around $\nu_c \approx 0.7$, a disorder-order transition is evidenced, where the ordered regime for $\nu > \nu_c$ is not described by Eq. (3.1). For some data and a more detailed discussion of $g(r)$, see Refs. 16 and 20.

For *bidisperse situations*, the correlation functions g_{11}, g_{22} and g_{12} are different for different species combinations [20]. The mixed correlation functions [10, 20], are here expressed in terms of $A_{1,2}$, R and ν:

$$g_{11} = \frac{1 - \nu\left(1 - \frac{9}{16}\frac{A_1}{A_2}\right)}{(1-\nu)^2}, \tag{3.2}$$

$$g_{22} = \frac{1 - \nu\left(1 - \frac{9}{16R}\frac{A_1}{A_2}\right)}{(1-\nu)^2}, \tag{3.3}$$

$$g_{12} = \frac{1 - \nu\left(1 - \frac{9}{8(1+R)}\frac{A_1}{A_2}\right)}{(1-\nu)^2}. \tag{3.4}$$

Note that all g_{ij} are identical to $g_2(\nu)$ in the monodisperse case with $R = 1$ and $A_1 = A_2 = 1$. Within statistical error, agreement between the theoretical predictions and simulation results is obtained [20].

The particle correlation functions from *polydisperse mixtures* are smooth functions with much less variety in magnitude than in the mono- and bidisperse situations. Interestingly, they resemble the distribution function of a gas or liquid with a smooth interaction potential [8]. Note that there is no indication of long range order even for the highest densities if the size distribution (bi- or polydisperse) is sufficiently wide, as discussed later in Sec. 4.

3.2. *Stress and the equation of state*

The stress tensor, defined for a test-volume V, has two contributions, one from the collisions and the other from the translational motion of the particles. Using a and b as indices for the cartesian coordinates, one has the components of the stress tensor

$$\sigma^{ab} = \frac{1}{V}\left[\sum_i m_i v_i^a v_i^b - \frac{1}{\Delta t}\sum_n \sum_{j=1,2} \Delta p_j^a \ell_j^b\right], \tag{3.5}$$

with ℓ_j^b, the components of the vector from the center-of-mass of the two colliding particles j to their contact points at collision n, where the momentum Δp_j^a is exchanged. The sum in the left term runs over all particles i, the first sum in the right term runs over all collisions n occuring in the time-interval Δt, and the second sum in the right term concerns the collision partners of collision n — in any case the corresponding particles must be inside the averaging volume V. The mean pressure $p = (\sigma_1 + \sigma_2)/2$, with the eigenvalues σ_1 and σ_2 of the stress tensor, is now obtained from the simulations for different volume fractions [20].

In the monodisperse system, we obtain crystallization around $\nu_c = 0.7$, and the data clearly deviate from $P_0 = pV/E - 1 = (1+r)\nu g_4(\nu)$, i.e. the pressure is strongly reduced due to crystallization and, thus, enhanced free volume. The monodisperse data diverge at the maximum packing fraction $\nu_{\rm max}^{\rm mono} = \pi/(2\sqrt{3})$ in 2D. The deviations of the polydisperse simulations from P_0 increase with the width of $w(a)$ and with increasing volume fraction. Note that there exists a deviation already for small $\nu \to 0$.

3.3. *Mixture pressure*

A more elaborate calculation in the style of Jenkins and Mancini [10], leads to the partial translational pressures $p_i^t = n_i E/V$ for the species i and to the collisional pressures $p_{ij}^c = \pi N_i N_j g_{ij} a_{ij}^2 (1 + r_{ij})T/(4V^2)$, with $a_{ij} = a_i + a_j$. In elastic simulations, the species temperatures are equal, so that the corresponding correction term can be dropped. Thus, the global mixture pressure is

$$\begin{aligned} p^{\rm m} &= p_1^t + p_2^t + p_{11}^c + 2p_{12}^c + p_{22}^c \\ &= \frac{E}{V}[1 + (1+r)\nu g_{\mathcal{A}}(\nu)] , \end{aligned} \tag{3.6}$$

with the effective correlation function

$$g_{\mathcal{A}}(\nu) = \frac{(1+\mathcal{A}) - \nu(1-\mathcal{A}/8)}{2(1-\nu)^2} , \tag{3.7}$$

dependent on $\mathcal{A} = \langle a \rangle^2 / \langle a^2 \rangle$. Note that $\mathcal{A}$ is well defined for any size distribution function, so that Eq. (3.7) can also be applied to polydisperse situations. However, we remark that Eq. (3.7) is not appropriate for very wide size distribution functions [24], when higher order corrections have to be taken into account. Using the effective particle correlations, one can define the dimensionless pressure $P_2(\nu) = \frac{p^{\rm m}V}{E} - 1 = (1+r)\nu g_{\mathcal{A}}(\nu)$, and compare it with simulation results P. An almost perfect agreement between P and $P_2(\nu)$ is obtained for $\nu < 0.4$ and even up to larger $\nu \approx 0.65$, the difference is always less than about two percent. Note that the quality factors for *all* simulations collapse and that the quality is perfect (within less than 0.5 percent for all $\nu < 0.65$) if $P_2(\nu)$ is multiplied by the empirical function $1 - \nu^4/10$, as fitted to the quality factor P/P_2. Thus, based on our simulation results, we propose the corrected, non-dimensional mixture pressure

$$P_4(\nu) = \frac{p^{\rm m}V}{E} - 1 = (1+r)\nu g_{\mathcal{A}}(\nu)[1 - a_g \nu^4] , \tag{3.8}$$

with the empirical constant $a_g \approx 0.1$, for all $\nu < 0.65$. Only in the monodisperse case, do we use $P_4(\nu) = (1+r)\nu g_4(\nu)$, since this form is of higher order in the expansion around small ν and needs no empirical correction. For larger ν the excluded volume effect becomes more and more important, leading to a divergence of P. In the high density regime, the behavior is strongly dependent on the width of the size distribution function.

3.4. *A global equation of state (monodisperse)*

Based on both kinetic theory for the disordered regime and hard sphere simulations, a mixture pressure was proposed in a simple form, dependent only on the width of the size distribution $\mathcal{A}$. For higher densities and small width, $1 - \mathcal{A}$, an ordered situation is obtained, whereas no order is evidenced for wide size distributions. The presence of an ordered situation is discussed below, in Sec. 4. The dense situations, however, deviate from the prediction in Eq. (3.8) due to the excluded volume in a dense packing.

The equation of state in the dense phase can be computed by means of a free volume theory [3, 12, 29], which leads in 2D to the reduced pressure $P_{\rm sf} = 1/(\sqrt{\nu_{\rm max}/\nu} - 1)$ [16], with the maximum volume fraction $\nu_{\rm max}$. Also slightly different functional forms $P_{\rm pm} = (\nu_{\rm max} + \nu)/(\nu_{\rm max} - \nu) - 1$ [6], and $P_{\rm fv} = 2\nu_{\rm max}/(\nu_{\rm max} - \nu)$ [20], do not lead to much better agreement. Note that the above functions are identical, in leading order for small $\nu_{\rm max} - \nu$, so that we use $P_{\rm fv}$. Based on our numerical data, we propose the corrected high density pressure

$$P_{\rm dense} = \frac{2\nu_{\rm max}}{\nu_{\rm max} - \nu} h_3(\nu_{\rm max} - \nu) - 1 \,, \tag{3.9}$$

where $h_3(x) = [1 + c_1 x + c_3 x^3]$ is a fit function with $c_1 = -0.04$ and $c_3 = 3.25$. To our knowledge, no theory exists, which combines the disordered and the ordered regime [14, 26]. Therefore, we propose a global equation of state

$$Q = P_4 + m(\nu)[P_{\rm dense} - P_4] \,, \tag{3.10}$$

with an empirical merging function $m(\nu) = [1 + \exp(-(\nu - \nu_c)/m_0)]^{-1}$ which selects P_4 for $\nu \ll \nu_c$ and $P_{\rm dense}$ for $\nu \gg \nu_c$ with the width of the transition m_0.

When the fit parameters $\nu_c = 0.701$ and $m_0 \approx 0.009$ are used, quantitative agreement between Q and the simulation results is achieved within about 1.5 percent (the agreement outside the transition region is much better than 0.2 percent). However, also a simpler version Q_0 without numerical corrections leads to reasonable agreement when $m_0 = 0.012$ and $\nu_c = 0.700$ is used. In the transition region, the function Q_0 has no negative slope but is continuous and differentiable, so that it allows for an easy and compact numerical integration of the pressure. We selected the parameters for Q_0 as a compromise between the quality of the fit on the one hand and the treatability of the function on the other hand.

The global equation of state can also be used for mixtures of particles with different sizes (data not shown here) if $\nu_{\rm max}$ is adjusted accordingly. For a more detailed comparison of the global equation of state with simulation data, see Refs. 16 and 20.

3.5. *Inhomogeneous systems*

Most of the data were obtained from elastic, homogeneous simulations. A few tests showed that also weakly inelastic situations lead to equivalent results as long as they are homogeneous. Data obtained from inhomogeneous systems which are not, in general, in global equilibrium due to *strong dissipation* cannot be compared to the results presented. Work is in progress to check how far subsystems in local equilibrium (or at least close to it) can be understood by the predictions valid for homogeneous, almost elastic systems. Note that the averaging in subsystems over short times introduces large statistical fluctuations, so that conclusions about one realization are not necessarily helpful. Furthermore, averaging is not possible over different realizations, since inhomogenous situations are incomparable from one run to the next and possibly even the global results differ.

4. How Much Disorder Is Necessary to Avoid Order?

In order to study the deviation of the pressure p in the simulations from the theoretical value P_4 it is helpful to examine the quality factor $q = P/P_4$ more closely. In Fig. 1, q is displayed for various densities and dimensionless distribution widths of bidisperse systems with $n_1 = 0.744$ and $R \in [0.5; 1]$. In the contour plot (left), q is plotted against the density, ν, and the width of the size distribution, $\sqrt{1-\mathcal{A}}$.

A valley in the q-surface corresponds to a reduction in pressure which we relate to order (which leads to a larger free path and thus to a reduced collision rate). The steep increase in the q-surface is due to the excluded volume of the particles. For every value of $\mathcal{A}$, one can obtain a specific value of the maximum packing density $\nu_{\max}$ by extrapolation.

It can be evidenced that P deviates from P_4 by less than 1% at densities $\nu < 0.65$ regardless of the type and width of the size distribution function. For smaller $\mathcal{A}$-values, the prediction from kinetic theory is correct up to even larger densities.

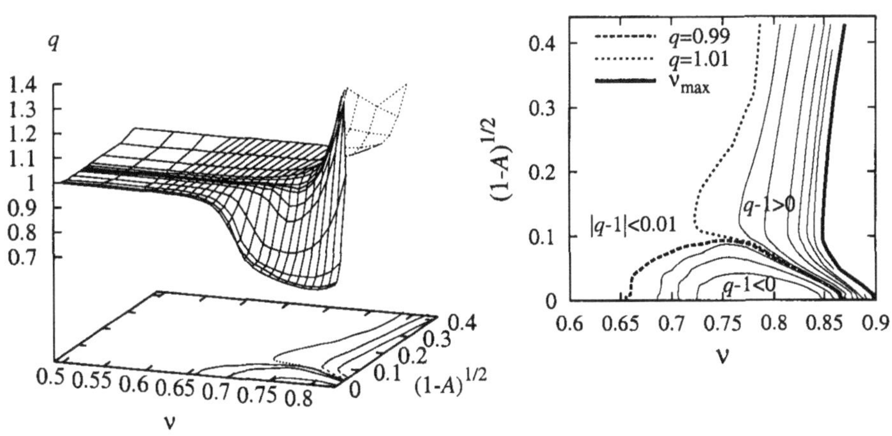

Fig. 1. Quality factor $q = P/P_4$, from simulations with different R- and ν-values, plotted against the volume fraction ν and $\sqrt{1-\mathcal{A}}$. The base of the left diagram and the right plot show the contour plot. Shown are the iso-lines for $q = 0.99$ (dashed), $q = 1.01$ (dotted) and $q = 0.75$, 0.85, 0.95, 1.05, 1.15, 1.5, 2, 3, and 5 (thin solid lines), as well as the maximum density (thick solid line). The maximum of the contour $q = 0.99$ is $\mathcal{A} = 1 - 0.093643^2 = 0.99123$, $\nu = 0.75199$, $R = 0.8143$.

In the domain of agreement between theory and simulations, we assume that the system is in the disordered, fluid state. Consequently, the system is ordered, if a pressure drop occurs, i.e. if $q - 1 < -0.01$. The increase in pressure is related to the dense packing of the system and thus to the solid state, i.e. a solid has a much lower compressibility than a fluid. Note that the corresponding plots for other polydisperse size distributions look very similar.

For broad enough size distributions, $\mathcal{A} < 0.99$, the disorder caused by the different particle sizes is sufficient to suppress the formation of an ordered structure. For our bidisperse systems with $n_1 = 0.744$, global disorder is predicted for $R < 0.814$. As a consequence, the pressure drop disappears and the validity of Eq. (3.8) is extended to densities up to $\nu \approx 0.75$ for very broad size distributions. For densities above $\nu \approx 0.75$ the pressure starts to increase strongly, exceeds the predicted value, and finally diverges when ν approaches $\nu_{\rm max}$.

The maximum packing density depends on the size distribution function and on the rate used to increase the density. Rapidly growing the particles increases the probability of trapping defects in the structure which results in a reduced $\nu_{\rm max}$. Here, we changed the density rather slowly, corresponding to the adiabatic limit, however, the rates of change are always finite in a numerical simulation.

5. Summary and Outlook

A *global equation of state* for almost elastic 2D granular gases with mono-, bi- and polydisperse size distributions was presented. For low and intermediate densities, the equation of state can be written in a closed form which only contains the width-correction $\mathcal{A} \le 1$ of the size-distribution function. A small, empirical correction can be added to the theories to raise the quality even further for medium densities. At high densities, the maximum packing density was obtained by extrapolation of the numerical data and a functional form was fitted to the high density, solid regime based on free volume arguments. Both the liquid and the solid regime were connected via an empirical merging function to give a *global equation of state* for *all* densities.

The simulations and the theories presented here were applied to homogeneous systems. The range of applicability may be reduced by the fact that dissipation can lead to strong inhomogeneities in density, temperature and pressure. In a freely cooling system, for example, clustering leads to all densities between $\nu \approx 0$ and $\nu \approx \nu_{\rm max}$ [17, 19]. The proposed *global equation of state* is a necessary first step to account for such strong inhomogeneities with very high densities above which the low-density theory fails. Since the results are based on a limited amount of data, it has to be checked whether it still makes sense in the extreme cases of narrow $w(a)$, where crystallization effects are rather strong, and for extremely broad, possibly algebraic $w(a)$, where $\mathcal{A}$ is not defined. What also remains to be done is to find similar expressions not only for pressure and energy dissipation rate but also for viscosity and heat-conductivity and to extend the present results to three dimensions.

Acknowledgments

We acknowledge the support by the Deutsche Forschungsgemeinschaft (DFG) and helpful discussions with B. Arnarson, D. Hong, J. Jenkins, M. Louge, B. Meerson, A. Santos and O. Strauß.

References

[1] Arnarson, B. Ö. and Willits, J. T., Thermal diffusion in binary mixtures of smooth, nearly elastic spheres with and without gravity, *Phys. Fluids* **10**, 1324–1328 (1998).

[2] Brey, J. J., Dufty, J. W. and Santos, A., Dissipative dynamics for hard spheres, *J. Stat. Phys.* **87** (5/6), 1051-1067 (1997).

[3] Buehler, R. J., Wentorf, R. H. Jr., Hirschfelder, J. O. and Curtiss, C. F., The free volume for rigid sphere molecules, *J. Chem. Phys.* **19** (1), 61–71 (1951).

[4] Chapman, S. and Cowling, T. G., *The Mathematical Theory of Nonuniform Gases* (Cambridge University Press, London, 1960).

[5] Dickinson, E., Molecular dynamics simulation of hard-disc mixtures. The equation of state, *Molecular Physics* **33** (5), 1463–1478 (1977).

[6] Grossman, E. L., Zhou, T. and Ben-Naim, E., Towards granular hydrodynamics in two-dimensions, *Phys. Rev.* **E55**, 4200 (1997).

[7] Haff, P. K., Grain flow as a fluid-mechanical phenomenon, *J. Fluid Mech.* **134**, 401–430 (1983).

[8] Hansen, J. P. and McDonald, I. R., *Theory of Simple Liquids* (Academic Press Limited, London, 1986).

[9] Herrmann, H. J., Hovi, J.-P. and Luding, S., eds., *Physics of Dry Granular Media*, NATO ASI Series E 350 (Kluwer Academic Publishers, Dordrecht, 1998).

[10] Jenkins, J. T. and Mancini, F., Balance laws and constitutive relations for plane flows of a dense, binary mixture of smooth, nearly elastic, circular discs, *J. Appl. Mech.* **54**, 27–34 (1987).

[11] Jenkins J. T. and Richman, M. W., Kinetic theory for plane flows of a dense gas of identical, rough, inelastic, circular disks, *Phys. Fluids* **28**, 3485–3494 (1985).

[12] Kirkwood, J. G., Maun, E. K. and Alder, B. J., Radial distribution functions and the equation of state of a fluid composed of rigid spherical molecules, *J. Chem. Phys.* **18** (8), 1040–1047 (1950).

[13] Landau, L. D. and Lifschitz, E. M., *Physikalische Kinetik* (Akademie Verlag Berlin, Berlin, 1986).

[14] Löwen, H., Fun with hard spheres, in *Statistical Physics and Spatial Statistics*, Mecke, K. R. and Stoyan, D., eds. (Springer, Berlin, 2000).

[15] Lubachevsky, B. D., How to simulate billards and similar systems, *J. Comp. Phys.* **94** (2), 255 (1991).

[16] Luding, S., Global equation of state of two-dimensional hard sphere systems, *Phys. Rev.* **E63**, 042201 (2001).

[17] Luding, S. and Herrmann, H. J., Cluster growth in freely cooling granular media, *Chaos* **9** (3), 673–681 (1999).

[18] Luding, S., Huthmann, M., McNamara, S. and Zippelius, A., Homogeneous cooling of rough dissipative particles: Theory and simulations, *Phys. Rev.* **E58**, 3416–3425 (1998).

[19] Luding, S. and McNamara, S., How to handle the inelastic collapse of a dissipative hard-sphere gas with the TC model, *Granular Matter* **1** (3), 113–128 (1998), cond-mat/9810009.

[20] Luding, S. and Strauß, O., The equation of state of polydisperse granular gases, in *Granular Gases*, Pöschel, T. and Luding, S., eds. (Springer, Berlin, 2001), pp. 389–409, cond-mat/0103015.
[21] Luding, S., Strauß, O. and McNamara, S., Segregation of polydisperse granular media in the presence of a temperature gradient, in *IUTAM Symposium on Segregation in Granular Flows*, Rosato, T., ed. (Kluwer Academic Publishers, Dordrecht, 1999), pp. 297–304.
[22] McNamara, S. and Luding, S., A simple method to mix granular materials, in *IUTAM Symposium on Segregation in Granular Flows*, Rosato, T., ed. (Kluwer Academic Publishers, Dordrecht, 1999), pp. 305–310.
[23] Pöschel, T. and Luding, S., eds., *Granular Gases*, Lecture Notes in Physics 564 (Springer, Berlin, 2001).
[24] Santos, A., Yuste, S. B. and Haro, M. L. D., Equation of state of a multicomponent *d*-dimensional hard sphere fluid, *Mol. Phys.* **96**, 1–5 (1999).
[25] Sela, N. and Goldhirsch, I., Hydrodynamic equations for rapid flows of smooth inelastic spheres, to burnett order, *J. Fluid Mech.* **361**, 41–74 (1998).
[26] Sengupta, S., Nielaba, P. and Binder, K., Elastic moduli, dislocation core energy, and melting of hard disks in two dimensions, *Phys. Rev.* **E61**, 6294 (2000).
[27] Sunthar, P. and Kumaran, V., Temperature scaling in a dense vibrofluidized granular material, *Phys. Rev.* **E60** (5), 1951–1956 (1999).
[28] Willits, J. T. and Arnarson, B. Ö., Kinetic theory of a binary mixture of nearly elastic disks, *Phys. Fluids* **11** (10), 3116–3122 (1999).
[29] Wood, W. W., Note on the free volume equation of state for hard spheres, *J. Chem. Phys.* **20**, 1334 (1952).
[30] Ziman, J. M., *Models of Disorder* (Cambridge University Press, Cambridge, 1979).

CHAPTER 11

Compaction and Density Fluctuations in Vibrated Granular Media

A. C. B. BARNUM, ARIF OZBAY and E. R. NOWAK*

Department of Physics and Astronomy, University of Delaware, Newark, Delaware 19716, USA

**nowak@udel.edu*

We report measurements of the density of a vibrated granular material as a function of time or taps. The material studied consists of monodisperse spherical glass beads confined to a long, thin cylindrical tube. Changes in vibration intensity are used to induce transitions between two steady state densities that depend on the intensity of the vibrations. We find a complex time evolution similar to previous work on the irreversible relaxation from a loose state toward a steady state. In addition, frequency dependent third order moments of the density fluctuations are measured. The data indicate a coupling between large variations in density on one time scale and noise power over a broad range of higher-frequency scales.

Keywords: Granular compaction; packing; noise; nonlinear dynamics; nonequilibrium thermodynamics.

1. Introduction

Recently, there has been renewed interest in applying the concepts of statistical mechanics to formulate a thermodynamic description of dense granular media [2, 4, 9]. In order to realize such a theory, it is necessary to know the phenomenology of how granular media respond to non-thermal excitations. Granular materials are not ordinary thermal systems since the thermal kinetic energy $\sim k_{\rm B}T$ per particle is negligible and gravity plays an unusually important role. Indeed, granular matter shares a number of features exhibited by glasses; such as slow relaxation, hysteresis, and a non-trivial fluctuation spectrum [6, 7]. Analogies between them were pointed out some time ago [5, 13]. One of the main differences that distinguishes granular matter from glasses is the non-thermal manner in which energy is supplied to the grains (vibration, shearing) and lost by them through inelastic collisions. In dynamical processes of granular media a crucial role is played by disorder and geometric frustration. This property can prevent an 'equilibrium' state from being reached over experimentally accessible time scales. In this paper, we address the question of how the packing density of such a granular system relaxes

First published in *Advances in Complex Systems*, Vol. 4, No. 4 (2001), pp. 389–396.

from an initial perturbed configuration toward its steady state density for a given vibration intensity. Previous experiments [8] focused on the irreversible relaxation process from a 'loose' configuration whereas our investigations concentrate on relaxation between two closely separated densities (the small perturbation limit). As in glasses, new methods are needed to investigate the properties of a dense state that lacks conventional order. Here, we import and apply some of the statistical tools developed to study electronic noise in spin glasses into the granular materials field. We are particularly interested in investigating whether fast relaxation mechanisms are affected by processes with much longer timescales. To study this we employ third and fourth moment correlations of the density fluctuations about a steady state density.

2. Experimental Procedure

Experiments were performed using 2 mm diameter, spherical soda lime glass beads. The beads were placed in a 1 m tall glass tube (22 mm inner diameter) mounted vertically on an electromagnetic vibration exciter. As in previous experiments [12], vertical vibrations were applied in the form of discrete 'taps' by driving the exciter with a single cycle of a 30 Hz sine wave. The vibration intensity is parameterized by $\Gamma \equiv a/g$, where a is the peak acceleration during a tap and g is the gravitational constant. The time interval between taps (about 0.5 s) was sufficiently long to ensure that the system was completely at rest before measurements of the density were taken. Measurements of the packing density were made using a capacitive technique described elsewhere [8]. Three capacitors allowed for independent measurements of the density relaxation and fluctuations near the bottom, middle and upper regions of the tube. A separate tube filled with identical beads, but which was not vibrated, was used to monitor and correct for any environmental changes or drifting in the instrumentation electronics.

The steady-state density as a function of Γ is determined by measuring the irreversible-reversible curve [12], as shown in Fig. 1. This was done by first 'fluffing' the system to a uniformly loose state by blowing dry nitrogen gas from the bottom of the tube. The irreversible branch of the compaction was determined by successively incrementing Γ by 0.6 and measuring the packing density after 10^4 taps at each value of Γ, up to $\Gamma = 8$. The reversible behavior was mapped out in a similar way except that Γ was decremented by 0.6. Subsequent changes to Γ result in the packing fraction moving reproducibly along the reversible branch and hence, it represents the steady-state density of the system as a function of the vibration intensity. For sufficiently low $\Gamma(< 2)$, the timescale for reaching the steady-state density will exceed 10^4 taps, and the system may get trapped in a low-density metastable state. This is a plausible explanation for the experimentally observed flattening of the reversible branch at the lowest Γ. The irreversible-reversible curve shown in Fig. 1 was determined from an average over the three sections of the tube. Although we have experimentally determined that the capacitance is a linear function of the

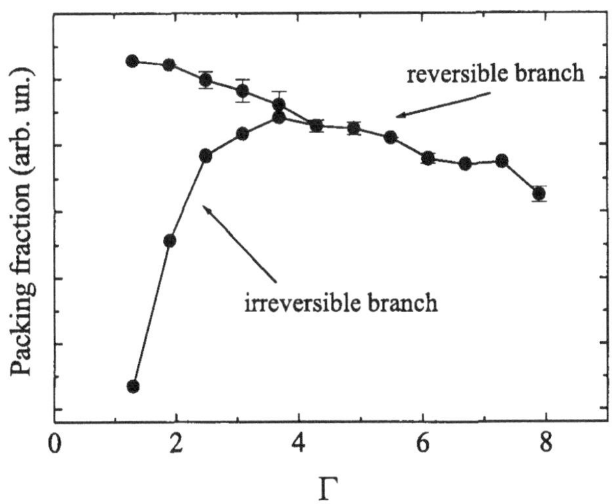

Fig. 1. The irreversible-reversible curve for 2 mm diameter spherical glass beads. The packing fraction represents an average over the bottom, middle and top regions of the tube and it was recorded after 10^4 taps at each vibration intensity Γ. The reversible branch is an average over several sweeps in Γ and the error bars represent the standard deviation.

packing fraction, a calibration curve relating the measured capacitance values to the packing fraction was not available in this case. Consequently, all the density data is shown having arbitrary units. Since our analysis concerns trends, this will not be a limitation. Nevertheless, we expect the maximum packing fraction in Fig. 1 to be at or above the random close packed limit (0.64) as previously observed [12].

Unlike the relaxation from a loosely packed configuration [8], the density relaxation can be quite small between two steady-state densities closely separated in Γ. Experimental measurements of such relaxations required a high-precision capacitance bridge (such as the Andeen–Hagerling, Model 2500A) and many ensemble averages (typically 10 to 15) were taken in order to attain a smooth curve that could be used to ascertain the functional dependence of the relaxation on taps. For practical reasons having to do with the duration of the relaxations, detailed investigations focused on intermediate to high vibration intensities. We report results on the density relaxation between two pairs of steady-state densities associated with the following vibration intensities: (4.3, 6.1) and (6.1, 7.9). For each relaxation measurement, the system was returned to a steady-state density corresponding to the chosen initial vibration intensity Γ_{initial}. The same initial steady state density could be obtained reproducibly simply by tapping the system between 10^4 and 10^5 times at Γ_{initial}. Γ was then changed and the packing fraction was measured over 10^3 to 10^4 taps. The relaxations over the three sections of the tube were summed and the process was then repeated. We note that beads near the top of the tube underwent larger changes in packing fraction in comparison to those at the middle and bottom of the tube. The density relaxations of many such identical trials were averaged in order to reduce the statistical variations in the shape of the relaxation curve.

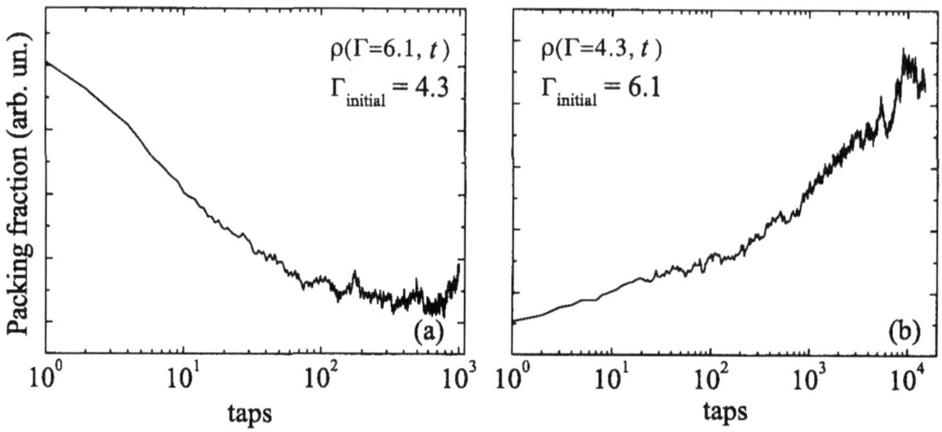

Fig. 2. The time evolution of the density between two steady state densities associated with vibration intensities, $\Gamma = 4.3$ and 6.1, along the reversible branch in Fig. 1. In panel (a) the granular pile relaxes rapidly to a lower steady state density whereas in (b) the relaxation back to the original (higher) density is slower.

3. Results and Discussion

Figure 2(a) shows the density relaxation, $\rho(t)$, from a higher steady-state density ($\Gamma_{\text{initial}} = 4.3$) to lower density ($\Gamma = 6.1$) along the reversible branch in Fig. 1. The opposite case ($\Gamma_{\text{initial}} = 6.1$) is depicted in Fig. 2(b). Here, time t is in units of taps. It is clear that for both cases the relaxation is slow. In that context, the behavior is similar to when the system relaxes from the irreversible regime towards a steady-state density. The relaxation in Fig. 2(b) ($\Gamma < \Gamma_{\text{initial}}$) is slower than in Fig. 2(a), indicating that the timescales for attaining a steady-state density are larger when the system is relaxing towards a higher density. Indeed, the relaxation timescale is roughly a factor of 10 slower depending on the functional form used to fit the data, as discussed below. Similar behavior was observed for another pair of vibration intensities, $\Gamma = 6.1$ and 7.9.

Several functional forms have been proposed to describe the density relaxation in granular media. In previous experiments [8], a logarithmic dependence was found empirically to best describe the relaxation from the irreversible branch. With our experimental data it is difficult to state conclusively what functional form best describes the relaxation along the reversible branch. For example, good fits to the relaxations in Fig. 2 can be obtained by using either a double exponential expression as in Ref. 1, a stretched exponential, and, in most cases, the logarithmic dependence. The data is, however, incompatible with a single exponential relaxation. In view of all the present data which spans different experimental realizations and vibration intensities, we find that the double exponential expression is systematically the best fit to the density relaxation along the reversible branch.

We have also begun detailed investigations of the density fluctuations in the steady state along the reversible branch, which tells us something about the low frequency behavior of the vibrated granular material. Recently, we introduced a non-Gaussian (fourth-order moment) noise analysis technique dubbed the 'second spectrum' [10]. That work showed that the variance in the high-frequency power does not come from independent pulses (which would give frequency independent second spectra) but rather from events on time scales much longer than those corresponding to the high-frequency noise power. Here, we introduce an additional noise analysis technique based on third-order moments which are called three-halves (1.5) spectra [14]. These are cross-spectra of the fluctuating density and the time series of the octave band powers. Stated another way, the Fourier transform of the 1.5 spectrum is the cross correlation function for the density and its own noise power (at a higher frequency band).

The time series of the octave band powers were obtained by repeatedly calculating the ordinary power spectrum, $S(f_1)$, from the measured time series of the density fluctuations, $\rho(t)$. $S(f_1)$ was calculated by taking a 1008-point segment of $\rho(t)$ and discrete Fourier transforming and squaring it. Frequency, f_1, has units of inverse taps. $S(f_1)$ was found to be similar to previous reports [11] and it can be loosely characterized as $1/f$-like. For broadband noise where the noise power does not vary strongly with frequency, we can reduce the numbers to be dealt with and the fractional uncertainty associated with sampling a random signal by summing each power spectrum into seven octaves, O_i with $i = 1, 2, \ldots, 7$. The lowest octave, O_1, is roughly 0.004–0.007 taps^{-1}, consisting of four power spectra points, and the highest octave, O_7, containing 192 points ranges from 0.19 to 0.38 taps^{-1}. This procedure was repeated on subsequent 1008-point segments of $\rho(t)$ to obtain a time series of noise powers in seven frequency bands, $O_i(t_2)$. The timescale t_2 also has units of taps with every increment in t_2 corresponding to a 1008 point segment of $\rho(t)$. The normalized 1.5 spectrum for octave i is defined as

$$S_{1.5}(f_2, f_1) \equiv \frac{F[\rho(t)](F[O_i(t_2)])^*}{\langle O_i(t_2)\rangle},$$

where F is the Fourier transform operator, $f_2 = 1/t_2$, and the asterisk denotes the complex conjugate. The 1.5 spectrum is useful for showing the connection between pulses on one time scale and power on higher-frequency scales.

Figure 3 shows the time series for $\rho(t_2)$ along with the seven octave band noise powers at a steady state density associated with $\Gamma = 5.5$ for a region near the middle of the tube. The most striking feature is the negative correlation between the density and the envelope of the octave noise powers in the latter half of the time series. This behavior carries through all seven octaves. An increase in noise power with decreasing density is consistent with independent measurements that show that the variance increases with decreasing density (increasing Γ) along the reversible branch; for example, see Ref. 11.

Models employing distributions of independent pulses, with the spectral shape of $S(f_1)$ determined by the distributions of pulse widths and heights would fail qualitatively to describe the higher order spectra. In any model in which each pulse

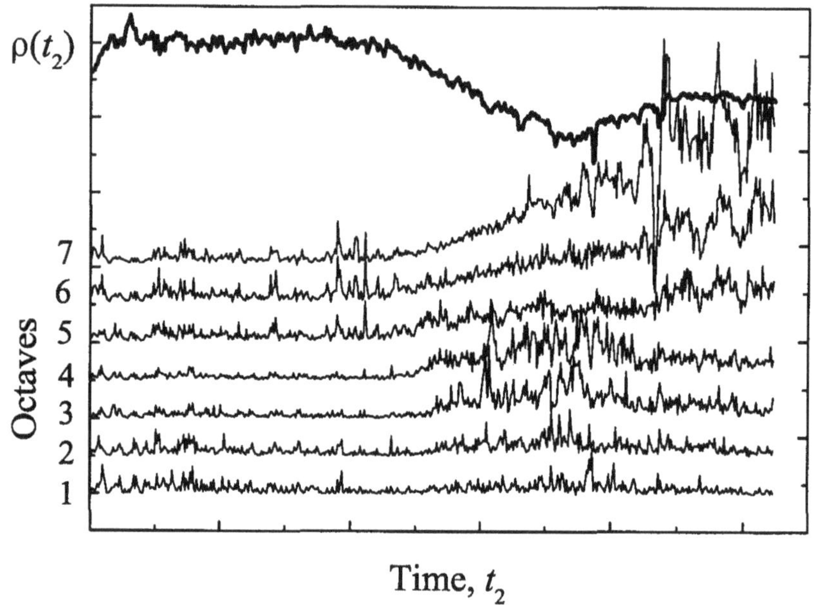

Fig. 3. The variation in density (top curve) and the octave noise powers as a function of time (taps) are plotted for a region near the middle of the tube at a vibration intensity of 5.5 g. The octave powers are obtained from repeated measurements of the power spectrum over time intervals of 1008 taps in the time series of the density. The time series shown here contain approximately 530 points which required in excess of 534,000 taps. The octave time series are typically characterized by alternating periods of large and small noise power. In this figure the large low-frequency variation in density is negatively correlated with an onset of large noise power across a broad range of frequencies (octaves).

has a narrow distribution (on a logarithmic scale) of characteristic rates and pulses occur independently of each other, the second spectrum will be nearly independent of f_2, since the pulse contributing to the noise at frequencies f_1 will not have any characteristic rates $f_2 \ll f_1$. For this set of data the second spectrum is not frequency independent but rather varies approximately as $f_2^{-0.8}$. Likewise, if $S(f_1)$ primarily comes from distinct short-duration pulses, the correlation between low-frequency density fluctuations and low-frequency fluctuations in high frequency noise power will be a strongly decreasing function of f_1/f_2, since the rapid pulses (duration about $1/f_1$) will be uncorrelated with the longer pulses (duration $1/f_2$). It is clear at least from the broad feature in Fig. 3 that this is not the case.

Another question that can be addressed using $S_{1.5}(f_2, f_1)$ is how much of the power at high frequencies in $S(f_1)$ comes from high-frequency structure on low-frequency pulses rather than distinct high-frequency pulses. For the data in Fig. 3, $S_{1.5}(f_2, f_1)$ falls off approximately as f_1^{-1} and $f_2^{-0.7}$. The relatively strong dependence on f_1 indicates that the noise power at high frequencies does not contain a large tail from pulses with duration much longer than $1/f_1$. This too can be seen by inspection of Fig. 3.

Up to now we have discussed the higher order moments of the noise near the middle of the tube. Qualitatively similar behavior was also observed at the top and bottom, particularly with regards to the frequency dependence of $S_{1.5}$. However, the dip in density in Fig. 3 was not evident in the other parts of the tube. When we consider data from three different experiments, a generic feature that emerges from the time series of the octave noise power data is the occurrence of alternating periods of relatively low noise power followed by a burst of higher noise power activity. Occasionally these bursts are correlated with prominent changes in the density, as in Fig. 3. Physically, what may be happening is that the reconfiguration of a small subset (cluster) of beads alters the geometrical frustration acting on its neighboring beads. As a consequence the granular assembly may be more 'free' to explore a greater range of configurations corresponding to different densities (larger noise) at the same vibration intensity. Interestingly, bursts in noise power are also seen in a spin model used to describe granular compaction [3]. In that model, the dynamics of spins on a random graph with finite connectivity show cascades of successive spin-flips. These cascades give rise to sudden changes in density and associated with them are fluctuations of the noise power in different frequency bands that are strongly correlated across many octaves.

4. Conclusions

We have observed that the relaxation between two packing densities on the reversible branch is highly complex. The time evolution is not well described by a single exponential function; instead, two or more time scales may be involved. We have also observed that the relaxation process is slower for the case of relaxing toward a higher density. How these relaxation timescales grow with increasing density is the subject of present investigations. In addition, third-order moments were used to show that large, low-frequency variations in the packing density are associated with corresponding changes in the variation in noise power at higher frequency scales. This property, and the non-trivial frequency dependence of the noise power fluctuations (see Ref. 10), rule out models of granular dynamics that are based on simple superpositions of independently fluctuating entities. Finally, the high frequency fluctuations that give rise to the ordinary power spectrum, $S(f_1)$, are presumably associated with the fast relaxation mechanisms in granular dynamics. If so, it is interesting to speculate how the slow and fast dynamics are coupled. Third and fourth moment correlations are a potential way of addressing this issue.

Acknowledgments

This work was supported by the Department of Physics and Astronomy at the University of Delaware. A. C. B. Barnum gratefully acknowledges the University of Delaware Research Foundation and the Petroleum Research Foundation (35861-G5) for support. We are also grateful to Norbert Mulders for lending us the high-precision capacitance bridge.

References

[1] Barker, G. C. and Mehta, A., *Phys. Rev.* **E47**, 184 (1993).
[2] Barrat, A., Kurchan, J., Loreto, V. and Sellitto, M., *Phys. Rev. Lett.* **85**, 5034 (2000).
[3] Berg, J. and Mehta, A., (to be published).
[4] Edwards, S. F. and Oakeshott, R. B. S., *Physica* **A157**, 1080 (1989).
[5] Edwards, S. F. and Mehta, A., *J. Phys.* **50**, 2489 (1989).
[6] Jaeger, H. M. and Nagel, S. R., *Science* **255**, 1523 (1992).
[7] Jaeger, H. M., Nagel, S. R. and Behringer, R. P., *Rev. Mod. Phys.* **68**, 1259 (1996).
[8] Knight, J. B., Fandrich, C. G., Lau, C. N.. Jaeger, H. M. and Nagel, S. R., *Phys. Rev.* **E51**, 3957 (1995).
[9] Makse, H. A. and Kurchan, J., cond-mat/0107163 (2001).
[10] Nowak, E. R., Grushin, A., Barnum, A. C. B. and Weissman, M. B., *Phys. Rev.* **E63**, 020301(R) (2001).
[11] Nowak, E. R., Knight, J. B., Ben-Naim, E., Jaeger, H. M. and Nagel, S. R., *Phys. Rev.* **E57**, 1971 (1998).
[12] Nowak, E. R., Knight, J. B., Povinelli, M., Jaeger, H. M. and Nagel, S. R., *Powder Technology* **94**, 79 (1997).
[13] Struik, L. C. E., in *Physical Ageing in Amorphous Polymers and Other Materials*, Chap. 7 (Elsevier, Houston, 1978).
[14] Weissman, M. B., *Rev. Mod. Phys.* **60**, 537 (1988).

CHAPTER 12

Kinetic Theory and Hydrodynamics for a Low Density Granular Gas

JAMES W. DUFTY
Department of Physics, University of Florida,
Gainesville, FL 32611, USA

Many features of real granular fluids under rapid flow are exhibited as well by a system of smooth hard spheres with inelastic collisions. For such a system, it is tempting to apply standard methods of kinetic theory and hydrodynamics to calculate properties of interest. The domain of validity for such methods is *a priori* uncertain due to the inelasticity, but recent systematic studies continue to support the utility of kinetic theory and hydrodynamics as both qualitative and quantitative descriptions for many physical states. The basis for kinetic theory and hydrodynamic descriptions is discussed briefly for the special case of a low density gas.

Keywords: Kinetic theory; hydrodynamics; granular gas; inelastic collisions; transport; fluctuations; correlations.

1. Introduction

Granular media in rapid, dilute flow exhibit a surprising similarity to ordinary fluids and the utility of a hydrodynamic description for such conditions has been recognized for many years [15]. This phenomenology has come under scrutiny in recent years with questions about the domain of its validity and the associated constitutive equations appearing in the hydrodynamic equations [13, 16]. Answers to such questions can be found in a more fundamental microscopic description where the tools of nonequilibrium statistical mechanics are available for critical analysis. An intermediate mesoscopic description between statistical mechanics and hydrodynamics is that of kinetic theory, whose applicability to granular matter also poses questions. The objective in this short communication is to provide an example of a precise context in which the relevance of kinetic and hydrodynamic descriptions can be assessed. A more general review of the current status for this problem with extensive references was prepared for this workshop [9] and should be consulted for completeness.

The system considered is a one component gas of N smooth hard spheres at low density. The inelastic collisions are characterized by a normal restitution coefficient

First published in *Advances in Complex Systems*, Vol. 4, No. 4 (2001), pp. 397–406.

$\alpha \leq 1$, where $\alpha = 1$ is the elastic limit. The state of the system at each time is specified by a point in the $6N$-dimensional phase space and the dynamics is given by uniform motion and binary inelastic velocity changes at contact. These are the necessary ingredients to construct a statistical mechanics for this idealized model of granular media. It is straightforward to write the Liouville equation for the probability density and from it obtain the BBGKY hierarchy for the reduced distribution functions [6, 7, 10, 17]. In the next section the condition of low density is exploited to obtain a complete, closed kinetic theory description for the gas from the BBGKY hierarchy. The low density expansion leading to this result is formally the same as that for a gas with elastic collisions. Consequently, it would appear that the kinetic description for the granular gas has the same level of validity as that for normal gases.

The derivation of hydrodynamics from the kinetic theory is discussed in the third section. Hydrodynamics is defined generally as the composition of exact balance equations for the density, energy density (or granular temperature), and the momentum density (or flow field) plus constitutive equations for the associated fluxes and cooling rate. Constitutive equations exist whenever there is a "normal" solution to the kinetic theory. Such a solution can be constructed explicitly for weakly inhomogeneous states, leading to a Navier–Stokes hydrodynamics for the granular gas [5, 11, 12]. The need to go beyond the Navier–Stokes approximation is rare for normal gases, but is not uncommon for many relevant granular gas states. It remains a challenge for granular gas theory to understand the form of constitutive equations for these more general conditions, but this uncertainty should not be interpreted as a signature of the failure for hydrodynamics to apply.

2. Kinetic Theory

Concerns about a kinetic theory description for granular gases are often based on a too restrictive concept of the prerequisites (e.g. the existence of an equilibrium state). It is useful therefore to provide a formal "derivation" of the kinetic description at low density to demonstate the close similarity between normal and granular gases, without unwarranted conceptual prejudices. In this section a small parameter expansion of the reduced distribution functions is shown to provide a formal solution to the entire BBGKY hierarchy, in parallel to the corresponding analysis for normal gases [8]. The analysis is similar to an expansion proposed by Grad for the hard sphere gas with elastic collisions [14]. In this approach, there is no reference to concepts such as "approach to equilibrium," Maxwellian distribution, or "molecular chaos," and the distinction between inelastic or elastic collisions plays no explicit role. Thus, superficially at least, it appears the basis for the low density kinetic theory is the same in both cases.

The s-particle reduced distribution functions, $f^{(s)}(x_1, \ldots, x_s, t)$, obey the BBGKY hierarchy, where $x_i = (\mathbf{q}_i, \mathbf{v}_i)$. A dimensionless form of this hierarchy is obtained by scaling the space and time with the mean free path $\ell \equiv 1/(n\sigma^2)$ and the mean free time $t_0 \equiv \ell/v_0$. Here, n is the density, σ is the hard sphere diameter, and v_0 is some characteristic velocity. Similarly, the reduced distribution functions are scaled with $(n/v_0^3)^s$. The resulting dimensionless BBGKY hierarchy has the form

$$\left(\partial_t + \sum_{i=1}^{s} v_i \cdot \nabla_i - \lambda^2 \sum_{i<j}^{s} \bar{T}(i,j)\right) f^{(s)}(x_1, \cdots, x_s, t)$$

$$= \sum_{i=1}^{s} \int dx_{s+1} \bar{T}(i, s+1) f^{(s+1)}(x_1, \ldots, x_{s+1}, t) . \tag{2.1}$$

The binary scattering operator $\bar{T}(i,j)$ for a pair of particles $\{i,j\}$ is defined by

$$\bar{T}(i,j) = \int d\hat{\sigma} \Theta(\mathbf{g}_{ij} \cdot \hat{\sigma})(\mathbf{g}_{ij} \cdot \hat{\sigma})[\alpha^{-2}\delta(\mathbf{q}_{ij} - \lambda\hat{\sigma}) b_{ij}^{-1} - \delta(\mathbf{q}_{ij} + \lambda\hat{\sigma})] , \tag{2.2}$$

where $\hat{\boldsymbol{\sigma}}$ is a unit vector along $\mathbf{q}_{ij} = \mathbf{q}_i - \mathbf{q}_j$ and $d\hat{\boldsymbol{\sigma}}$ denotes a two-dimensional solid angle integration over the sphere for particles at contact. Also, $\mathbf{g}_{ij} \equiv \mathbf{v}_i - \mathbf{v}_j$ is the relative velocity, and b_{ij}^{-1} is the scattering operator defined for any function $X(\mathbf{v}_i, \mathbf{v}_j)$ by $b_{ij}^{-1} X(\mathbf{v}_i, \mathbf{v}_j) \equiv X(\mathbf{v}_i', \mathbf{v}_j')$. The "restituting" velocities are

$$\mathbf{v}_i' \equiv \mathbf{v}_i - \frac{1+\alpha}{2\alpha}(\mathbf{g}_{ij} \cdot \hat{\sigma})\hat{\sigma} , \qquad \mathbf{v}_j' \equiv \mathbf{v}_j + \frac{1+\alpha}{2\alpha}(\mathbf{g}_{ij} \cdot \hat{\sigma})\hat{\sigma} . \tag{2.3}$$

The α dependence of $\bar{T}(i,j)$ contains all aspects of the inelasticity, and plays no explicit role in the following expansion.

In this dimensionless form the BBGKY hierarchy depends on the single dimensionless parameter $\lambda \equiv \sigma/\ell = n\sigma^3$, the ratio of the "force range" to the mean free path. This parameter is small at low density, suggesting an expansion for a solution to the entire hierarchy as a power series in λ. The dependence on λ occurs explicitly as shown on the left side of Eq. (2.1) and implicitly through the finite separation of the colliding particles in $\bar{T}(i,j)$. The structural features of the expansion in λ are simplest if it is performed at fixed $\bar{T}(i,j)$. The s-particle reduced distribution functions are taken to have the representation

$$f^{(s)}(x_1, \ldots, x_s, t) = f_0^{(s)}(x_1, \ldots, x_s, t) + \lambda^2 f_1^{(s)}(x_1, \ldots, x_s, t) + \cdots . \tag{2.4}$$

It is then readily shown that the hierarchy is solved exactly to order λ^2 in the form

$$f_0^{(s)}(x_1, \ldots, x_s, t) = \prod_{i=1}^{s} f_0^{(1)}(x_i, t) , \tag{2.5}$$

$$f_1^{(s)}(x_1, \ldots, x_s, t) = \sum_{j=1}^{s} \prod_{i \neq j}^{s} f_0^{(1)}(x_i, t) f_1^{(1)}(x_j, t)$$

$$+ \sum_{i<j}^{s} \prod_{k \neq i,j}^{s} f_0^{(1)}(x_k, t) G(x_i, x_j, t) , \tag{2.6}$$

where the expression for $f_1^{(s)}$ holds for $s \geq 2$. Thus, the reduced distribution functions for any number of particles are determined as a sum of products of the single particle functions $f_0^{(1)}(x_1,t)$ and $f_1^{(1)}(x_1,t)$, and pair function $G(x_1,x_2,t)$. These are determined from the set of three fundamental kinetic equations

$$\left(\frac{\partial}{\partial t}+\mathbf{v}_1\cdot\nabla_1\right)f_0^{(1)}(x_1,t)=J(x_1,t|f_0^{(1)}), \tag{2.7}$$

$$\left(\frac{\partial}{\partial t}+\mathbf{v}_1\cdot\nabla_1-I_1+\mathbf{v}_2\cdot\nabla_2-I_2\right)G(x_1,x_2,t)=\bar{T}(1,2)f_0^{(1)}(x_1,t)f_0^{(1)}(x_2,t), \tag{2.8}$$

$$\left(\frac{\partial}{\partial t}+\mathbf{v}_1\cdot\nabla_1-I_1\right)f_1^{(1)}(x_1,t)=\int dx_2\,\bar{T}(1,2)G(x_1,x_2,t). \tag{2.9}$$

Here $J(x_1,\ t|f_0^{(1)})$ is the Boltzmann–Bogoliubov collision operator and I_1, defined over functions of x_1, is its linearized form:

$$J(x_1,t|f_0^{(1)})=\int dx_2\,\bar{T}(1,2)f_0^{(1)}(x_1,t)f_0^{(1)}(x_2,t), \tag{2.10}$$

$$I_1h(x_1)\equiv\int dx_2\,\bar{T}(1,2)\left(f_0^{(1)}(x_1,t)h(x_2)+h(x_1)f_0^{(1)}(x_2,t)\right). \tag{2.11}$$

These low density results, Eqs. (2.7)–(2.9), provide the kinetic theory description for the gas. They are remarkably rich. As expected, the leading order distribution function $f_0^{(1)}$ is the solution to the Boltzmann equation. The two particle correlations are generated from the uncorrelated product of Boltzmann solutions through inelastic binary collisions $\bar{T}(1,2)f_0^{(1)}(x_1,t)f_0^{(1)}(x_2,t)$ on the right-hand side of Eq. (2.8). Finally, corrections to the Boltzmann solution due to correlations are given by a coupling of the distribution function to the correlations in Eq. (2.9) (the so-called "ring" recollision effects). The solutions and implications of these kinetic equations can be quite different for elastic and inelastic collisions. But these differences come from the equations themselves and should not be interpreted as signatures of their failure to apply. For example, at $\alpha=1$ a possible solution for an isolated system is $f_0^{(1)}(x_1,t)\to f_M(v_1)$, $G(x_1,x_2,t)=0=f_1^{(1)}(x_1,t)$, where f_M is the Maxwell–Boltzmann distribution. Equation (2.10) supports f_M because energy is conserved, and $G(x_1,x_2,t)=0$ because $T(1,2)f_M(\mathbf{v}_1)f_M(\mathbf{v}_2)=0$ for the same reason. Since energy conservation no longer holds with $\alpha<1$ it is not surprising that the isolated system does not approach equilibrium, the Maxwellian is not a stationary solution, and that finite correlations exist. Indeed, the extent to which such predicted differences agree with observations from molecular dynamics provides *support* for the kinetic theory, not *limitations* on it as is sometimes implied.

Clearly, the above derivation has not restricted this kinetic description to isolated systems or near-equilibrium states. In fact, the most interesting cases of practical interest are response to boundary conditions and/or external fields. The similarities between normal and granular fluids is closest for such "nonequilibrium" conditions. Too often, properties of granular gases are contrasted only to those of the equilibrium state for normal gases.

It is important to note that practical access to the solutions to the above kinetic equations is possible for a wide range of conditions by direct simulation Monte Carlo

(DSMC) [1]. Much attention has been given to the special "homogeneous cooling state" (HCS) which is a solution to the Boltzmann equation with the scaling form

$$f_0^{(1)}(\mathbf{v},t) = v_0^{-3}(t)n\phi(v/v_0(t)), \qquad v_0^2(t) = 2T(t)/m. \tag{2.12}$$

The form of this HCS distribution is known to good approximation by analytic methods, which have been confirmed by DSMC [17]. The correlations for this state have also been studied in some detail [4] by both analytic and simulation methods. Finally, studies of more complex states (e.g. shear flow) also have been given [2]. In summary, the kinetic theory description appears to describe well a wealth of new phenomena peculiar to inelastic collisions.

3. Hydrodynamics

Consider now a spatially inhomogeneous state, created either by initial preparation or by boundary conditions. The local balance equations for the density $n(\mathbf{r},t)$, granular temperature $T(\mathbf{r},t)$ (or energy density), and flow velocity $\mathbf{U}(\mathbf{r},t)$ follow directly by taking moments of the Boltzmann equation (2.7) with respect to 1, $\mathbf{v}$ and v^2:

$$D_t n + n\nabla\cdot\mathbf{U} = 0, \tag{3.1}$$

$$D_t T + \frac{2}{3nk_B}(P_{ij}\partial_j U_i + \nabla\cdot\mathbf{q}) = -T\zeta, \tag{3.2}$$

$$D_t U_i + (mn)^{-1}\partial_j P_{ij} = 0. \tag{3.3}$$

Here $D_t = \partial_t + \mathbf{U}\cdot\nabla$ is the material derivative, $P_{ij}(\mathbf{r},t)$ is the pressure tensor and $\mathbf{q}(\mathbf{r},t)$ is the heat flux. The form of these balance equations is the same as for fluids with elastic collisions except for the source term on the right-hand side of Eq. (3.2) due to the dissipative collisions, where $\zeta \propto (1-\alpha^2)$ is identified as the cooling rate. The fluxes P_{ij}, $\mathbf{q}$ and the cooling rate ζ are given as explicit low degree moments of the distribution function $f^{(1)}(x_1,t)$:

$$P_{ij} = \int d\mathbf{v} m V_i V_j f(\mathbf{r},\mathbf{v},t), \qquad \mathbf{q} = \int d\mathbf{v}\, \frac{1}{2} m V^2 \mathbf{V} f(\mathbf{r},\mathbf{v},t), \tag{3.4}$$

$$\zeta = (1-\alpha^2)\frac{\beta m\sigma^2}{12nT}\int d\mathbf{v}_1 d\mathbf{v}_2 d\hat{\sigma}\Theta(\hat{\sigma}\cdot\mathbf{g})(\hat{\sigma}\cdot\mathbf{g})^3 f(\mathbf{r}_1,\mathbf{v}_1,t,)f(\mathbf{r}_2,\mathbf{v}_2,t). \tag{3.5}$$

The utility of these balance equations is limited without further specification of P_{ij}, $\mathbf{q}$ and ζ, which, in general, have a complex space and time dependence. However, for a fluid with elastic collisions this dependence "simplifies" on sufficiently large space and time scales, where it is given entirely through a functional dependence on the fields n, T and $\mathbf{U}$. The resulting functional dependencies of P_{ij} and $\mathbf{q}$ on these fields are called constitutive equations and their discovery can be a difficult many-body problem. The above balance equations, together with the constitutive equations, become a closed set of equations for n, T and $\mathbf{U}$ called *hydrodynamic equations*. This is the most general and abstract notion of hydrodynamics, which encompasses both the Navier–Stokes form for small spatial gradients and more general forms for nonlinear rheological transport. The primary feature of a hydrodynamic description is the reduction of the description from many microscopic degrees of freedom to a set of equations for only five local fields.

The critical problem for a hydrodynamic description is therefore to determine the existence and form of the constitutive equations. It is clear from Eqs. (3.4) and (3.5) that they can be obtained if the Boltzmann equation admits a "normal" solution, whose space and time dependence occurs entirely through its functional dependence on the fields:

$$f(\mathbf{r}, \mathbf{v}, t) = F(\mathbf{v}|n, T, \mathbf{U}) . \tag{3.6}$$

The fluxes and cooling rate then inherit this space and time dependence and become constitutive equations. The space and time dependence of the fields follows from the solution to the resulting hydrodynamic equations to complete the self-consistent description of F. An example is given by the HCS distribution in Eq. (2.12), where there is no space dependence and all of the time dependence occurs through $T(t)$. The latter is determined from the hydrodynamic equation (3.2), which reduces to $\partial_t T = -T\zeta$. The use of Eq. (2.12) in Eq. (3.5) gives the constitutive equation $\zeta = \zeta_0 T^{1/2}$, where ζ_0 is a constant.

For gases with elastic collisions the prototypical hydrodynamics is that of the Navier–Stokes equations. The corresponding constitutive equations, Newton's viscosity law and Fourier's heat law, follow from a normal solution to the Boltzmann equation obtained from an expansion in the spatial gradients. The reference state is the *local* Maxwellian whose parameters are the exact density, temperature and flow velocity for the nonequilibrium state being described. Deviations from this reference state are proportional to the spatial gradients of the temperature and flow field. The systematic expansion for the normal solution to the Boltzmann equation is generated by the Chapman–Enskog method. There have been three main objections/reservations/concerns regarding application of this method for granular gases: (1) the absence of an equilibrium state as the basis for the local Maxwellian reference state, (2) the inherent time dependence of any reference state due to collisional cooling, and (3) the inclusion of the energy (temperature) as a hydrodynamic field when it is not associated with a conserved density and does not have a time scale solely characterized by the degree of spatial inhomogeneity.

The first two concerns are primarily technical rather than conceptual issues that can be answered by direct application of the Chapman–Enskog method to see if it indeed generates a normal solution to the granular Boltzmann equation. Consider a state for which the spatial variations of n, T and $\mathbf{U}$ are small on the scale of the mean free path, $\ell\nabla \ln n \ll 1$, where $\ell = 1/n\sigma^2$ is the mean free path. Then, it is expected that the functional dependence of the normal solution on the hydrodynamic fields can be made local in space through a Taylor series expansion about a point $\mathbf{r}$ for which the distribution function is being evaluated. Let ϵ denote a formal small "uniformity" parameter measuring the small spatial gradients in the fields. It is worth emphasizing that there is no restriction on the cooling rate ζ; only the spatial gradients are being ordered by the uniformity parameter. The distribution function, collision operator, and time derivatives are given by the representations

$$\begin{aligned} F &= F^{(0)} + \epsilon F^{(1)} + \cdots , \\ J_E &= J^{(0)} + \epsilon J^{(1)} + \cdots , \\ \partial_t &= \partial_t^{(0)} + \epsilon \partial_t^{(1)} + \cdots . \end{aligned} \tag{3.7}$$

The coefficients in the time derivative expansion are identified from the balance equations with a similar expansion for P_{ij}, $\mathbf{q}$ and ζ generated through their definitions (3.4) and (3.5) as functionals of F. As for elastic collisions the leading term $F^{(0)}$ is further constrained to have the same moments with respect to 1, v^2 and $\mathbf{v}$ as the full distribution F. This assures that the hydrodynamic fields occuring in $F^{(0)}$ are exactly those for the nonequilibrium state, and hence $F^{(0)}$ is normal. To zeroth order in ϵ the Boltzmann kinetic equation becomes

$$\frac{1}{2}\zeta^{(0)} \nabla_V \cdot \left(\mathbf{V} F^{(0)}\right) = J^{(0)}\left[F^{(0)}, F^{(0)}\right], \tag{3.8}$$

where $\mathbf{V} = \mathbf{v} - \mathbf{U}$.

The first concern above regarding the reference distribution now can be addressed. The distribution $F^{(0)}$ is not free to be chosen, but rather is determined by the kinetic equation itself. For the granular gas it is *not* the local Maxwellian, but rather the local (normal) HCS solution (i.e. Eq. (2.12) with the density, temperature and flow field replaced by their nonequlibrium values). There is no *a priori* requirement of an equilibrium state for the Chapman–Enskog method to apply, and in fact early applications based on a local equilibrium state are inconsistent and lead to incorrect transport coefficients even for weak dissipation. The concept of "approach to equilibrium" is no longer relevant for granular gases.

Mathematically, the changes in this method for granular gases arise from the fact that the time derivative of the temperature does not vanish to lowest order in ϵ, as it does for a gas with elastic collisions. The reference state $F^{(0)}$ incorporates this zeroth order time dependence of the temperature even for strong dissipation. This is the origin of the second concern noted above. However, since $F^{(0)}$ is normal, it necessarily has the exact time dependence of all hydrodynamic fields. The only difference for granular gases is the introduction of a new time scale $1/\zeta^{(0)}$ in the reference state. There is nothing *a priori* inconsistent with a description of slow spatial decay towards a time dependent reference state. In fact, this is an important feature of the Chapman–Enskog scheme for both granular and normal fluids alike — use of a time independent reference state would limit the derivation to only *linear* hydrodynamics.

Implementation of the Chapman–Enskog method to the first order in ϵ is now straightforward and has been carried out in detail and without approximation recently for the Boltzmann equation [5] (and for its dense fluid generalization, the Enskog equation [11]); the case of a two component mixture is considered in

Ref. 12. The constitutive equations for the one component fluid found to this order are

$$P_{ij} \to p\delta_{ij} - \eta\left(\partial_j U_i + \partial_i U_j - \frac{2}{3}\delta_{ij}\nabla\cdot\mathbf{U}\right) - \gamma\delta_{ij}\nabla\cdot\mathbf{U}\,, \tag{3.9}$$

$$\mathbf{q} \to -\kappa\nabla T - \mu\nabla n \tag{3.10}$$

(the bulk viscosity γ vanishes at low density).

The form of the pressure tensor is the same as that for fluids with elastic collisions, where $\eta(\alpha)$ is the shear viscosity depending on the restitution coefficient. The heat flux is similar to Fourier's law, where $\kappa(\alpha)$ is the thermal conductivity. There is a new transport coefficient $\mu(\alpha)$ coupling the heat flux to a density gradient, which vanishes at $\alpha = 1$. The transport coefficients are given in terms of solutions to inhomogeneous integral equations involving the linearized Boltzmann collision operator. Solubility conditions for the existence of solutions have been proven and approximate solutions in terms of polynomial expansions have been obtained for practical purposes, just as for the case of elastic collisions [11]. Consequently, the transport coefficients are known as explicit functions of α. The Chapman–Enskog method places no explicit restriction on α so the results are not limited to weak dissipation.

It remains to discuss the third concern, the use of the temperature as a hydrodynamic field which is no longer associated with a conserved density. Analysis of the Navier–Stokes approximation described above shows there are two classes of inherent time scales to this hydrodynamic description. The first scales with the wavelength of the phenomena considered and can be made large (long time scales) by considering sufficiently smooth disturbances. This is the case for gases with elastic collisions. For granular gases there is a second time scale in the temperature equation, the inverse cooling rate, which is bounded for a given value of the restitution coefficient. The concern is that this new time scale may be shorter than that required for validity of the normal solution to the kinetic equation.

Qualitatively, a wide class of solutions for spatially inhomogeneous states evolves in two stages. During a short transient period of the order of the mean free time (the kinetic stage), the velocity distribution approaches closely the local distribution $F^{(0)}$ and becomes normal as in Eq. (3.6). Subsequently, on a longer time scale the space and time dependence of the distribution occurs only through the fields that are governed by hydrodynamic equations. This notion of a two stage relaxation is similar to that for gases with elastic collisions and is confirmed for granular gases by DSMC. The question, therefore, is whether the inverse cooling rate is smaller than the mean free time. Certainly, this is the case for weak dissipation since $\zeta^{(0)}$ is proportional to $\left(1-\alpha^2\right)$. For strong dissipation it is necessary to study in detail the spectrum of non-hydrodynamic (kinetic) modes to find the slowest mode and compare it with the inverse cooling rate. This has been done for the diverse examples of Brownian motion, kinetic models and uniform shear flow (see Ref. 9). In all cases it is found that there is a separation of time scales between the slowest kinetic mode and the inverse cooling rate, even at strong dissipation. Based on these examples, it appears that inclusion of the temperature in the hydrodynamic description is justified for a wide range of degrees of dissipation.

The validity of Navier–Stokes hydrodynamics and the dependence of the transport coefficients on the restitution coefficient has been verified in a number of simulations, both DSMC and MD, giving excellent agreement with the predictions from the Chapman–Enskog method. The tests at low density based on the Boltzmann equation have been reviewed recently by Brey and Cubero [3].

The Navier–Stokes equations associated with Eqs. (3.9) and (3.10) are also known as Newtonian hydrodynamics. Although the Chapman–Enskog method can be carried out to second order in ϵ (Burnett order), it is likely that failure of the Navier–Stokes approximation signals more complex non-Newtonian behavior for which other methods to construct the normal solution are required that are not based on a small gradient expansion.

4. Discussion

Two questions have been addressed here: (1) Can kinetic theory provide a valid mesoscopic description of rapid flow (fluidized) granular media? and (2) Can a hydrodynamic description be formulated and justified for a macroscopic description? The origin of the kinetic description for a low density gas has been shown to follow from an expansion that makes no reference as to whether the collisions are elastic or inelastic. The latter affects only the solution to these equations. It appears that many of the questions, concerns and objections raised regarding the validity of a kinetic theory description are removed in this way. Furthermore, kinetic theory provides a powerful tool for analysis and predictions of rapid flow gas dynamics when combined with the method of DSMC for numerical solution. Its full potential for both conceptual and practical questions has not yet been exploited.

An important use of the kinetic theory is to determine if a hydrodynamic description applies and to define its domain of validity. Equations (2.7)–(2.9) leave no room for speculation. For given initial and boundary conditions the solution either approaches a normal form on some space and time scale or it does not. The issue of a hydrodynamic description is now a precise mathematical question. This is particularly important if the hydrodynamic description is more complex than that of Navier–Stokes. Non-Newtonian hydrodynamics is rare or unphysical for simple atomic systems with elastic collisions. In contrast, it is more common for granular gases in steady states where the gradients are strongly correlated to the coefficient of restitution. Kinetic theory is perhaps the only systematic means to determine the constitutive equations in these cases.

It is clear from recent studies that granular media exhibit a wide range of interesting phenomena for which a Navier–Stokes hydrodynamics is an accurate and practical tool (see Ref. 9 for specific examples and references). Although the context here has been limited to low density gases, a similar kinetic theory basis has been developed for Navier–Stokes hydrodynamics in moderately dense fluids [11]. More generally, non-Newtonian behavior puts granular fluids in a class of complex materials with mysterious, as yet unexplained, properties [13, 16]. Kinetic theory and hydrodynamics (in the broader sense) can be expected to provide much of this explanation.

Acknowledgments

This research was supported in part by National Science Foundation grant PHY 9722133.

References

[1] Bird, G. A., *Molecular Gas Dynamics and the Direct Simulation of Gas Flows* (Clarendon, Oxford, 1994).

[2] Brey, J. J., Model kinetic equations for rapid granular flow, in *Dynamics: Models and Kinetic Methods for Non-Equilibrium Many Body Systems*, Karkheck, J., ed. (Kluwer Academic, Boston, 2000).

[3] Brey J. J. and Cubero, D., Hydrodynamic transport coefficients of granular gases, in *Granular Gases*, Poschel, T. and Luding, S., eds. (Springer, New York, 2001).

[4] Brey, J. J., Moreno, F. and Ruiz-Montero, M. J., Spatial correlations in dilute granular flows: A kinetic model study, *Phys. Fluids* **10**, 2965 (1998).

[5] Brey, J. J., Dufty, J. W., Kim, C. S. and Santos, A., Hydrodynamics for granular flow at low density, *Phys. Rev.* **E58**, 4638 (1998).

[6] Brey, J., Dufty, J. W. and Santos, A., Dissipative dynamics for hard spheres, *J. Stat. Phys.* **87**, 1051 (1997).

[7] Dufty, J. W., Statistical mechanics, kinetic theory and hydrodynamics for rapid granular flow, *J. Phys. Condens. Matter* **A12**, 47 (2000).

[8] Dufty, J. W., Lee, M. and Brey, J. J., Kinetic model for pair correlations, Appendix A, *Phys. Rev.* **E51**, 297 (1995); see also Marchetti, M. C., Ph. D. thesis, University of Florida, Appendix D, (1982).

[9] Dufty, J., Kinetic theory and hydrodynamics for rapid granular flow — A perspective, cond-mat/0108444 v1 (2001).

[10] Ernst, M. H., Kinetic theory of granular fluids: Hard and soft inelastic spheres, in *Dynamics: Models and Kinetic Methods for Non-Equilibrium Many Body Systems*, Karkheck, J., ed. (Kluwer Academic, Boston, 2000).

[11] Garzo, V. and Dufty, J., Dense fluid transport for inelastic hard spheres, *Phys. Rev.* **E59**, 5895 (1999).

[12] Garzo, V., and Dufty, J., Hydrodynamics for a granular mixture at low density, *Phys. Fluids* (to appear); cond-mat/0105395 v1 (2001).

[13] Goldhirsh, I., Granular gases — probing the boundaries of hydrodynamics, in *Granular Gases*, Poschel, T. and Luding, S., eds. (Springer, New York, 2001).

[14] Grad, H., Principles of the kinetic theory of gases, in *Encyclopedia of Physics*, Flugge, S., ed. (Springer, New York, 1958).

[15] Haff, P. K., Grain flow as a fluid mechanical phenomenon, *J. Fluid Mech.* **134**, 187 (1983).

[16] Kadanoff, L. P., Built upon sand: Theoretical ideas inspired by the flow of granular materials, *Sand Review* V2.2, 1 (1997); *Rev. Mod. Phys.* **71**, 435 (1999).

[17] van Noije, T. P. C. and Ernst, M. H., Kinetic theory of granular gases, in *Granular Gases*, Poschel, T. and Luding, S., eds. (Springer, New York, 2001).

CHAPTER 13

Surface Granular Flows: Two Related Examples

D. V. KHAKHAR and ASHISH V. ORPE
Department of Chemical Engineering, Indian Institute of Technology – Bombay, Powai, Mumbai 400076, India

J. M. OTTINO
Department of Chemical and Mechanical Engineering, Northwestern University, Evanston, IL 60208, USA

Granular surface flows are common in industrial practice and natural systems, however, theoretical description of such flows is at present incomplete. Two prototype systems involving surface flow are compared: heap formation by pouring at a point and rotating cylinders. Continuum models for analysis of these flows are reviewed, and experimental results for quasi-2D systems are presented. Experimental results in both systems are well described by continuum models.

Keywords: Complex systems; granular flow; rotating cylinder; heap flow; continuum model.

1. Introduction

Granular flows have been the subject of considerable, recent work [7, 11, 16, 26, 31] driven by both technological needs [6, 14] and the recognition that many aspects of the basic physics are poorly understood [9]. Surface flows of granular materials, that is flows confined to a surface layer on a static granular bed, are important in industrial practice and nature. Industrial examples appear in the transportation, processing and storage of materials in systems such as rotary kilns, tumbling mixers, and feeding and discharge of silos. Examples in nature include the formation of sand dunes, lava flow, avalanches, and transport of sediments in rivers. Although considerable progress has been made, theoretical description of surface flows is incomplete at present. Several approaches, based on different assumptions about the physics of the flows, have been proposed [2, 4, 5, 10, 13, 18, 19, 23, 28, 30, 33]. A few experimental studies are also available [1, 3, 8, 12, 15, 17, 19–22, 24, 25, 27–29, 32]. Most work is focussed on two systems: heap flow and rotating cylinder flow shown schematically in Fig. 1.

First published in *Advances in Complex Systems*, Vol. 4, No. 4 (2001), pp. 407–417.

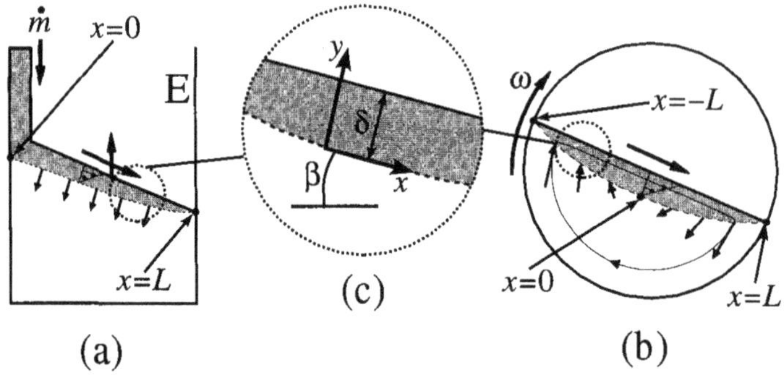

Fig. 1. Schematic view of surface flow systems: (a) Heap flow, (b) Rotating cylinder flow, (c) Co-ordinate system used in the analysis.

An important feature of surface granular flows is the interchange of particles between the flowing layer and the fixed bed. In the case of a rotating cylinder the interchange rate is determined by kinematics since the velocity of the fixed bed at the bed-layer interface is known. The situation in the case of heap flow is more complicated. Bouchaud *et al.* [4] proposed a phenomenological model (BCRE model) in which the interchange rate is determined by the local surface angle. A variation of this model proposed by Boutreux *et al.* [5] (BRdG model) has been broadly validated by continuum models [10, 19] and experiments [19], as we show below. Continuum models developed previously, for both heaps and rotating cylinders, are all based on depth-averaged hydrodynamic equations and differ primarily in the constitutive equations used. All the models contain parameters which must be evaluated from experiments, but in most cases, these parameters have not been determined.

Here we present a common continuum based framework for the analysis of both heap flow and rotating cylinder flow. The treatment closely follows that given in Refs. 19 and 25. Model predictions are compared to experimental results and to predictions of previous models. The general continuum model is presented first. Results for the heap formation and rotating cylinder flow are given next followed by conclusions.

2. General Continuum Model

Consider a flowing layer on the surface of a granular bed assuming the flow is nearly uni-directional in the layer and curvature effects are small. The depth averaged continuity equation and the x-momentum balance equation are simplified using the following assumptions. The bulk density in the layer (ρ) is nearly constant (since the dilation of the flowing particles is not too large in the relatively slow flows being considered). The velocity profile in the layer is linear and of the form [3, 17] $v_x = 2u(y/\delta)$, where $u(x,t)$ is the depth averaged velocity in the layer and δ is the layer thickness. Slow plastic deformation [21] is neglected. The shear stress at the interface is taken to be [25]

$$\tau_{xy}|_{y=0} = \rho d^2 f(\rho)\left(\frac{\partial v_x}{\partial y}\right)^2 - \rho g \delta \cos\beta \tan\beta_s\,, \tag{2.1}$$

where d is the particle diameter and $\tan\beta_s$ is the effective coefficient of *dynamic* friction, with β_s taken to be the static angle of repose. The stress is sensitively dependent on the local bulk density, and based on recent empirical evidence [25] we take $f(\rho) = c\delta/d$ with $c \approx 1.5$. The governing equations then reduce to

$$\frac{\partial \delta}{\partial t} + \frac{\partial}{\partial x}(\delta u) = -\Gamma\,, \tag{2.2}$$

$$\frac{\partial}{\partial t}(\delta u) + \frac{4}{3}\frac{\partial}{\partial x}(\delta u^2) = -4cd\frac{u^2}{\delta} + g\delta\frac{\sin(\beta - \beta_s)}{\cos\beta_s}\,, \tag{2.3}$$

where Γ is the flux from the layer into the bed. Further, assuming the static friction forces at the heap-layer interface to be fully mobilized, the Mohr–Coulomb criterion yields

$$\tau_{xy}|_{y=0} = -\rho g \delta \cos\beta \tan\beta_m\,, \tag{2.4}$$

where $\tan\beta_m$ is the effective coefficient of *static* friction. Using Eq. (2.1) and the assumptions given above, Eq. (2.4) yields

$$u = \frac{\dot\gamma\delta}{2}\,, \tag{2.5}$$

with the shear rate given by

$$\dot\gamma = \left[\frac{g\cos\beta\sin(\beta_m - \beta_s)}{cd\cos\beta_m\cos\beta_s}\right]^{1/2}. \tag{2.6}$$

A similar analysis is given by Douady *et al.* [10] with the difference that no stress constitutive equation such as Eq. (2.1) is used and instead the shear rate in the flowing layer is assumed to be constant.

3. Heap Formation

Consider a quasi-steady flow ($\partial\delta/\partial t, \partial u/\partial t \approx 0$) and a slowly varying interface angle ($\partial\beta/\partial x \approx 0$) during heap formation. The continuity equation (Eq. (2.2)) together with Eq. (2.5) then reduces to

$$\dot\gamma\delta\frac{\partial\delta}{\partial x} = -\Gamma\,, \tag{3.1}$$

and the momentum balance equation (Eq. (2.3)) together with Eq. (2.4) simplifies to

$$\dot\gamma^2\delta\frac{\partial\delta}{\partial x} = -\frac{g\sin(\beta_m - \beta)}{\cos\beta_m}\,. \tag{3.2}$$

Combining Eqs. (3.1) and (3.2) yields $\Gamma = g\sin(\beta_m - \beta)/\dot{\gamma}\cos\beta_m$, which, for the case when $\beta_m \approx \beta$, reduces to

$$\Gamma \approx V(\beta_m - \beta)\,, \tag{3.3}$$

where $V = g/\dot{\gamma}\cos\beta_m$. Thus, the continuum model yields a source term similar to the BRdG model; the scaling of V is also similar to the BRdG model.

We further simplify the above equations for two different geometries of heap formation: *closed*, as shown in Fig. 1(a), and *open* in which the end wall (E, Fig. 1(a)) is removed. In the open system at steady state, all the material entering the system leaves at the far edge of the heap and no particles are absorbed or eroded. This implies that $\Gamma = 0$, which on substituting into Eq. (3.3) gives $\beta = \beta_m \equiv$ constant. Using these results in Eqs. (3.1) and (3.2) shows that the average velocity (u) and thickness (δ) of the flowing layer are also constant in open systems. The mass flow rate in the system is given by $\dot{m} = \rho u \delta T$, where T is the width of the layer. This expression, together with Eq. (2.5), gives the following relationship between the layer thickness and mass flow rate

$$\delta = \left[\frac{2\dot{m}}{T\rho\dot{\gamma}}\right]^{1/2}. \tag{3.4}$$

Experimental results [19] based on flow visualization studies validate the above predictions, and sample results are given below. Fig. 2(a) shows the variation of the maximum angle of repose with mass flow rate in the system for 2 mm steel balls in an open heap system (filled symbols). The data indicate that β_m, and thus the coefficient of static friction at the heap-layer interface ($\tan\beta_m$), is not a constant but increases with the local flow rate. An increase in surface angle with flow rate was also reported by Lemieux and Durian [22]. Figure 2(b) shows the variation of the layer thickness (δ) with mass flow rate. The solid line is a fitted curve of the form $\delta \propto \dot{m}^{1/2}$. This indicates agreement with theoretical predictions (Eq. (3.4)) if the product $\rho\dot{\gamma}$ is independent of mass flow rate.

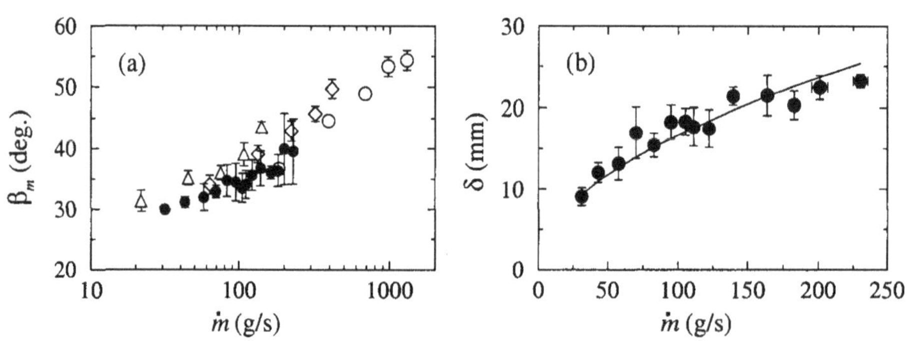

Fig. 2. Variation of the (a) maximum angle of repose (β_m) and (b) layer thickness (δ) with mass flow rate ($\dot{m}$) for 2 mm steel balls. Filled symbols: open heap system [19]. Open symbols: rotating cylinder system for three cylinder sizes [25]. The solid line in (b) is a best fit of the form $\delta \propto \dot{m}^{1/2}$.

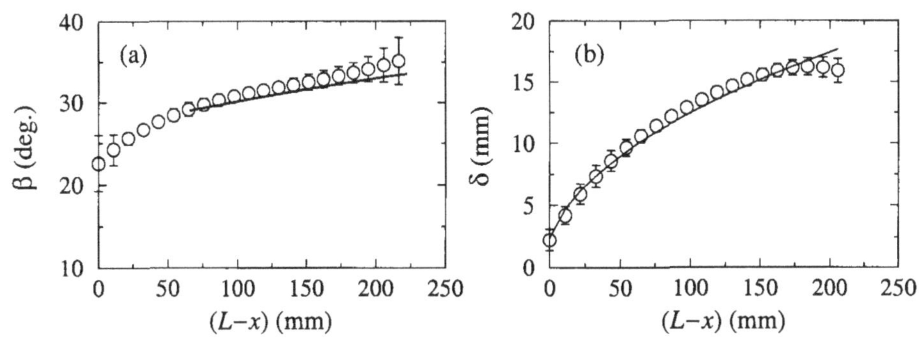

Fig. 3. Variation of (a) surface angle (β) and (b) layer thickness (δ) with distance from the edge of the heap ($L - x$) for flow of 2 mm steel balls in a closed system. Symbols are experimental data and error bars indicate the standard deviation over six measurements. Solid line in (a) is a fit of Eq. (3.3) and in (b) is the prediction of Eq. (3.5).

In a closed system (Fig. 1(a)), at steady state we must have $\Gamma \equiv$ constant for the heap to rise uniformly. Integrating Eq. (3.1), the layer thickness profile is obtained as

$$\delta = \left[\delta_L^2 + \frac{2\Gamma(L-x)}{\dot{\gamma}}\right]^{1/2}, \tag{3.5}$$

where δ_L is the layer thickness at the end of the layer, $x = L$, and L is the length of the interface (Fig. 1(a)). The rise velocity is related to the mass flow rate by

$$\Gamma = \frac{\dot{m}}{(TL\rho)}, \tag{3.6}$$

and the interface angle is calculated from Eq. (3.3).

Experimental results [19] for closed systems show that the rise velocity (Γ) varies nearly linearly with mass flow rate in agreement with Eq. (3.6), and the bulk density, which is found to be constant, is $\rho = 3.2$ g/cm^3. Figure 3 shows the variation of both interface angle (β) and layer thickness (δ) with length along the bed-layer interface ($L - x$) for a fixed mass flow rate. The solid line in Fig. 3(b) is a fit of Eq. (3.5). There is a good match between the fitted line and the experimental data, which suggests that the shear rate, $\dot{\gamma}$, is constant. Similar results are obtained for all flow rates studied. Using experimental results for the rise velocity (Γ) and the interface length (L), we obtain $\dot{\gamma} = 20 \pm 2$ s^{-1} from Eq. (3.5), where the standard deviation indicated is calculated for all ten flow rates studied. Using the value of the bulk density obtained above, we find from Eq. (3.4) that the shear rate for the open system is $\dot{\gamma} = 22 \pm 3$ s^{-1}. The value of the shear rate predicted from Eq. (2.6) is $\dot{\gamma} = 20 \pm 5$ s^{-1} for the range of mass flow rates considered. Thus, the shear rates for the open and closed systems are the same within experimental error, and predictions of theory are in reasonable agreement with experimental values.

4. Rotating Cylinder

The simplest case corresponds to rotating cylinder flow for 50% fill fraction. Assuming a nearly flat interface, the source term is given by $\Gamma = \omega x$. Substituting into the continuity equation (Eq. (2.2)) and integrating we obtain

$$u\delta = \frac{\omega}{2}(L^2 - x^2)\,. \tag{4.1}$$

Using Eq. (2.2) the momentum balance equation (Eq. (2.3)) simplifies to

$$u\frac{du}{dx} = \frac{3g}{4}\frac{\sin(\beta - \beta_s)}{\cos\beta_s} - 3\frac{cdu^2}{\delta^2} + \frac{\omega x u}{\delta}\,. \tag{4.2}$$

We consider two different limiting solutions to Eqs. (4.1) and (4.2) below.

Firstly, consider the case when shear rate ($\dot{\gamma}$) is nearly constant. Using Eq. (2.5), the flux equation (Eq. (4.1)) gives the layer thickness profile as

$$\delta = \left(\frac{\omega}{\dot{\gamma}}\right)^{1/2}(L^2 - x^2)^{1/2}\,, \tag{4.3}$$

which is symmetric for all rotational speeds (ω). Equation (4.3) corresponds to the model of Makse [23], in which the shear rate is assumed to be a fitting parameter. In the present case the shear rate is obtained from Eq. (2.6) and the mean velocity is given by $u = \dot{\gamma}\delta/2$. Substituting these results in Eq. (4.2), and using the Mohr–Coulomb condition (Eq. (2.4)) yields Eq. (3.3) with $\Gamma = \omega x$ and $\beta_m \approx \beta$. This allows for calculation of the angle (β) along the interface. Thus, the assumption of a constant shear rate is consistent with the model, and gives a complete description of the flow. However, it is not apparent from the analysis, under what conditions the solution is valid.

Consider next the case when the shear rate is not constant along the layer, but when the acceleration (du/dx) is small. Eliminating δ using Eq. (4.1), the scaled momentum balance becomes

$$\bar{u}\frac{d\bar{u}}{d\xi} = \frac{3}{4Fr}\frac{\sin(\beta - \beta_s)}{\cos\beta_s} - 12cs\frac{\bar{u}^4}{(1-\xi^2)^2} + \frac{2\xi\bar{u}^2}{1-\xi^2}\,, \tag{4.4}$$

where $\bar{u} = u/\omega L$, $\xi = x/L$ and the dimensionless parameters are the Froude number, $Fr = \omega^2 L/g$, and the size ratio, $s = d/L$. The first term on the right-hand side of Eq. (4.4) is the net driving force, that is, the gravitational force less the frictional resistance to flow, and is independent of the flow velocity ($\bar{u}$). The second term is the 'viscous' resistance due to collisional stresses, and the third term arises as a result of in-flow and out-flow of particles from the layer. Both these terms depend on the flow velocity. Typical experimental Froude numbers for experiments in rotating cylinders are in the range $O(10^{-3})$ to $O(10^{-2})$. In these cases the driving force term ($O(1/Fr)$) is much larger than the acceleration term ($O(\xi/\sqrt{sFr})$ based on Eq. 2.5), particularly near the midpoint of the layer ($\xi = 0$). The collisional stress term is of the same magnitude as the net driving force term since the flow velocity increases to balance the two. Thus, for $\xi\sqrt{Fr/s} \ll 1$ the acceleration term may be neglected.

For negligible acceleration ($d\bar{u}/d\xi \approx 0$), the scaled mean flow velocity is obtained from Eq. (4.4) as

$$\bar{u} = \left(\frac{1-\xi^2}{12cs}\right)^{1/2}\left[\xi + \left(\xi^2 + \frac{9csA}{Fr}\right)^{1/2}\right]^{1/2}, \tag{4.5}$$

where $A = \sin(\beta - \beta_s)/\cos\beta_s$. Using Eq. (4.1), the scaled layer thickness profile is

$$\bar{\delta} = \left[\frac{3cs(1-\xi^2)}{\xi + (\xi^2 + 9csA/Fr)^{1/2}}\right]^{1/2}, \tag{4.6}$$

where $\bar{\delta} = \delta/L$. The above solution is valid only if $A > 0$, that is, if $\beta > \beta_s$. For $\beta \leq \beta_s$, we have $\bar{u} = \bar{\delta} = 0$, thus there is no steady flow possible if the interface angle is less than the static angle of repose. This is consistent with the definition of the static angle of repose. Note that the layer profile is not symmetric about $\xi = 0$, and for any $\xi > 0$ we have $\bar{\delta}(-\xi) > \bar{\delta}(\xi)$, that is, the upper part of the layer ($\xi < 0$) is thicker than the lower part. The source of the asymmetry is the in-flow/out-flow term in the momentum balance (third term on the right-hand side of Eq. 4.4). In the upper part of the layer ($\xi < 0$) the flow is retarded by material entering the layer from the bed ($\Gamma < 0$) and the reverse is true in the lower part of the layer. Thus, the layer is thicker in the upper part because of the lower velocity relative to the lower part of the layer ($\xi > 0$), resulting in a skewed profile. Further, Eq. (4.6) indicates that the profile becomes more skewed with increasing Froude number (Fr) and decreasing size ratio (s). In the limit, $Fr/s \ll 1$, the scaled layer thickness profile becomes $\bar{\delta} = (csFr/A)^{1/4}(1-\xi^2)^{1/2}$, which is identical to the result obtained assuming a constant shear (Eq. (4.3)) when Eq. (2.6) is used to calculate the shear rate. This implies that a profile symmetric about the layer midpoint ($\xi = 0$) is obtained at very low Froude numbers and relatively high size ratios, and in this limit the shear rate is nearly constant.

The interface angle profile ($\beta(\xi)$) is obtained from Eq. (3.3), using $\Gamma = \omega x$ and Eq. (2.6) as

$$\beta(\xi) = \beta_m - \frac{Fr\cos\beta_m}{3cs}\left[\xi + \left(\xi^2 + \frac{9csA}{Fr}\right)^{1/2}\right]^{1/2}\xi. \tag{4.7}$$

In simplifying the preceding equation we assume $\beta_m \approx \beta \approx \beta_s$. Equation (4.7) indicates that the interface angle decreases monotonically with distance along the interface and at $\xi = 0$, $\beta = \beta_m$. Thus, in the rotating cylinder flow the maximum angle of repose can be experimentally obtained by measuring the interface angle at the midpoint of the layer. For $(Fr/s)\xi$ sufficiently large and $\xi > 0$, we get $\beta < 0$, that is, for small size ratios and large Froude numbers the layer profile may turn up at the end. Conversely, when $(Fr/s) \ll 1$, Eq. (4.7) yields $\beta \approx \beta_m$, and the interface profile is nearly flat. Neglecting terms $O(\xi Fr/s)$, which

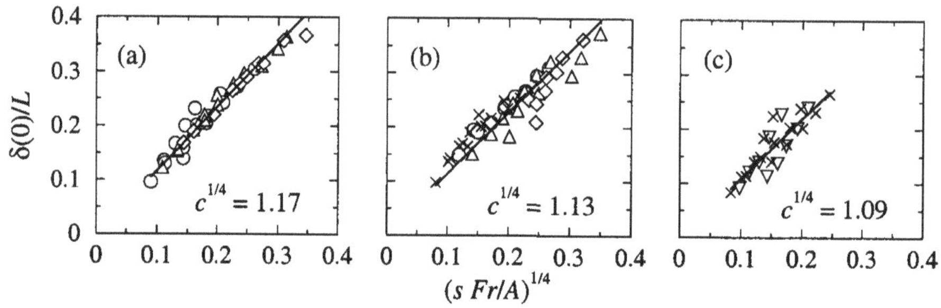

Fig. 4. Variation of the layer thickness at the midpoint ($\delta(0)$) with $(sFr/A)^{1/4}$ for (a) steel balls, (b) glass beads, and (c) sand particles in cylinders of different sizes and at different rotational speeds. Symbols are experimental data for different sized particles: $d = 1$ mm (○), $d = 2$ mm (△), $d = 4$ mm (◇), $d = 0.4$ mm (▽) and $d = 0.8$ mm (×). The solid line is a fit of Eq. (4.6) and the values of the parameter $c^{1/4}$ are indicated.

is consistent with the approximation in the momentum balance equation, we get $A = (\beta_m - \beta_s)/\cos\beta_s$.

Consider next a comparison of the theoretical results to experimental data. A few key numbers are reported, as they convey a sense of qualitative agreement. However, for full details the reader is referred to [25]. The model parameters required are β_s, β_m and c. Data of Orpe and Khakhar [25] for the first two parameters are shown in Fig. 2(a) (open symbols) for 2 mm steel balls in rotating cylinders of three sizes and for different rotational speeds of the cylinders. The data correlates reasonably well with the mass flow rate at the midpoint of the layer calculated from $\dot{m} = \rho\omega L^2 T/2$, where T is the cylinder length and the same density as in the heap experiments ($\rho = 3.2$ g/cm^3) is used. Data spanning nearly two decades of flow rate fall on a single curve, although with some scatter. The maximum angle of repose increases with mass flow rate, and the measured values are similar to those from heap experiments which are also shown in the same figure. The static angle of repose is the angle at $\dot{m} = 0$.

Orpe and Khakhar [25] had obtained $c \approx 1.5$ by fitting the theory of Khakhar *et al.* [18] to experimental layer thickness profiles. We obtain a new estimate of the parameter based on the layer thickness at the midpoint of the layer ($\xi = 0$), which, from Eq. (4.6), is $\bar{\delta}(0) = (csFr/A)^{1/4}$. Figure 4 shows experimental data for $\bar{\delta}(0)$ versus $(sFr/A)^{1/4}$ for experimental data for 90 experiments comprising steel balls, glass beads and sand of different sizes in cylinders of different sizes and for different rotational speeds. The data falls on a straight line for each material (although with some scatter) and a least squares fit gives $c = 1.9$ for steel balls, $c = 1.6$ for glass beads and $c = 1.4$ for sand. Since the model is essentially exact at $\xi = 0$, the good fit implies that the proposed constitutive equation for stress is reasonable, and the shear rate in the layer is well-described by Eq. (2.6) at $\xi = 0$.

Predictions of the model for the layer thickness profile are compared to experimental data in Fig. 5 for sand particles and steel balls for different Froude numbers

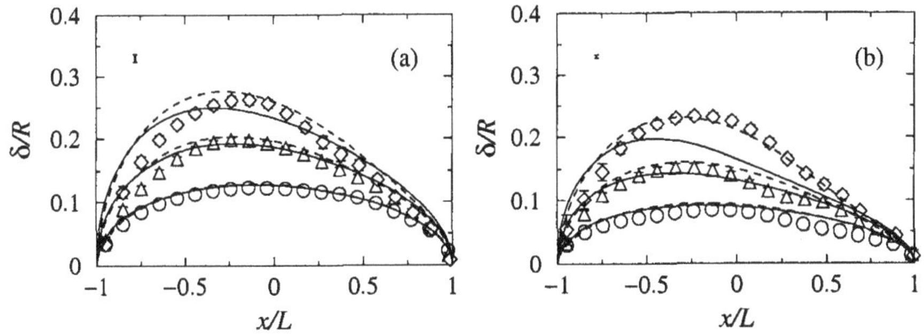

Fig. 5. Layer thickness profiles for (a) 2 mm steel balls and (b) 0.8 mm sand. Symbols denote experimental data for three different froude numbers (Fr): $Fr = 2 \times 10^{-3}$ (○), $Fr = 22 \times 10^{-3}$ (△) and $Fr = 64 \times 10^{-3}$ (◇). Solid lines are predictions of Eq. (4.6) and dashed lines are the predictions of the model of Khakhar *et al.* [18]. The error bars give the standard deviation over six measurements and the bar indicates the scaled diameter of a particle ($s = d/R$).

in a cylinder of radius 16 cm, using the value of c obtained above and experimental values for β_m and β_s. The agreement is good except at the highest Fr and low s, and all the qualitative features of the data are reproduced. At low Fr and relatively high s studied, the profile is nearly symmetric (steel balls at the lowest Fr), and the profiles become more skewed with increasing Fr and decreasing s. The deviation at the high Froude numbers and low size ratio are due to neglect of the acceleration term. Similar agreement is obtained for the other cases studied as well. The predictions of the model of Khakhar *et al.* [18] are shown in the figure as dashed lines. These nearly coincide with the results from the present model, except for the highest Fr for sand, indicating that the approximations made are reasonable for the parameter values of interest. It is remarkable that such a simple theory is able to describe the behavior of the system over such a wide range of parameters: materials include steel balls, glass beads and sand; varying shapes with steel balls being spherical, glass beads, nearly spherical and sand being irregularly shaped; size ratios in the range $s \in (0.005, 0.05)$ and Froude numbers in the range $Fr \in (2 \times 10^{-3}, 64 \times 10^{-3})$. Model predictions of the interface angle profile are in reasonable agreement with experiments [25].

5. Conclusions

A theoretical framework serves to unify the behaviour of surface flows for two prototypical systems: heap flow and rotating cylinder flow. The model is based on a stress constitutive equation and failure criterion which contain three material parameters: β_s, β_m and c. Analytical results for both systems give a complete description of the systems in terms of the layer thickness profiles ($\delta(x)$), average velocity of flow ($u(x)$) and the interface angle profile ($\beta(x)$). In open heap systems a layer of uniform thickness with a uniform flow velocity is obtained, whereas in the closed heap system $\delta^2 \propto x$. The interface angle is constant and equal to the maximum angle of repose in the open system, whereas it decreases with distance from the pouring point in the closed system. Results for the rotating cylinder are obtained for the case when the acceleration of particles in the layer is small ($\xi\sqrt{Fr/s} \ll 1$). The layer profile is found to be asymmetric about the midpoint of the layer ($\xi = 0$) with the upper part of the layer ($\xi < 0$) being thicker. The skewness increases with increasing Froude numbers and decreasing size ratios. The scaled shear rate ($\dot{\gamma}/\omega$) decreases with increasing Froude number and size ratio. The layer interface angle decreases with distance in the flow direction. For high $\xi Fr/s$ and $\xi > 0$ the layer turns up, whereas when $\xi Fr/s$ is small a nearly flat interface is obtained.

Quasi-2D experiments carried out for open and closed heaps and rotating cylinders of different sizes, by and large, validate the predictions of the theory. The three material parameters of the model (β_s, β_m and c) are all obtained from relatively simple measurements. The model equation can thus be applied to more complex geometries. Deviations of the model from experimental data appear in the interface angle profile in the rotating cylinder flow. This is most likely due to end wall effects which are discussed in Ref. 25.

Acknowledgments

D. V. Khakhar acknowledges the financial support of the Department of Science and Technology, India, through the Swarnajayanti Fellowship project (DST/SF/8/98) for part of this work. This work was supported in part by grants to J. M. Ottino from the Division of Basic Energy Sciences of the Department of Energy, the National Science Foundation, Division of Fluid and Particulate Systems, and the Donors of the Petroleum Research Fund, administered by the American Chemical Society.

References

[1] Bagnold, R. A., *Proc. R. Soc. London Ser.* **A255**, 49 (1954).
[2] Boateng, A. A. and Barr, P. V., *J. Fluid Mech.* **330**, 233 (1997).
[3] Bonamy, D., Faucherand, B., Planelle, M., Daviaud F. and Laurent, L., in *Powders and Grains*, Kishino, Y. ed. (Swets and Zeitlinger, Lisse, 2001), p. 463.
[4] Bouchaud, J. P., Cates, M., Ravi Prakash, J. and Edwards, S., *J. Phys. France* **I4**, 1383 (1994).
[5] Boutreux, T., Raphaël, E. and de Gennes, P. G., *Phys. Rev.* **E58**, 4692 (1998).
[6] Bridgewater, J., *Chem. Eng. Sci.* **50**, 4081 (1995).
[7] Campbell, C. S., *Annu. Rev. Fluid Mech.* **22**, 57 (1990).
[8] Daerr, A. and Douady, S., *Nature* **399**, 241 (1999).
[9] de Gennes, P. G., *Rev. Mod. Phys.* **71**, S374 (1999).
[10] Douady, S., Andreotti, B. and Daerr, A., *Eur. Phys. J.* **11**, 131 (1999).
[11] Duran, J., *Powder and Grains* (Springer-Verlag, New York, 2000).
[12] Dury, C. M., Ristow, G. H., Moss, J. L. and Nakagawa, M., *Phys. Rev.* **E57**, 4491 (1998).
[13] Elperin, T. and Vikhansky, A., *Europhys. Lett.* **42**, 619 (1998).
[14] Ennis, B. J., Green, J. and Davis, R., *Chem. Eng. Prog.* **90**, 32 (1994).
[15] Henein, H., Brimacombe, J. K. and Watkinson, A. P., *Metall. Trans.* **B14**, 191 (1983).
[16] Jaeger, H. M., Nagel, S. R. and Behringer, R. P., *Rev. Mod. Phys.* **68**, 1259 (1996).
[17] Jain, N., Ottino, J. M. and Lueptow, R. M., *Phys. Fluids* (2001) (in press).

[18] Khakhar, D. V., McCarthy, J. J., Shinbrot, T. and Ottino, J. M., *Phys. Fluids* **9**, 31 (1997).

[19] Khakhar, D. V., Orpe, A. V., Andresén, P. and Ottino, J. M., *J. Fluid Mech.* **441**, 255 (2001).

[20] Khosropour, R., Valachovic, E. and Lincoln, B., *Phys. Rev.* **E62**, 807 (2000).

[21] Komatsu, T. S., Inagaki, S., Nakagawa N. and Nasuno, S., *Phys. Rev. Lett.* **86**, 1757 (2001).

[22] Lemieux, P.-A. and Durian, D. J., *Phys. Rev. Lett.* **85**, 4273 (2000).

[23] Makse, H. A., *Phys. Rev. Lett.* **83**, 3186 (1999).

[24] Nakagawa, M., Altobelli, S. A., Caprihan, A., Fukushima, E. and Jeong, E. K., *Exp. Fluids* **16**, 54 (1993).

[25] Orpe, A. V. and Khakhar, D. V., *Phys. Rev.* **E64**, 031302 (2001).

[26] Ottino, J. M. and Khakhar, D. V., *Annu. Rev. Fluid Mech.* **32**, 55 (2000).

[27] Rajchenbach, J., Clément, E. and Duran, J., in *Fractal Aspects of Materials*, Family, F., ed., MRS Symposium Proceedings No. 367 (Materials Research Society, Pittsburgh, 1995), p. 525.

[28] Rajchenbach, J., *Phys. Rev. Lett.* **65**, 2221 (1990).

[29] Rajchenbach, J., in *Physics of Dry Granular Media*, Hermann, H., ed. (Kluwer Academic, Dordrecht, 1998), p. 421.

[30] Rao, S. J., Bhatia, S. K. and Khakhar, D. V., *Powder Technol.* **67**, 153 (1991).

[31] Ristow, G. H., *Pattern Formation in Granular Materials* (Springer, Berlin, 2000).

[32] Yamane, K., Nakagawa, M., Altobelli, S. A., Tanaka, T. and Tsuji, Y., *Phys. Fluids* **10**, 1419 (1998).

[33] Zik, O., Levine, D., Lipson, S. G., Shtrikman, S. and Stavans, J., *Phys. Rev. Lett.* **73**, 644 (1994).

CHAPTER 14

Rheology of Dense Granular Flow

THOMAS C. HALSEY and DENIZ ERTAŞ
ExxonMobil Research and Engineering,
Route 22 East, Annandale, NJ 08801, USA

GARY S. GREST and LEONARDO E. SILBERT
Sandia National Laboratories, Albuquerque, NM 87185, USA

DOV LEVINE
Department of Physics, Technion, Haifa, 32000 Israel

We have performed numerical studies of dense granular flows on an incline with a rough bottom in two and three dimensions. This flow geometry produces a constant density profile that satisfies scaling relations of the Bagnold, rather than the viscous, kind. No surface-only flows were observed. The bulk and the surface layer differ in their rheology, as evidenced by the change in principal stress directions near the surface; a Mohr–Coulomb type failure criterion is seen only near the surface. In the bulk, normal stress anomalies are observed both in two and in three dimensions. We do not observe iso-staticity in static frictional piles obtained by arresting the flow.

Keywords: Granular flow; rheology; non-Newtonian fluid dynamics; granular packings; powder mechanics; iso-staticity.

1. Introduction

Despite their importance in many areas of science and technology, most aspects of the physics of hard granular systems remain obscure. Not least among these is the dynamics of dense flow. Despite the wide attention, from a variety of points of view, given to the subject of avalanches over the last decade and a half [6], the phenomenology of unconfined dense granular flows on inclined planes (chute flows) remains controversial. Recent experimental work of Pouliquen has suggested that the overall rheology of these systems is of the type proposed by Bagnold a half century ago, in which the shear rate varies as the square root of the distance from a free surface [1, 12].

In this contribution we report on a series of computational studies that have confirmed this rheology, and which raise additional questions about more subtle aspects of the rheology as well as the role of the upper and lower surfaces in such

First published in *Advances in Complex Systems*, Vol. 4, No. 4 (2001), pp. 419–428.

flows [3, 14, 15]. We have performed two- and three-dimensional studies of up to $N = 20,000$ monodisperse particle systems. Our approach is fundamentally a molecular dynamics approach; we forward integrate the mechanical equations of motion for the particle assembly. We include Coulomb friction between the particles, which interact with viscously damped forces of either the linear spring (Hookean) or the more realistic Hertz–Mindlin elastic type [7]. We use rough bottom surfaces to reduce the likelihood of plug flow. Our primary results are:

- The angle of repose depends significantly on the height of the pile, as reported experimentally.
- The kinematics of chute flows is of the Bagnold type, with shear rates varying as the square root of depth from the free surface. Alternatively, the magnitude of the stress tensor is proportional to the square of the strain rate.
- The overall amplitude of the flow velocity goes to zero at the angle of repose in three dimensions, but not in two dimensions.
- The structure of the stress tensor is non-Newtonian, with normal stresses in the directions of the flow, of the perpendicular to the free surface, and (in three dimensions) the vorticity, all differing from one another.
- In the bulk of the flow, there is only a small difference between the normal stress in the flow direction and in the direction normal to the free surface, even as one approaches the angle of repose. Classical analyses of failure, such as the Mohr–Coulomb criterion (discussed below), predict that this difference should be appreciable if the pile is at the angle of repose.
- Nevertheless, in the vicinity of the free surface, the normal stresses appear to approach a Mohr–Coulomb type criterion at the angle of repose, despite the fact that this is not the case deeper in the flow. (This is true only in three dimensions, and not in two dimensions.)
- A large fraction of contacts are at plastic yield in any flowing state; however, as soon as the flow arrests the vast majority of contacts are elastic, i.e. below Coulomb yield.
- The number of contacts in the static states obtained are significantly larger than needed to insure static equilibrium of the pile (isostaticity).

These results hold for a variety of specific values of parameters such as viscous damping rate (particle inelasticity), particle stiffness, friction coefficient, etc.

In Sec. 2, we give the dimensional argument for Bagnold kinematics, and discuss its appearance in our numerical results. In Sec. 3, we report on the structure of the stress tensor observed, and discuss the normal stress anomalies seen, in the bulk and at the surface. In Sec. 4, we discuss the distribution of plastic versus elastic contacts, in both flowing and static states, as well as the behavior of the particle coordination number. Finally, in Sec. 5, we conclude.

The work reported in this contribution is discussed in more detail in Refs. 3, 14 and 15. Details of our simulation procedures are found in Ref. 14, which we wrote with S. J. Plimpton of Sandia National Laboratories.

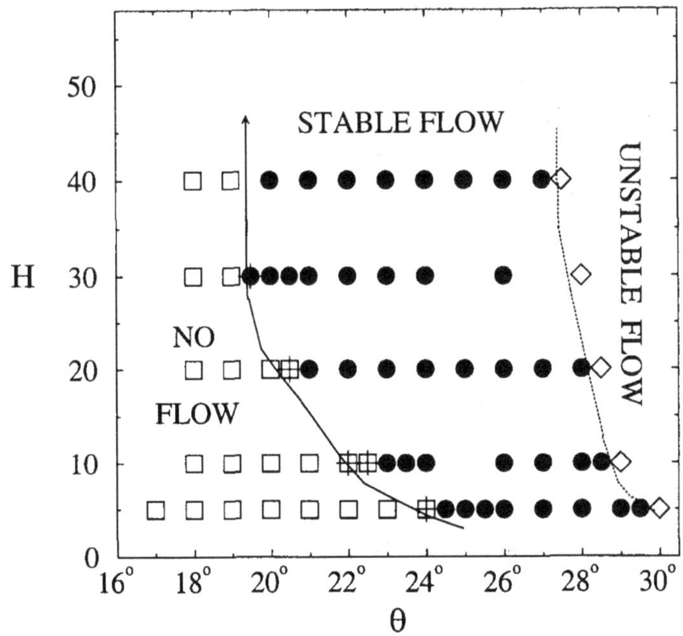

Fig. 1. Phase behavior of Hookean granular particles in chute flow geometry in three dimensions, as a function of inclination angle θ and height of pile H [14]. A regime of steady state flow is observed for angles θ obeying $\theta_r < \theta < \theta_{max}$. The angle of repose θ_r depends upon the height of the pile. Squares with crosses indicate hysteretic angles and heights for which both no flow and stable flow states can be observed, depending on preparation.

2. Bagnold Kinematics

2.1. *Phase behavior*

The overall behavior of these flows is exhibited as a function of inclination angle θ and height of the pile H in Fig. 1. Below an angle of repose θ_r, which decreases with increasing height of the pile, there is a static regime. Above a maximum angle θ_{max} the flows accelerate and run away, failing to reach steady state, while for $\theta_r < \theta < \theta_{max}$ steady state flows are observed. A small hysteresis can be seen in the value of θ_r. These results are in agreement with the observations of Pouliquen [12].

2.2. *Shear stress scaling*

Let us define axes for our chute flow: the z-axis is normal to the free surface of the flow, the x-axis is in the direction of flow, and the y-axis is parallel to the vorticity. We take $z = 0$ at the base of the flow, so that $z = H$ at the free surface. As above, the free surface is oriented at an angle θ with respect to the horizontal. If the shear stress σ_{xz} is determined locally by the strain rate $\dot{\gamma} \equiv \partial_z v_x$, then there are strong dimensional constraints on this local relation. In particular, suppose that the particles are rigid; i.e. that their elastic moduli are much greater than the stress tensor, so that the particles are hardly deformed by the stresses to which they are subjected. This is typically the case in powder mechanics and in geophysical flows. We would anticipate that in this case the stress-strain rate relation will be independent of the elastic moduli. Furthermore, the other parameters controlling the rheology of this system: the coefficient of restitution, and the microscopic coefficient of friction, are both dimensionless. Thus, the only parameters left that can contribute to a local rheology are the mass M of the particles and their radius R. We immediately conclude that

$$\sigma_{xz} \propto \frac{M}{R}\dot{\gamma}^2 , \tag{2.1}$$

as originally proposed by Bagnold [1].

The stress tensor itself also satisfies the equations of motion

$$\partial_z \sigma_{zz} + \partial_x \sigma_{xz} = \rho g \cos\theta , \tag{2.2}$$

$$\partial_z \sigma_{zx} + \partial_x \sigma_{xx} = \rho g \sin\theta , \tag{2.3}$$

where ρ is the (presumed constant) density of the pile. These equations have the x, y independent solutions

$$\sigma_{xz} = \rho g(H - z) \sin\theta , \tag{2.4}$$

$$\sigma_{zz} = \rho g(H - z) \cos\theta . \tag{2.5}$$

Combined with the results above, we conclude that

$$\dot{\gamma} \propto \sqrt{H - z} . \tag{2.6}$$

Our numerical results are displayed in Fig. 2, and are in excellent agreement with this scaling. In addition, except for a dilated surface layer, we find that the packing fraction is independent of depth for this system.

Finally, we can also examine the amplitude of the velocity; i.e. the amplitude of the proportionality between the shear stress and the strain rate. Defining the parameter A_{Bag} by

$$\dot{\gamma} = A_{\mathrm{Bag}} \sqrt{\sigma_{xz}} , \tag{2.7}$$

we see from Fig. 3 that while this goes to zero seemingly continuously in $d = 3$ at the angle of repose, there appears to be a finite jump in this quantity at this angle in $d = 2$.

3. Structure of the Stress Tensor

We have also studied the structure of the normal stress anomalies in our chute flows, both in the bulk and near the surface. It is a characteristic of Newtonian fluids that all normal stresses are equal, i.e.

$$\sigma_{xx} = \sigma_{yy} = \sigma_{zz} . \tag{3.1}$$

This is also the prediction of most theories of granular rheology [8].

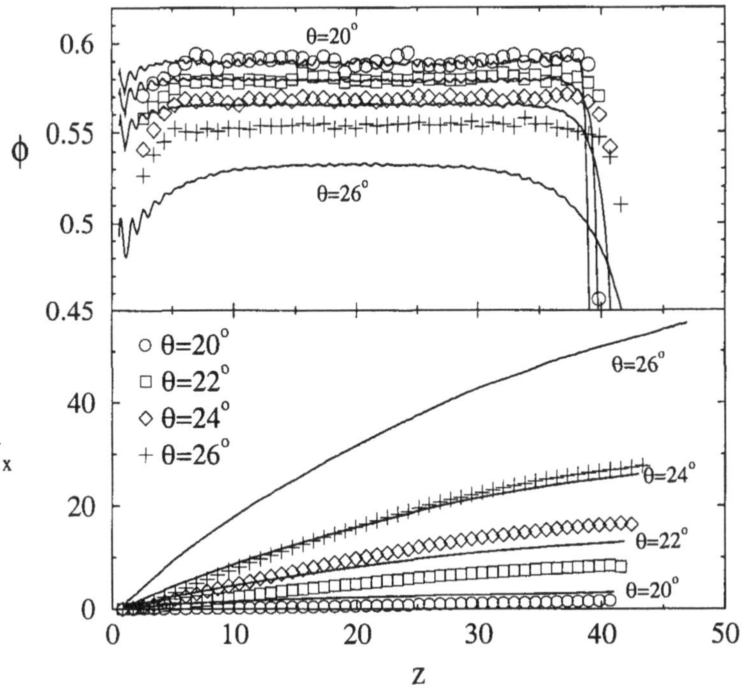

Fig. 2. Packing fraction ϕ and velocity profile $v_x(z)$ for three dimensional piles [14]. The velocity obeys Bagnold scaling, for which the shear rate $\partial_z v_x(z) \propto \sqrt{z}$. Points indicate results for Hookean inter-particle forces, while lines indicate results for Hertzian interactions.

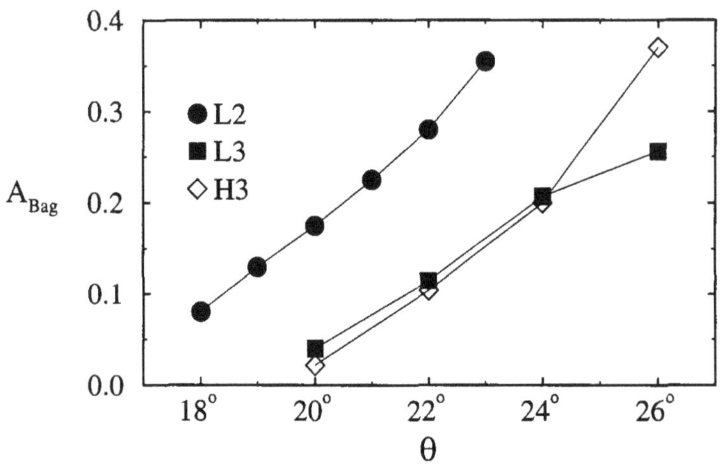

Fig. 3. The amplitude A_{Bag} of the velocity profile as a function of inclination angle θ in $d = 2, 3$ [14]. H3 and L3 refer respectively to three-dimensional Hertzian and Hookean sphere systems, while L2 refers to Hookean spheres in two dimensions. While this amplitude seems to go continuously to zero in $d = 3$, a discontinuity appears in $d = 2$.

However, it is already clear that there is a contradiction between this expectation and the classical analysis of granular failure due to Mohr and Coulomb [11]. In this analysis, failure occurs because the ratio of the shear stress to the normal stress across some plane internal to the pile, likely close to parallel to the surface, exceeds the coefficient of internal friction. This implies that at failure,

$$\sigma_{xx} > \sigma_{zz} \,. \tag{3.2}$$

Thus, if the stress state in a decelerating pile continuously approaches that characteristic of a static pile near failure, we would expect non-Newtonian normal stress behavior near the threshold of arrest.

The actual behavior seen is more subtle than either of these scenarios. First of all, the behavior of the stresses is non-Newtonian for all angles of inclination studied [13]. The non-Newtonian behavior is especially pronounced for σ_{yy}, which is typically about 10% less than σ_{zz}. The lowering of the normal stress parallel to the axis of vorticity is also typical of polymeric non-Newtonian rheologies.

On the other hand, we find that in the bulk of the flow $\sigma_{xx} \approx \sigma_{zz}$. The deviations from this equality, while small, are systematic and reproducible, both in $d = 2$ and in $d = 3$. One way of parameterizing the non-Newtonian behavior is through the "normal stress anomaly" χ defined by

$$\chi \equiv \frac{\sigma_{zz} - \sigma_{xx}}{\sigma_{zz}} \,. \tag{3.3}$$

$\chi = 0$ would correspond to Newtonian behavior in the $x - z$ plane. Results for χ as a function of inclination angle θ are shown for Hertzian and Hookean systems in $d = 3$ (H3 and L3 respectively) in Fig. 4. This figure also shows results for $d = 2$.

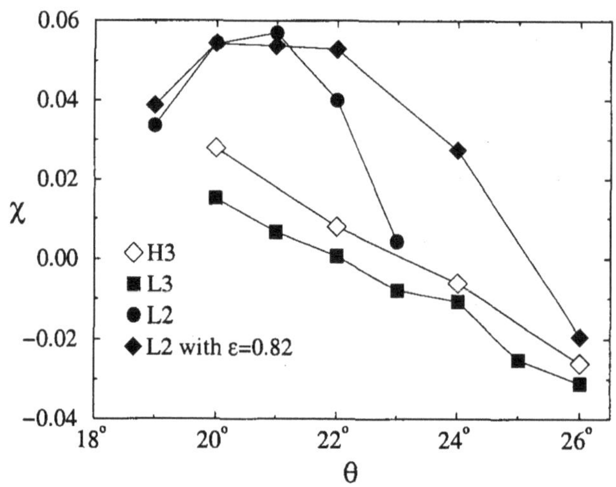

Fig. 4. The normal stress anomaly χ as a function of inclination angle for various two- and three-dimensional systems [14]. The $d = 2$ result with coefficient of restitution $\epsilon = 0.82$ is indicated specifically, by contrast with the more elastic standard L2 model with $\epsilon = 0.92$.

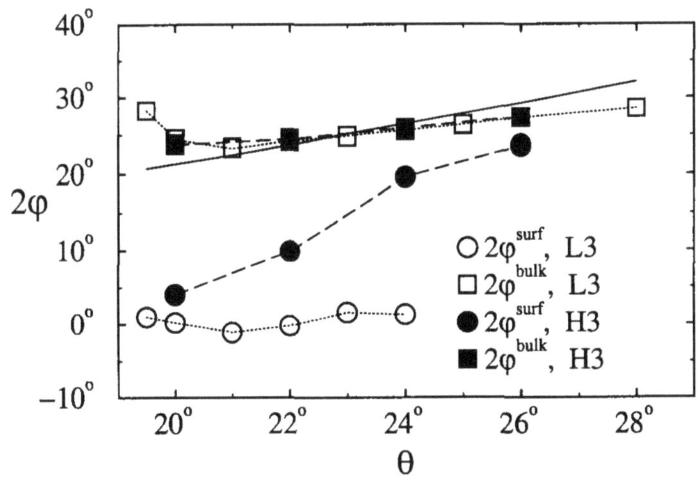

Fig. 5. The deviation φ of the stress state from Mohr–Coulomb criticality as a function of inclination angle θ, both in the bulk and at the surface of a three-dimensional flow [14]. $\varphi = 0$ corresponds to Mohr–Coulomb criticality, which is seen only near the surface. H3 and L3 refer to Hertzian and Hookean models, as discussed in the caption to Fig. 3. The solid line shows Newtonian behavior (no normal stress anomaly).

The situation near the surface is quite different. For inclination angles far from the angle of repose θ_r, the normal stresses near the surface show a similar behavior to the bulk case, at least for Hertzian forces. However, as $\theta \to \theta_r$, σ_{xx} near the surface becomes significantly larger than σ_{zz}, and approaches a value characteristic of a Mohr–Coulomb failure criterion. We can indicate this by a variable φ, which measures the angle between the principal axes of the stress tensor and those that would characterize Mohr–Coulomb failure; surface and bulk values of φ are displayed versus θ in Fig. 5. This measure is obviously equivalent to χ as defined above; where χ measures the deviation of the stress state from that of a Newtonian fluid, φ measures its deviation from Mohr–Coulomb criticality.

A natural interpretation of this result is that the failure of a dry granular pile is actually controlled by the instability of its free surface, so that the Mohr–Coulomb failure criterion actually applies only to the surface, and not to the bulk stress state. This contrasts with the case of failure with cohesive forces between the particles (obtained, e.g. with capillary wetting due to small levels of humidity), for which the failure seems to be a bulk phenomenon [4].

4. Grain Contact States

In addition to macroscopic measurements, we can also examine "mesoscopic" properties of our chute flow systems. On such property is the friction state at the individual grain-grain contacts. This state can be "elastic," for contacts at which the tangential force $t_i < \mu n_i$, where μ is the local coefficient of friction and n_i, t_i

are the normal and tangential forces at the ith contact, or "plastic," for frictionally saturated contacts at which $t_i = \mu n_i$.

Such measurements are interesting not only in the flowing, but also in the static state. Many authors have argued that static granular packings should typically be "isostatic," i.e. that the number of independent forces should equal the number of constraints imposed by the requirement that each grain be in mechanical equilibrium [2, 10]. For frictionless particles, this implies that each particle should, on average, have six neighbors with which it is in contact. For perfectly frictional particles ($\mu \to \infty$), this argument implies that the average coordination number z_c should instead be $z_c^{\rm iso} = 4$. For imperfectly frictional particles of the type we consider, an intermediate result is obtained,

$$z_c^{\rm iso} = \frac{12}{3 - n_c}, \tag{4.1}$$

where $0 \leq n_c \leq 1$ is the fraction of plastic contacts.

Alternatively, we can define a "staticity index" s_c by

$$s_c \equiv (3 - n_c) z_c / 2, \tag{4.2}$$

where $s_c = 6$ is the isostatic case, with $s_c > 6$ implying "hyperstaticity," i.e. an excess of contacts over the bare minimum necessary for a static packing. In hyperstatic cases, the precise nature of the elasticity of the particles controls the values of the forces in the packing; this is not the case for isostatic packings [5].

The coordination number, staticity index and fraction of plastic contacts are shown for a variety of inclination angles in Fig. 6. Flowing piles, even at relatively large coefficients of friction, have significant fractions of plastic contacts. This fraction drops to nearly zero for static piles. In addition, the static piles seem to be significantly hyperstatic, with the staticity index showing an apparent jump at the angle of repose.

One consequence of isostaticity is that the network of forces is, at least in principle, uniquely determined by the geometrical packing of the particles and the external (i.e. gravititational) loads placed on the particles. By contrast, as already remarked, the force network for a hyperstatic packing is influenced by the elasticity of the particles and the detailed contact mechanics between the particles. Many observers have seen convincing evidence that packings of frictionless particles are commonly isostatic [9]; thus, a fundamental difference seems to be emerging between the granular mechanics of frictionless and frictional particles.

Of course, our particles are not perfectly rigid, and the arguments advanced for isostaticity only apply in the ideal, rigid limit. Some authors have speculated the non-isostaticity in frictional piles disappears quite slowly as the modulus of the particles is taken to infinity [9]. However, the hyperstaticity we see seems to persist to quite realistic values of the elastic moduli of the particles, which suggests that the potential isostaticity of perfectly rigid particles may be of only academic interest.

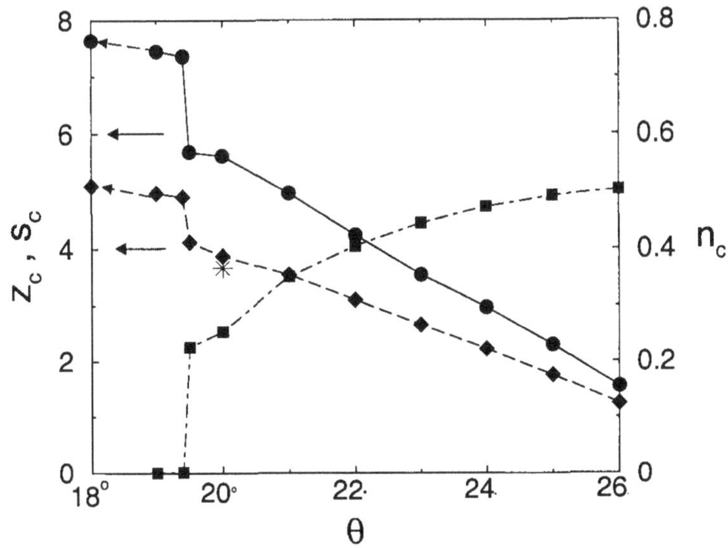

Fig. 6. The fraction of plastic contacts n_c (squares), the coordination number z_c (diamonds), and the staticity index s_c (circles) as functions of inclination angle θ [15]. All of these quantities appear to undergo discontinuous changes at the angle of repose, which is $\theta_r \approx 19.5°$ for the system displayed. The arrows mark the values obtained in isostatic packings of perfectly frictional particles.

5. Conclusion

Overall, our most important result is the observation of the Bagnold state in the steady-state flowing regime. We do not observe surficial flows; nor, with sufficiently rough bottom surfaces, do we observe any plug flow behavior in $d = 3$ (although such phenomena can be observed with smooth bottom surfaces.) The potential for crystallization in two dimensions somewhat moderates these statements, since hysteresis and plug flow corresponding to crystallization are relatively easy to observe in $d = 2$. Nevertheless, in $d = 3$, these observations, combined with experimental results, support the centrality of Bagnold scaling in the rheology of dense flows, despite the puzzle offered by the normal stress anomalies seen.

Our results raise some interesting questions regarding, not only the nature of the flowing state, but about the transition to a static state at the angle of repose. This transition offers a peculiar combination of continuous with discontinuous elements. Continuous elements include:

- Velocity amplitudes for the Bagnold scaling regime,
- Stress state at the surface, compared with the Mohr–Coulomb failure criterion;

while discontinuous elements include:

- Plastic versus elastic grain contact states,
- Staticity index and coordination number of particles.

We expect that understanding the nature of the surface control of the underlying failure will elucidate this combination of elements.

Acknowledgments

S. J. Plimpton parallelized the molecular dynamics code used in these studies. D. Levine was supported by the U.S.–Israel Binational Science Foundation grant 1999235. G. S. Grest and L. E. Silbert were supported by Sandia National Laboratories, a multiprogram laboratory operated by Sandia Corporation, a Lockheed Martin Company, for the United States Department of Energy under Contract DE-AC04-94AL85000.

References

[1] Bagnold, R. A., *Proc. Roy. Soc. London* **A225**, 49 (1954); *ibid.* **295**, 219 (1966).
[2] Edwards, S. F. and Grinev, D. V., *Physica* **A263**, 545 (1999).
[3] Ertaş, D., Grest, G. S., Halsey, T. C., Levine, D. and Silbert, L. E., *Europhys. Lett.*, in press.
[4] Halsey, T. C. and Levine, A. J., *Phys. Rev. Lett.* **80**, 3141 (1998); Mason, T. G., Levine, A. J., Ertaş, D. and Halsey, T.C., *Phys. Rev.* **E60**, R5044 (1999).
[5] Halsey, T. C. and Ertaş, D., *Phys. Rev. Lett.* **83**, 5007 (1999).
[6] Jaeger, H. M., Nagel, S. R. and Behringer, R. P., *Rev. Mod. Phys.* **68**, 1259 (1996).
[7] Johnson, K. L., *Contact Mechanics* (Cambridge University Press, New York, 1985).
[8] Losert, W., Bocquet, L., Lubensky, T. C. and Gollub, J. P., *Phys. Rev. Lett.* **85**, 1428 (2000).
[9] Makse, H. A., Johnson, D. L. and Schwartz, L. M., *Phys. Rev. Lett.* **84**, 4160 (2000).
[10] Moukarzel, C. F., *Phys. Rev. Lett.* **81**, 1634 (1998).
[11] Nedderman, R. M., *Statics and Kinematics of Granular Materials* (Cambridge University Press, New York, 1992).
[12] Pouliquen, O., *Phys. Fluids* **11**, 542 (1999).
[13] Savage, S. B., *J. Fluid Mech.* **92**, 53 (1979).
[14] Silbert, L. E., Ertaş, D., Grest, G. S., Halsey, T. C., Levine, D. and Plimpton, S. J., *Phys. Rev. E*, in press; cond-mat/0105071.
[15] Silbert, L. E., Ertaş, D., Grest, G. S., Halsey, T. C. and Levine, D., unpublished.

CHAPTER 15

Glassy States in a Shaken Sandbox

PETER F. STADLER
Institut für Theoretische Chemie und Molekulare Strukturbiologie, Universität Wien, Währingerstraße 17, A-1090 Wien, Austria
and
The Santa Fe Institute, 1399 Hyde Park Rd., Santa Fe NM 87501, USA

ANITA MEHTA
S N Bose National Centre for Basic Sciences, Block JD Sector 3, Salt Lake, Calcutta 700098, India
and
ICTP, Strada Costiera 11, I-34100 Trieste, Italy

JEAN-MARC LUCK
Service de Physique Théorique (URA 2306 of CNRS), CEA Saclay, 91191 Gif-sur-Yvette cedex, France

Our model of shaken sand, presented in earlier work, has been extended to include a more realistic 'glassy' state, i.e. when the sandbox is shaken at very low intensities of vibration. We revisit some of our earlier results, and compare them with our new results on the revised model. Our analysis of the glassy dynamics in our model shows that a variety of ground states is obtained; these fall into two categories, which we argue are representative of regular and irregular packings.

Keywords: Vibrated granular media; glassy dynamics; lattice models.

1. Introduction

The test of a good lattice model of a complex system is whether it succeeds in capturing the essential physics of a real system in its bid to reduce its technical complexity. Areas as diverse as traffic flow [10], plate tectonics [12, 13] and granular flow [27] are examples where lattice models have been used rather successfully, despite their apparent simplicity, to describe at least a few of the salient features of some genuinely complex systems.

In this spirit, we present two versions of a model of shaken sand in the following; both models exhibit behavior that is representative of shaken sand between the

First published in *Advances in Complex Systems*, Vol. 4, No. 4 (2001), pp. 429–439.

fluidized and the glassy regimes. While the second model includes rather complex interactions between the 'grains' in the frozen state, in contrast to the first, the latter is nevertheless surprisingly successful in replicating at least some of the qualitative features associated with the glassy regime. One of the aims of this contribution is then to identify some of the essential features that are needed for a (discrete) minimal model of the system in question.

The earlier model ('old model') [31] is the generalization of a cellular-automaton (CA) model [4, 27] of an avalanching sandpile. This model shows *both* fast and slow dynamics in the appropriate regimes; in particular, it reduces to an exactly solvable model of noninteracting grains in the frozen ('glassy') regime, and provides one with a toy model for ageing in vibrated sand [9, 24].

Next, we present a more realistic version of the model ('new model'), which is the topic of ongoing research. While identical in the fluidized regime, the latter model is less of a toy model for the glassy state, in that the grains are no longer non-interacting, but are coupled to each other based on their orientations. Our analysis shows that a rich variety of ground states is obtained, which we analyze in terms of a particular parameter that has an interpretation in terms of the irregularity of the grains.

2. The Model

We consider a rectangular lattice of height H and width W with $N \leq HW$ grains located at the lattice points, shaken with vibration intensity Γ. Each grain is a rectangle with sides 1 and $a \leq 1$, respectively. Consider a grain (i,j) in row i (counting from the bottom) and column j whose height at any given time is given by $h_{ij} = n_{ij-} + a n_{ij+}$, where n_{ij-} is the number of vertical grains and n_{ij+} is the number of horizontal grains below (i,j).

The dynamics of the model are described by the following rules:

(i) If lattice sites $(i+1, j-1)$, $(i+1, j)$ or $(i+1, j+1)$ are empty, grain (i,j) moves there with a probability $\exp(-1/\Gamma)$, in units such that the acceleration due to gravity, the mass of a grain, and the height of a lattice cell all equal unity.

(ii) If the lattice site $(i-1, j)$ below the grain is empty, it will fall down.

(iii) If lattice sites $(i-1, j \pm 1)$ are empty, the grain at height h_{ij} will fall to either lower neighbor, provided the height difference $h_{ij} - h_{i-1,j\pm1} \geq 2$.

(iv) The grain flips from horizontal to vertical with probability $\exp(-m_{ij}(\Delta H + \Delta h)/\Gamma)$, where m_{ij} is the mass of the pile (consisting of grains of unit mass) above grain (i,j). For a rectangular grain, $\Delta H = 1 - a$ is the height difference between the initial horizontal and the final vertical state of the grain. Similarly, the *activation energy* for a flip reads $\Delta h = b - 1$, where $b = \sqrt{1 + a^2}$ is the diagonal length of a grain.

(v) The grain flips from vertical to horizontal with probability $\exp(-m_{ij}\Delta h/\Gamma)$.

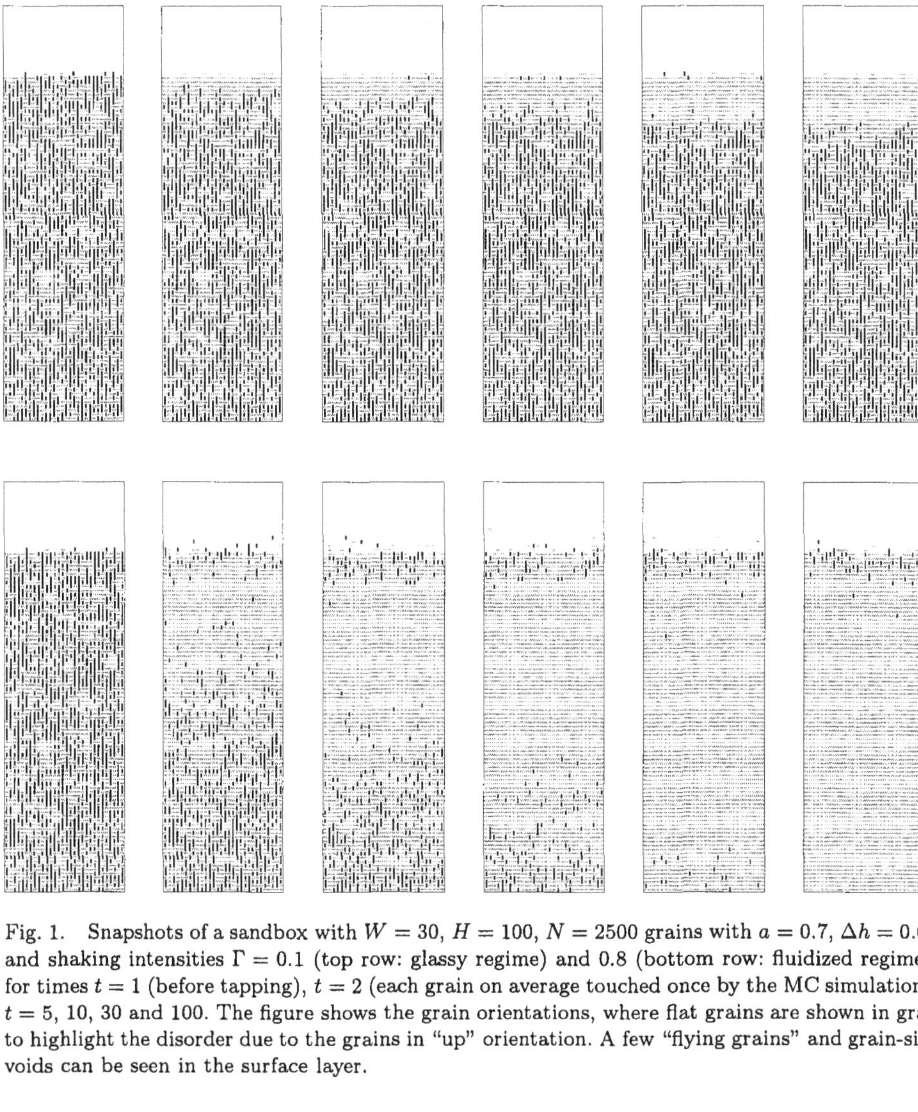

Fig. 1. Snapshots of a sandbox with $W = 30$, $H = 100$, $N = 2500$ grains with $a = 0.7$, $\Delta h = 0.05$ and shaking intensities $\Gamma = 0.1$ (top row: glassy regime) and 0.8 (bottom row: fluidized regime), for times $t = 1$ (before tapping), $t = 2$ (each grain on average touched once by the MC simulation), $t = 5$, 10, 30 and 100. The figure shows the grain orientations, where flat grains are shown in gray to highlight the disorder due to the grains in "up" orientation. A few "flying grains" and grain-size voids can be seen in the surface layer.

We use a Monte-Carlo update scheme, i.e. in each elementary step a position (i, j) in the box is picked at random; then the rules (i) to (v) are applied in order. The time unit is defined as HW such update attempts.

These rules yield a rich variety of dynamical behavior, depicted in Fig. 1, which shows the time evolution of the sandbox under the effect of different vibrational intensities. All the boxes are started in the same initial state; the observed behaviors correspond to the glassy regime (top row) and the fluidized regime (bottom row). In the intermediate regime separating the two [31], individual-particle relaxations are the mechanism for each particle to find local stability. These take place via a *fast* dynamics, which is non-equilibrium and non-ergodic in nature, at the end of which each particle is in its locally optimal configuration; the threshold at which this happens is called [7, 8] the *single particle relaxation threshold* (SPRT).

In line with recent investigations of compaction [1–3, 11, 16, 19, 20, 26, 29, 30], we have examined the behavior of the packing fraction of our model, as a function of the vibration intensity Γ. Let N^- and N^+ be the numbers of vertical and horizontal grains in the box. The packing fraction[a] ϕ is:

$$\phi = \frac{N^+ - aN^-}{N^+ + aN^-}, \tag{2.1}$$

which we use as an order parameter reflective of the behavior of the compactivity [19, 20]. Since the vertical orientation of a grain thus wastes space proportional to $1 - a$, relative to the horizontal one, this is a measure of the disorder in the packing. Throughout the contribution we use the value $a = 0.7$.

The advantage of our current model is that at least conceptually it can be extended to non-rectangular grain shapes. In this sense, we can and will consider a, ΔH and Δh as phenomenological parameters in the following.

3. A Spin-Model for the Ordered Regime

The frozen regime is characterized by an absence of holes within the sandbox, and negligible surface roughness. Here, the earlier model [31] reduces to an exactly solvable model of W independent columns of H noninteracting grain orientations $\sigma_n(t) = \pm 1$, with $\sigma = +1$ denoting a horizontal grain, and $\sigma = -1$ denoting a vertical grain. The orientation of the grain at depth n, measured from the top of the system, evolves according to a continuous-time Markov dynamics, with rates

$$\begin{cases} w(- \to +) = \exp\left(-\dfrac{n\Delta h}{\Gamma}\right), \\ w(+ \to -) = \exp\left(-\dfrac{n(\Delta H + \Delta h)}{\Gamma}\right), \end{cases} \tag{3.1}$$

as $m_{ij} = n = H+1-i$. The parameters ΔH and Δh correspond to two characteristic lengths of the model,

$$\xi_{\text{eq}} = \frac{\Gamma}{\Delta H}, \qquad \xi_{\text{dyn}} = \frac{\Gamma}{\Delta h}. \tag{3.2}$$

The equilibrium length ξ_{eq} is the typical depth below which all grains are frozen into their horizontal ground-state orientation, while the dynamical length ξ_{dyn} is characteristic of the divergence of the relaxation time with depth, $\tau \sim \exp(n/\xi_{\text{dyn}})$. As a consequence, any perturbation propagates only logarithmically slowly down the system, over an ordering length $\Lambda(t) \approx \xi_{\text{dyn}} \ln t$.

[a]The "packing fraction" defined here is strictly valid in a system without voids, i.e. where grain (mis)orientations are the only way of wasting space. The analogue in the case of a fluidized system is the 'voids ratio' used in engineering applications.

The detailed analysis of this model in earlier work, despite its 'toy model' nature, was remarkably successful in reproducing certain features of the glassy state. In this work, we present an improved version where the spins are no longer noninteracting, in the frozen regime. They are in fact coupled in a rather complex way; these more realistic interactions lead to a rich variety of ground states depending on the analogue of the aspect ratio a, which we argue is representative of the shape of the grains. We present this model in the next section.

4. A Generalization

The generalization of our earlier model involves the insertion of Eq. (2.1) into the transition rates of the system. We thus require that, for a given value of a, the transitions are such that the packing fraction of the system is locally minimized. We now allow a to take arbitrary values (while always remaining positive): $1-a$ can then be visualized as the size of the 'void' associated with the 'wrong' orientation of the grain. For convenience, we write these rules in a more general form below.

Thus, the dynamical rules (iv) and (v) of the rectangular grain model may be regarded as a special case of a more general model with transition rates

$$\begin{cases} w(- \to +) = \exp\left(-\dfrac{(\lambda_{ij} - \eta_{ij})}{\Gamma}\right), \\ w(+ \to -) = \exp\left(-\dfrac{(\lambda_{ij} + \eta_{ij})}{\Gamma}\right), \end{cases} \tag{4.1}$$

where

$$\eta_{ij} = Am_{ij+} + Bm_{ij-}, \qquad \lambda_{ij} = Cm_{ij+} + Dm_{ij-}, \tag{4.2}$$

are, respectively, the ordering field and the activation energy felt by grain (ij). In these expressions, $m_{ij\pm}$ is the number of horizontal and vertical grains above the grain (i,j), respectively. Note that grains are only counted from (i,j) to the first grain-sized void (empty lattice site) above level j. Thus, $m_{ij} = m_{ij+} + m_{ij-}$.

The earlier model is recovered by setting

$$A = B = \frac{\Delta H}{2}, \qquad C = D = \frac{\Delta H}{2} + \Delta h. \tag{4.3}$$

In the general situation where the equalities $A = B$, $C = D$ are not obeyed, the rates (4.1) depend on the orientations of all the grains above the grain under consideration. Our new model is therefore a fully directed model of interacting grains, where causality acts both in time and in space, as the orientation of a given grain only influences the grains below it and at later times.

The key parameter which governs the statics and dynamics of the model at low shaking intensity turns out to be the dimensionless ratio $\varepsilon = A/B$. Consider the ordered regime, where there are no holes, in the zero-temperature limit ($\Gamma \to 0$). In this regime, one of the rates defined in Eq. (4.1) is exponentially large with respect to the other one. As a consequence, the grains have well-defined steady-state orientations, which are given by the following deterministic equation:

$$\sigma_n = \mathrm{sign}(\varepsilon n_+(n) + n_-(n))\,, \tag{4.4}$$

where $n_\pm(n)$ is the number of horizontal and vertical grains at depths $0, \ldots, n-1$, so that $n_+(n) + n_-(n) = n = 1, 2, \ldots$. Equation (4.4) is recursive, because its right-hand side only involves the orientations of grains numbered $0, \ldots, n-1$.

Assume $\sigma_0 = +1$. As long as $\varepsilon \geq 0$, (4.4) leads to the trivial ground state where all the grains are horizontal ($\sigma_n = +1$), generalizing thus the case of the earlier model ($\varepsilon = 1$). In the frustrated regime ($\varepsilon < 0$), ground states have a richer structure. They contain non-trivial fractions $f_+ = 1/(1-\varepsilon)$, $f_- = -\varepsilon/(1-\varepsilon)$ of horizontal and vertical grains. The grain orientations are distributed in a way which depends, rather unexpectedly, on whether ε is rational or not.

(1) *Rational case*

If $\varepsilon = -p/q$ is a negative rational number (in irreducible form), Eq. (4.4) determines the grain orientation σ_n whenever the depth n is not a multiple of $r = p + q$. The orientations of the latter grains are left free. We thus obtain an extensively degenerate set of ground states, each of them being a random sequence of two types of r-mers, i.e. clusters of r grains. The associated configurational entropy per grain reads $\Sigma = (\ln 2)/r$. The simplest example is $\varepsilon = -1$, hence $r = 2$ and $f_- = 1/2$, where the clusters are $+-$ and $-+$, so that the ground states are all the dimerized grain configurations. Two examples consist of trimers ($r = 3$), namely $\varepsilon = -1/2$, with $f_- = 1/3$ and clusters $+-+$ and $-++$, and $\varepsilon = -2$, with $f_- = 2/3$ and clusters $+--$ and $-+-$.

(2) *Irrational case*

If ε is a negative irrational number, Eq. (4.4) determines all the grain orientations, so that the model admits a unique, non-degenerate ground state, where the grain orientations are distributed in a quasiperiodic fashion. The rule (4.4) is indeed equivalent to the cut-and-project algorithm used to build quasiperiodic binary chains, which are one-dimensional analogues of perfect quasicrystals [15, 17, 18, 21–23, 25]. For instance, for $-\varepsilon = (\sqrt{5}-1)/2 \approx 0.618033$ (the inverse golden mean), the unique ground state of the model is given by the well-known Fibonacci sequence: $+-+-++-+-++-++-+-++-+-++-++-+-+\cdots$.

5. Numerical Results

In Fig. 2 we show the behavior of the packing fraction as a function of Γ, the shaking intensity, for both models described above for a square box of side 120, containing 10,000 grains of unit mass, each of which has an aspect ratio of 0.7.

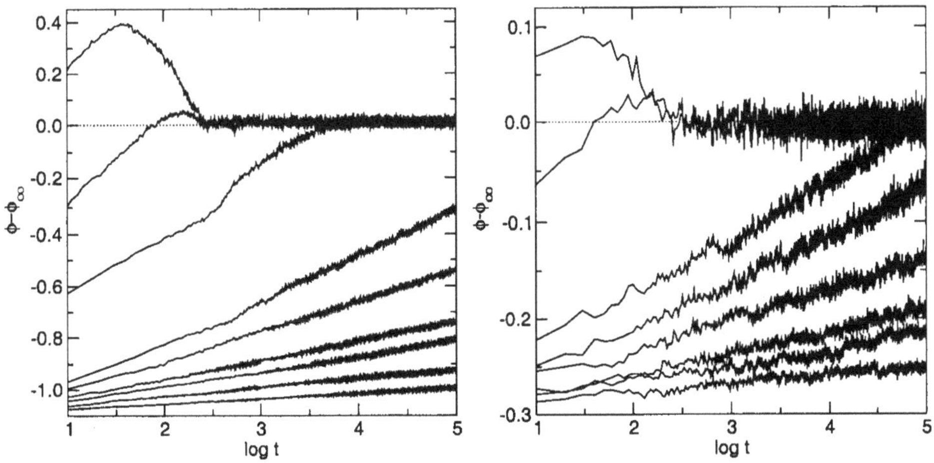

Fig. 2. Behavior of packing fraction for both models described in text, with an aspect ratio of 0.7 in each case. For the old model (left), $A = B = 0.35$, $C = D = 0.4$, while for the new model (right) $A = -0.2$, $B = C = D = 0.4$. The shaking intensities are, from bottom to top, $\Gamma = 0.3$ (old model only), 0.5, 0.8, 1, 1.5, 2, 5 and 10.

We note that the overall behavior of the two models is rather similar, although the complexity of the second makes it far more prone to fluctuations even after the steady state has been reached. In each case, we observe:

(1) *Fluidized region.* For $\Gamma \gg 1$, we observe an initial increase (caused by a *non-equilibrium* and transient 'ordering' of grains in the boundary layer) of the packing fraction that quickly relaxes to the equilibrium values ϕ_∞ in each case. This over-shooting effect in Fig. 2 increases with Γ, since grains ever deeper in the sandbox can now overcome their activation energy to relax to the horizontal.

(2) *Intermediate region* (for $\Gamma \approx 1$). Here the packing fraction remains approximately constant in the bulk, while the surface equilibrates via the fast dynamics of *single-particle relaxation.* The specific ϕ_∞ at which this occurs, is the *single-particle relaxation threshold density* observed in Ref. 8; non-equilibrium, non-ergodic, fast dynamics allows single particles locally to find their equilibrium configurations at this density. Analogous effects have been observed in recent experiments on colloids [34], where the correlated dynamics of *fast* particles was seen to be responsible for most relaxational behavior before the onset of the glass transition.

(3) *Frozen region* (for $\Gamma \ll 1$). Here the slow dynamics of the system results in a *logarithmic growth* of packing fraction with time:

$$\phi - \phi_\infty = b(\Gamma)\ln t + a\,, \tag{5.1}$$

where $b(\Gamma)$ increases with Γ, in good agreement with experiment [29, 30]. The slow dynamics has been identified [8] with a cascade process, where the free volume released by the relaxation of one or more grains allows for the ongoing relaxation of other grains in an extended neighbourhood. It includes the phenomenon of *bridge collapse*, which, for low vibration intensities, has been seen to be a major mechanism of compaction [2, 3, 26]. As Γ decreases, the corresponding ϕ_∞ increases asymptotically towards the jamming limit ϕ_{jam}, identified with a *dynamical phase transition* in related work [6].

While we have presented in earlier work [31] a full analysis of two-time correlation functions for the old model, work is currently in progress to investigate this in the rather more complex new model.

We finally investigate the analogue of 'annealed cooling,' where Γ is increased and decreased cyclically, and the response of the packing fraction observed [29, 30]. The results obtained here are similar to those [5] seen using more realistic models of shaken spheres, but the simplicity of the present lattice-based models allows for a greater transparency.

Starting with the sand in a fluidized state, as in the experiment [29, 30], we submit the sandbox to taps at a given intensity Γ for a time t_{tap} and increase the intensity in steps of $\delta\Gamma$; at a certain point, the cycle is reversed, to go from higher to lower intensities. The entire process is then iterated twice. Figure 3 shows the

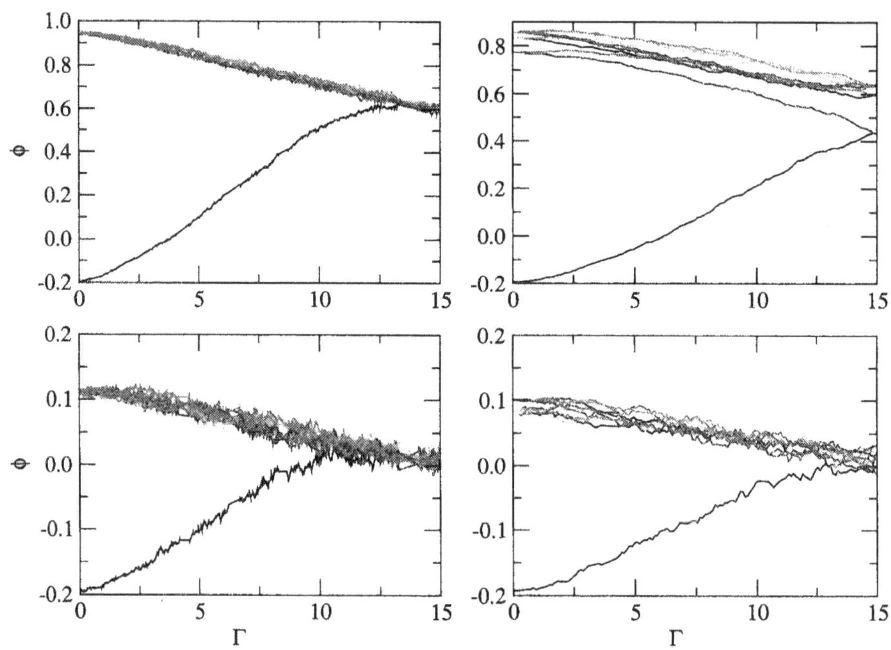

Fig. 3. Hysteresis curves. Top row: old model. Bottom row: new model. Left: $\delta\Gamma = 0.1$, $t_{\mathrm{tap}} =$ 2000 time units. Right: $\delta\Gamma = 0.001$, $t_{\mathrm{tap}} = 10^5$ time units. Note the approach of the irreversibility point Γ^* to the 'shoulder' Γ_{jam}, as the ramp rate $\delta\Gamma/t_{\mathrm{tap}}$ is lowered. The packing fraction tends to its close-packing limit in the limit of low intensities for the old model, while it asymptotes towards the jamming limit for the new model (see text). Note that negative values of phi correspond to a preponderance of disordered grains.

resulting behavior of the volume fraction ϕ as a function of Γ, where an 'irreversible' branch and a 'reversible' branch of the compaction curve are seen, which meet at the 'irreversibility point' Γ^* [29, 30]. The left- and right-hand side of Fig. 3 correspond respectively to high and low values of the 'ramp rate' $\delta\Gamma/t_{\mathrm{tap}}$ [29, 30], while the upper and lower panels correspond respectively to the old and new versions of our model. As the ramp rate is lowered, we note that:

(1) The width of the hysteresis loop in the so-called reversible branch decreases, in both cases. The 'reversible' branch is thus not reversible at all; more realistic simulations of shaken spheres [2, 3, 26] confirm the first-order, irreversible nature of the transition, which allows the density to attain values that are substantially higher than random close packing, and quite close to the crystalline limit [28]. Precisely such a transition has also recently been observed experimentally in the compaction of rods [33].

(2) In both panels, the 'irreversibility point' Γ^* approaches Γ_{jam} (the shaking intensity at which the jamming limit ϕ_{jam} is approached), in agreement with results on other discrete models [14]. However, in the upper panel, the packing fraction at low intensities tends towards close-packing (the so-called dynamical transition referred to in Ref. 8); this is at odds with the results of real experiment, which models with a greater degree of complexity [7] have been able to replicate. In the lower panel, which corresponds to our new model, we see tentative indications of an improvement in this respect vis-a-vis the old model; the packing fraction here asymptotes towards the jamming limit, rather than rising indefinitely towards the close-packing limit [7, 29].

6. Discussion

We have presented two models of shaken sandboxes; while their design was such that they would show identical behavior in regimes where there were a finite density of voids, the modelling of the densely packed regime was completely distinct in each case. In the first case, the model reduced to a model of noninteracting spins, while in the second, the insistence that all allowed transitions minimized a suitably defined local packing fraction led in fact to an intricate coupling between the grains. This physically motivated interaction was extremely nonlocal as well as directional. In this way we were able to generate a model that, despite being one-dimensional, has an extremely complex ground-state structure, depending on the *regularity* of the grain shape. Our present investigations, to be published elsewhere, concern the effect of zero-temperature and finite-temperature tapping of this system; our preliminary studies indicate that for regularly shaped grains, strong metastability in the achievable ground states is observed. For irregularly shaped grains, as in reality, a far better packing is achievable, since orientations of irregularly shaped grains are much better able to fill space [32].

It is, however, a rather salutary exercise to see that despite the relative sophistication of the new model in its inclusion of non-trivial interactions, most of the qualitative behavior of the packing fraction as a function of steady as well as annealed tapping, remains unchanged. We expect that quantitative features such as two-time correlation functions will be far more non-trivial in the second model than the first, although we expect their overall features to be rather similar. It is tempting to speculate that the directionality due to gravity (which leads to strongly non-Hamiltonian behavior, since grain couplings are propagating down the pile) which unites both first and second models might well be the most important ingredient that is needed to describe such lattice-based models of shaken sand.

References

[1] Aradian, A., Raphaël, E. and de Gennes, P. G., Thick surface flows of granular materials: Effect of the velocity profile on the avalanche amplitude, *Phys. Rev.* **E60**, 2009–2019 (1999).

[2] Barker, G. C. and Mehta, A., Vibrated powders — structure, correlations and dynamics, *Phys. Rev.* **A45**, 3435–3446 (1992).

[3] Barker, G. C. and Mehta, A., Transient phenomena, self-diffusion and orientational effects in vibrated powders, *Phys. Rev.* **E47**, 184–188 (1993).

[4] Barker, G. C. and Mehta, A., Avalanches at rough surfaces *Phys. Rev.* **E61**, 6765–6772 (2000).

[5] Barker, G. C. and Mehta, A., Inhomogeneous relaxation in vibrated granular media: Consolidation waves (2001); Phase Transitions, to appear; cond-mat/0010268.

[6] Berg, J. M. and Mehta, A., see contribution in this volume.

[7] Berg, J. M. and Mehta, A., *Phys. Rev.* **E65**, 031305 (2002), cond-mat/0108225.

[8] Berg, J. M. and Mehta, A., On random graphs and the statistical mechanics of granular matter (2001); *Europhys. Lett.* **56**, 784–791 (2001); cond-mat/0012416.

[9] Berthier, L., Cugliandolo, L. F. and Iguain, J. L., Glassy systems under time-dependent driving forces: Application to slow granular rheology, *Phys. Rev.* **E63**, 051302 (2001).

[10] Biham, O., Middleton, A. A. and Levine, D., Self-organization and a dynamical transition in traffic-flow models, *Phys. Rev.* **A46**, R6124–6127 (1992).

[11] Boutreux, Th., Raphaël, E. and de Gennes, P. G., Surface flows of granular materials: A modified picture for thick avalanches, *Phys. Rev.* **E58**, 4692–4700 (1998).

[12] Burridge, R. and Knopoff, L., Model and theoretical seismicity, *Bull. Seis. Soc. Am.* **57**, 341–371 (1967).

[13] Carlson, J. M. and Langer, J. S., Mechanical model of an earthquake fault, *Phys. Rev.* **A40**, 6470–6484 (1989).

[14] Coniglio, A. and Nicodemi, M., The jamming transition of granular media, *J. Phys. Cond. Matt.* **12**, 6601–6610 (2000).

[15] de Bruijn, N. G., Sequences of zeros and ones generated by special production rules, *Nederl. Akad. Wetensch. Proc.* **A84**, 27–37 (1981).

[16] de Gennes, P. G., Tapping of granular packs: A model based on local two-level systems, *J. Coll. Int. Sci.* **226**, 1–4 (2000).

[17] Duneau, M. and Katz, A., Quasiperiodic patterns, *Phys. Rev. Lett.* **54**, 2688–2691 (1985).

[18] Duneau, M. and Katz, A., Quasiperiodic patterns and icosahedral symmetry, *J. Phys. (France)* **47**, 181–196 (1986).

[19] Edwards, S. F., The role of entropy in the specification of a powder, in *Granular Matter: An Interdisciplinary Approach*, Mehta, A., ed. (Springer-Verlag, New York, 1994).
[20] Edwards, S. F. and Grinev, D. V., Statistical mechanics of vibration-induced compaction of powders, *Phys. Rev.* **E58**, 4758–4762 (1999).
[21] Elser, V., Indexing problems in quasicrystal diffraction, *Phys. Rev.* **B32**, 4892–4898 (1985).
[22] Kalugin, P. A., Kitayev, A. Yu. and Levitov, L. S., $Al_{0.86}Mn_{0.14}$: A six-dimensional crystal, *J.E.T.P. Lett.* **41**, 145–149 (1985).
[23] Kalugin, P. A., Kitayev, A. Yu. and Levitov, L. S., Six-dimensional properties of $Al_{0.86}$ $Mn_{0.14}$ alloy, *J. Phys. Lett. (France)* **46**, L601–L607 (1985).
[24] Kurchan, J., Emergence of macroscopic temperatures in systems that are not thermodynamical microscopically: Towards a thermodynamical description of slow granular rheology, *J. Phys. Cond. Matt.* **12**, 6611–6617 (2000).
[25] Luck, J.-M., Godrèche, C., Janner, A. and Janssen, T., The nature of the atomic surfaces of quasiperiodic self-similar structures, *J. Phys.* **A26**, 1951–1999 (1993).
[26] Mehta, A. and Barker, G. C., Vibrated powders — a microscopic approach, *Phys. Rev. Lett.* **67**, 394–397 (1991).
[27] Mehta, A. and Barker, G. C., Disorder, memory and avalanches in sandpiles, *Europhys. Lett.* **27**, 501–506 (1994).
[28] Mehta, A. and Barker, G. C., Glassy dynamics of granular compaction, *J. Phys. Cond. Matt.* **12**, 6619–6628 (2000).
[29] Nowak, E. R., Knight, J. B., Ben-Naim, E., Jaeger, H. M. and Nagel, S. R., Density fluctuations in vibrated granular materials, *Phys. Rev.* **E57**, 1971–1982 (1998).
[30] Nowak, E. R., Knight, J. B., Povinelli, M., Jaeger, H. M. and Nagel, S. R., Reversibility and irreversibility in the packing of vibrated granular material, *Powder Technology* **94**, 79–83 (1997).
[31] Stadler, P. F., Luck, J.-M. and Mehta, A., *Europhys. Lett.* **57**, 46–53 (2002), cond-mat/0103076.
[32] Torquato, S. and Chaikin, P., (unpublished).
[33] Villarruel, F. X., Lauderdale, B. E., Mueth, D. M. and Jaeger, H. M., Compaction of rods: Relaxation and ordering in vibrated, anisotropic granular material, *Phys. Rev.* **E61**, 6914 (2000).
[34] Weeks, E. R., Crocker, J. C., Levitt, A. C., Schofield, A. and Weitz, D. A., Three-dimensional direct imaging of structural relaxation near the colloidal glass transition, *Science* **287**, 627–631 (2000).

CHAPTER 16

Slow Dense Granular Flows as a Self-Induced Process

OLIVIER POULIQUEN* and YOËL FORTERRE
IUSTI, Université de Provence, 5 rue Enrico Fermi,
13453 Marseille Cedex 13, France
** olivier@iusti.univ-mrs.fr*

STÉPHANE LE DIZES
IRPHE, Université de Provence, 49 rue Joliot Curie,
B.P. 146 13384 Marseille Cedex 13, France

A simple model is presented for the description of steady uniform shear flow of granular material. The model is based on a stress fluctuation activated process. The basic idea is that shear between two particle layers induces fluctuations in the media that may trigger a shear at some other position. Based on this idea, a minimum model is derived and applied to different configurations of granular shear flow.

Keywords: Granular media; rheology; friction; fluctuations.

1. Introduction

The description of the flow of cohesionless granular material still represents a challenge [23]. In a collisional regime, when the medium is dilute and strongly agitated, hydrodynamic equations have been proposed by analogy with a molecular gas [7, 14]. Assuming the collisions between particles to be instantaneous and inelastic, one can derive constitutive equations for the density, velocity and granular temperature (a measure of the velocity fluctuations). However, in many cases energy injected is not sufficient and the dissipation due to the inelastic collisions is so efficient that the medium does not stay in a collisional regime and flows in a so-called dense regime. The particles experience multibody interactions and long-lived contacts. The material can no longer be seen as a granular gas and there is a need for another description. Experiments on dense granular flows have been carried out in different configurations including shear cells [13, 16, 27], silos [3, 18–20], flow down inclined planes [4, 5, 21], and flow at the surface of a heap [10, 11]. For these configurations precise information is now available about the velocity profiles, the

First published in *Advances in Complex Systems*, Vol. 4, No. 4 (2001), pp. 441–450.

fluctuations, and stresses that developed during the flows. However, there is a lack of unified theoretical description.

From the theoretical point of view, several approaches have been proposed to describe dense granular flows. A first attempt has been made to extend the kinetic theory of granular matter by introducing a shear rate independent term in order to incorporate friction [9, 24]. Another approach has been proposed recently by Bocquet *et al.* [2]. They suggest changing the density dependence of the viscosity in order to take into account the fact that particles at high volume fraction are trapped in cages. However, it is not clear that such approaches are still valid when interactions are not collisional. Savage [25] proposed a hydrodynamic model based on a fluctuating plasticity model. He obtained a hydrodynamic description close to the kinetic theory where the viscosity decreases with the temperature. Aranson *et al.* [1] developed an empirical theory based on the idea of a mixture of a fluid and a solid, the relative concentration of both constituents being driven by a Landau equation. This model seems to be efficient in describing non-stationary effects like avalanche triggering. Mills *et al.* [15] and Jenkins and Chevoir [8] described the granular fluid as a classical viscous fluid in which collective objects such as arches or columns are present, propagating the stresses in a non-local way. They provide a non-local description of the flow down inclined planes.

In this paper we present an alternative approach for describing slow dense granular flows. The model is based on a stress fluctuation activated process. The idea is that a shear somewhere in the material induces stress fluctuations in the medium which can in turn help shearing somewhere else. A crude model of a fluctuation activated process was previously proposed in the context of flow in a silo [20]. More recently, Debregeas *et al.* [6] developed a fluctuation model to describe velocity profiles and correlations observed in simple shear flow experiments on foams. In this paper we will present a simple model where the stress fluctuations are induced by the shear itself. We apply the model to different configurations.

2. Description of the Model

The main idea of the model is sketched out in Fig. 1. If a shear motion occurs at a level z' between two layers, local rearrangements will occur and induce stress fluctuations in the assembly of particles. These stress fluctuations can help the material to yield somewhere else in z. The motion in z is then related to the fluctuations created by the level z'.

In order to apply this idea, we consider a stationary parallel shear flow of particles with a mean velocity $u(z)$ along x which depends only on the vertical coordinate z (Fig. 1). We call $P(z)$ and $\tau(z)$ respectively the pressure and the shear stress at the level z. The particle diameter is d. For simplicity, the model is written in terms of a stack of layers and only the direction z is considered.

The yield criterion in z can be expressed in terms of a Coulomb criterion by introducing a friction coefficient μ; the material will spontaneously shear in z if

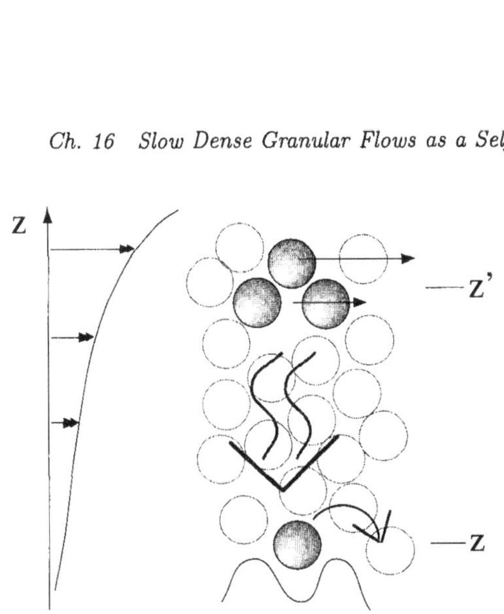

Fig. 1. Sketch of the self-activated process: shear at a level z' induces fluctuations which can induce yielding in z.

the local shear stress $|\tau(z)|$ reaches the value $\mu P(z)$, otherwise nothing happens. The friction coefficient μ takes into account both the friction between grains and the geometric entanglement. However, if stress fluctuations exist in addition to the mean stresses, the yield criterion can be locally reached even if the mean values of the shear and normal stresses do not verify the yield criterion.

This is, for example, the case when shear occurs at the level z' in the medium. If $\delta\sigma_{z'\to z}$ is the amplitude of the stress fluctuation induced by a shear in z' and measured at the level z, then this fluctuation will be sufficient to induce yielding in z if $\delta\sigma_{z'\to z} > (\mu P - |\tau|)$.

We then write that each time z' sends a fluctuation, the particle in z will jump to the next hole on the right with a probability equal to the probability that the stress fluctuation is higher than the threshold: $\mathcal{P}[\delta\sigma_{z'\to z} > (\mu P - \tau)]$. By symmetry we can write that the probability of jumping to the next hole on the left is $\mathcal{P}[\delta\sigma_{z'\to z} > (\mu P + \tau)]$.

We then write that the frequency at which z' sends fluctuations is simply the shear rate amplitude at z': $\left|\frac{du}{dz}(z')\right|$. Stipulating that the event coming from different levels z' are uncorrelated, we can write the shear rate in z as a sum over z':

$$\frac{du}{dz}(z) = \sum_{z'} \left|\frac{du}{dz}(z')\right| \left(\mathcal{P}[\delta\sigma_{z'\to z} > (\mu P - \tau)] - \mathcal{P}[\delta\sigma_{z'\to z} > (\mu P + \tau)]\right). \qquad (2.1)$$

It is important to keep in mind that in Eq. (2.1) several assumptions have been made. First, it is assumed that there is no memory in the process; if the stress fluctuation received at some point does not reach the threshold, the system returns to its initial state and the next fluctuation will have to reach the same threshold to induce yielding. There is no stress accumulation.

A second assumption which is made is that the jumps are instantaneous. The time necessary for the particle to jump to the next hole (of the order of $d\sqrt{\rho/P}$) is negligible compared to the elapsed time between two successive fluctuations

(du/dz^{-1}). This means that the model developed here will only apply for slow shear rates and quasi-static deformations for which $du/dz\, d\sqrt{\rho/P} \ll 1$.

In order to express the probabilities in Eq. (2.1) we have to choose a probability distribution for the stress fluctuation $\delta\sigma_{z'\to z}$ induced by z' on z. A plausible assumption is that the amplitude of fluctuation is a decreasing function of the distance $(z'-z)$ and that it is maximum for $z = z'$ equal to some value $\delta\sigma_0(z')$. We choose the following:

$$\delta\sigma_{z'\to z} = \frac{\delta\sigma_0(z')}{1+\beta(z-z')^2/d^2},$$

where β is a dimensionless parameter of the order of unity measuring the characteristic length in particle diameters over which the fluctuation decays. We tried other decreasing functions and the results described in this paper were qualitatively unchanged.

The amplitude $\delta\sigma_0(z')$ is a random variable with a mean which has to be of the order of the local pressure $P(z')$. Recent studies of the distribution of forces in granular piles suggest that the distribution is exponential over a large range of forces [12, 17, 22]. We then assume in the following that $\delta\sigma_0(z')$ follows an exponential distribution, i.e. the probability $pd(\delta\sigma_0)$ of having the value $\delta\sigma_0$ within a range $d(\delta\sigma_0)$ is equal to

$$pd(\delta\sigma_0) = \frac{1}{P(z')}\exp\left(-\frac{\delta\sigma_0}{P(z')}\right)d(\delta\sigma_0)\,.$$

We can then easily show that the probability of jumps in Eq. (2.1) can also be expressed in term of an exponential:

$$\begin{aligned}\mathcal{P}[\delta\sigma_{z'\to z} > (\mu P - \tau)] &= \int_{(\mu P-\tau)(1+\beta(z-z')^2/d^2)}^{\infty} pd(\delta\sigma_0)\\ &= \exp\left(-\frac{(\mu P(z)-\tau(z))(1+\beta(z-z')^2/d^2)}{P(z')}\right). \end{aligned} \quad (2.2)$$

It is interesting to note that the above expression is similar to an activated process where the role of the energy barrier would be played by $\mu P(z) - \tau(z)$ and the role the temperature by the mean stress fluctuation.

Finally, transforming the discrete sum in Eq. (2.1) to an integral, we can write the final expression for the shear rate as a function of the shear stress:

$$\begin{aligned}\frac{du}{dz}(z) = \frac{1}{d}\int\left|\frac{du}{dz}(z')\right| &\left(\exp\left(-\frac{(\mu P(z)-\tau(z))(1+\beta(z-z')^2/d^2)}{P(z')}\right)\right.\\ &\left.-\exp\left(-\frac{(\mu P(z)+\tau(z))(1+\beta(z-z')^2/d^2)}{P(z')}\right)\right)dz'\,. \end{aligned} \quad (2.3)$$

This integral equation gives a rheological law relating the stresses and shear rate. There are only two parameters in the model, μ the friction coefficient at the grain level, and β the dimensionless extension of the fluctuations. In the following we solve this equation in different configurations. Note that if the shear keeps the same sign across the layer, Eq. (2.3) is linear with respect to the shear rate. This means that a non-zero shear rate profile will correspond to specific values of the stress distribution. This eigenvalue problem is solved numerically by discretizing the integral in Eq. (2.3). Once the shear rate profile is known, the velocity profile is obtain by integration with the condition of zero velocity on the walls.

3. Uniform Shear

Let us first consider the simple rheological test of an infinite medium without gravity with an imposed constant shear: $\frac{du}{dz} = \Gamma = cte$. The pressure P is fixed to a constant. Equation (2.3) then indicates that a non-zero shear rate exists only if the shear stress τ satisfies

$$\frac{e^{-(\mu-\tau/P)}}{\sqrt{\beta(\mu-\tau/P)}} - \frac{e^{-(\mu+\tau/P)}}{\sqrt{\beta(\mu+\tau/P)}} = \sqrt{\frac{\beta}{\pi}}\,. \tag{3.1}$$

According to the model, a simple shear is then obtained for a specific value of the ratio τ/P. The material behaves like a frictional material with a coefficient of friction depending on the two parameters μ and β. For example, for the typical values $\mu = 0.7$ and $\beta = 2$, one finds that $\tau/P = 0.42$. The macroscopic friction coefficient τ/P is then less than the microscopic one μ.

4. Silo

In this section we address the problem of granular material flowing in a silo consisting of two parallel rough walls. Experiments have been carried out to measure the velocity profile in two-dimensional and three-dimensional configurations. The main result is that the shear is localized close to the rough walls in two shear bands whose thickness is between five and ten particles diameters, not very sensitive to the distance between the walls [3, 19, 20]. However, it has been shown that the thickness of the shear zones is affected when the silo is inclined from the vertical. The shear zone close to the bottom wall is larger and the other one thinner [20].

With the model presented in this paper it is possible to solve the problem for the flow between two rough planes. In this configuration the material is bounded by two walls at a distance L. In the model the walls are simply the first and last row of particles at $-L/2$ and $L/2$ which becomes the limits of the integral in Eq. (2.3). According to the equilibrium relations the stresses satisfy

$$\frac{dP}{dz} = -\rho g \cos(\theta)\,, \qquad \frac{d\tau}{dz} = -\rho g \sin(\theta)\,, \tag{4.1}$$

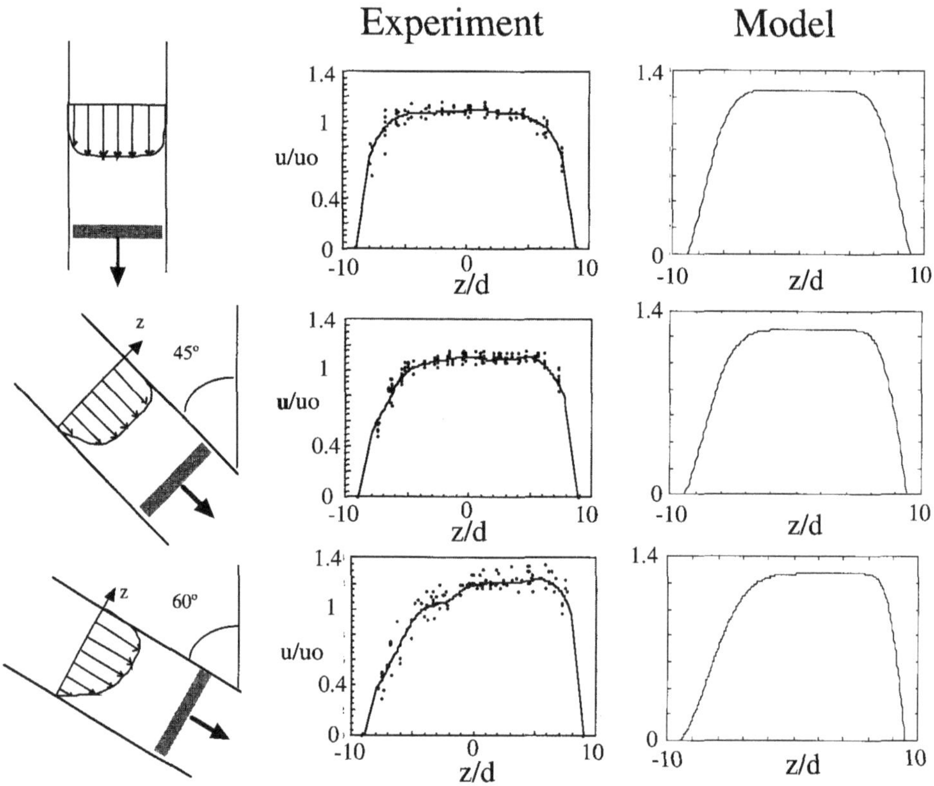

Fig. 2. Velocity profiles observed in an inclined silo in 2D experiments (Data from Ref. 20) and predicted by the model ($\mu = 0.7$, $\beta = 2$).

where θ is the angle between the walls and the horizontal, ρ is the density of the grain packing and g the gravity. Hence, $P(z) = -\rho g \cos(\theta) z + P_0$ and $\tau(z) = -\rho g \sin(\theta) + \tau_0$, where P_0 and τ_0 are constant which are *a priori* unknown. The constants are determined by the eigenvalue problem of Eq. (2.3). As we expect the shear rate to change sign in the silo, we have to solve Eq. (2.3) for both signs of the absolute value $\left|\frac{du}{dz}(z')\right|$ and then match the two solutions to get the correct solution of Eq. (2.3). This matching condition allows one to find the two constants P_0 and τ_0, and the associated shear rate profile.

The predicted velocity profiles are presented in Fig. 2 for three different inclinations of the silo. We have plotted both the velocity profile observed in the experiments and predicted by the model. For the vertical case, the model predicts the existence of shear zones localized close to the walls as observed in the experiments. The width of the shear zone varies slightly with the distance between the rough walls but is of the order of five particle diameters as shown in Fig. 3. When the silo is inclined, the velocity profile is asymmetric and the shear zone is larger on the bottom side than on the upper side as observed in the experiments.

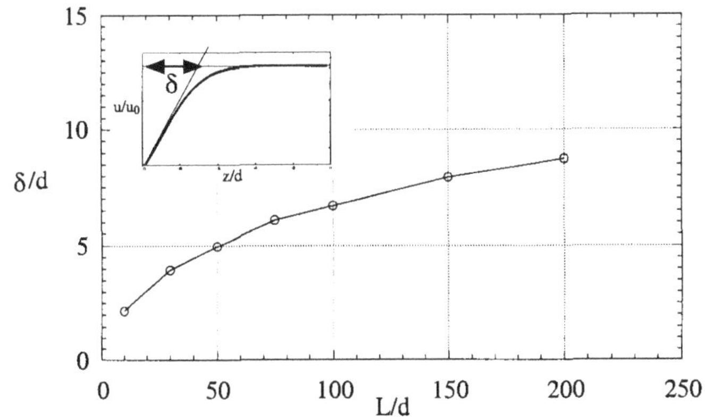

Fig. 3. Thickness of the shear zone as a function of the distance between the walls ($\mu = 0.7$, $\beta = 2$).

5. Flow Down Inclined Planes

We then address the problem of the flow down a rough plane. The two control parameters are, in this case, the inclination θ of the plane and the thickness of the flow h (Fig. 4). Numerical and experimental studies have been carried out giving information about the properties of steady uniform flows [4, 21, 26]. The linear model developed in this paper is only valid for quasi-static deformations, which means that we will not be able to describe the fully developed flow down inclines. A non-linear extension of the model is under development and will be the subject of another study. However, from the present model we can get information about the onset of flow. It has been shown experimentally and numerically that for a given inclination θ, there exists a minimum thickness $h_{\text{stop}}(\theta)$ below which no steady uniform flow is observed. The present model is able to predict such behavior.

In this configuration the material is bounded by a rough bottom which in the model is simply the first row of particles, and a free surface at $z = h$. In order to use the same definition for the layer thickness as in the experimental work, i.e. where the fixed layer is not counted, the integration domain in Eq. (2.3) is from $-d$ up to h. The stress distribution is derived from the equilibrium and from the no stress boundary condition at the free surface: $P = \rho g \cos(\theta)(h - z)$, and $\tau = \rho g \sin(\theta)(h - z)$.

From Eq. (2.3) it appears that for a given thickness h, a non-zero shear rate profile exists only for a given value of the inclination. This critical inclination is a function of the thickness of the layer h as shown in Fig. 4. The model predicts that in order to flow, a thin layer has to be more inclined than a thick layer. This result is reminiscent of the onset of flow observed in experiments. This behavior can be easily understood in the present model: for a thin layer, the integral in Eq. (2.3) extends over a narrower region, which means that there are fewer sources of fluctuation to help yielding. The inclination has then to be higher in order to be closer to the yielding threshold.

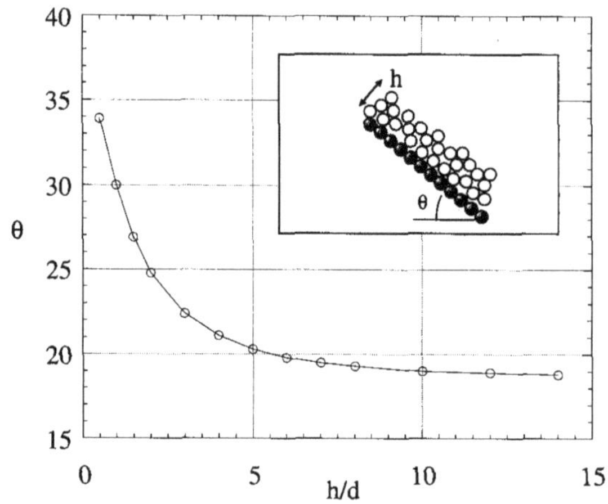

Fig. 4. Critical angle at which a flow down an inclined plane is possible as a function of the thickness of the layer h ($\mu = 0.7$, $\beta = 2$).

6. Shear Between Plates

The last example we want to address is the simple shear experiment. Recently, careful measurements of the velocity profile have been performed in a Taylor Couette cell where the granular material is sheared between two concentric cylinders in 2D or 3D geometry [13, 16, 27]. The shear is localized close to the moving walls and extends up to five or ten particle diameters in the bulk. The shear zone is associated with a slightly lower solid fraction.

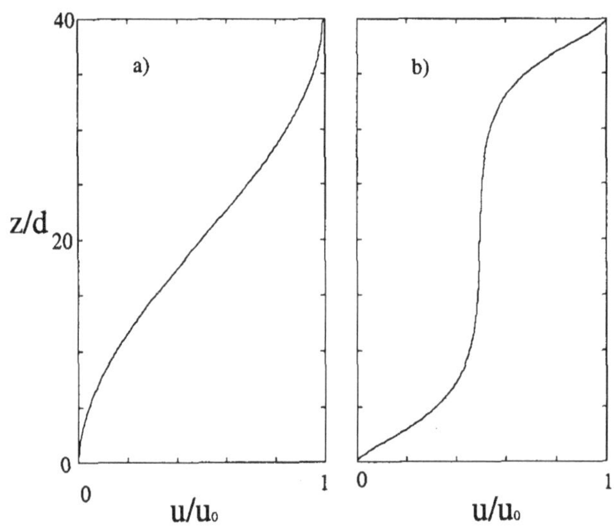

Fig. 5. Velocity profile predicted for the shear between two plates. (a) simple model (b) model with image sources of fluctuation.

We have solved our model for the shear flow between two plates, one being fixed, the other one moving at a velocity of unity. The pressure P and the shear stress τ are constant across the layer. Again, the linearity of the problem imposes that a solution is found only for a specific value of the ratio τ/P. The corresponding velocity profile is plotted in Fig. 5(a) for a gap between the plates equal to $40d$. The velocity profile we obtained is not localized close to the walls; the shear is distributed over the whole material. The phenomena of localization in simple shear flow is then not predicted with the simple model presented here.

However, several modifications of the model give rise to a localization. First, our configuration is plane whereas most of the experiments are carried out in cylindric geometry. A small asymmetry introduced, for example, in the pressure distribution to take into account of the cylindric geometry, gives rise to a localized shear band. However, in this case the thickness of the shear zone depends on the asymmetry which does not seem to be observed in experiments.

A more convincing improvement of the model consists of taking into account boundary effects in the perturbation function $\delta\sigma_{z'\to z}$. The stress fluctuation induced by a shear close to the wall is not the same as the fluctuation induced when the shear occurs far from the wall in the bulk. The wall being rigid, we have a zero displacement condition. Qualitatively, we then expect that a shear close to a wall induces higher fluctuation than a shear in the bulk, as part of the fluctuation is reflected by the wall. A simple way to take this wall effect into account is to add an additional source of fluctuation on the other side of the wall, an image of the real one. In the case of the shear cell where we have two rigid walls at position 0 and H, we then write the perturbation $\delta\sigma_{z'\to z}$ as the sum of three terms corresponding to a source in z', a source in $z'' = -z'$ and one in $z''' = 2H - z'$.

Using the image sources does not qualitatively change any of the results presented above for the silo or the inclined plane. However, for the shear between two walls, the solution shows two shear zones localized close to the walls as shown in Fig. 5(b). Therefore, more dangerous fluctuations close to the walls induce a localization of the shear bands.

7. Discussion and Conclusions

In this paper we have presented a simple model for slow sheared granular flows which is based on stress fluctuations. The process we describe is self-induced as the shear induces fluctuations which in turn induce shear. As a result we obtain an integral rheological law, where the shear rate at a position is related to the shear rate in the vicinity. Quasi-static flow, like flows between vertical plates or in shear cells, are well described by the model. It is also able to predict the onset of flow for the inclined plane configuration.

However, the model is still very crude. First, it is one-dimensional; the stress fluctuations are described by the z coordinate. A careful 3D analysis is certainly necessary to describe the fluctuations and shear induced motions better. A second improvement we are working on is the finite duration of the jump. When yielding is activated at some place by a fluctuation, it takes a finite time for the particle to go to the next hole. Taking into account this time delay should allow one to describe more rapid flows like inclined chute flows.

Acknowledgments

This work is supported by the French Ministry of Research and Education (ACI Blanche #2018). We would like to thank G. Debregeas and C. Caroli for fruitful discussions.

References

[1] Aranson, I. S. and Tsimring, L. S., *Phys. Rev.* **E64**, 020301 (2001).
[2] Bocquet, L. *et al.*, to be published in *Phys. Rev. E*, cond-mat/0012356 (2001).
[3] Chevoir, F. *et al.*, in *Powders and Grains*, Kishino, Y., ed. (Lisse, Swets and Zeitlinger, 2001), pp. 399–402.
[4] Chevoir, F. *et al.*, in *Powder and Grains 2001*, Kishino, Y., ed. (Lisse, Swets and Zeitlinger, 2001), pp. 373–376.
[5] Daerr, A. and Douady, S., *Nature* **399**, 241 (1999).
[6] Debregeas, G., Tabuteau, H. and di Meglio, J. M., to be published in *Phys. Rev. Lett.*, cond-mat/0103440.
[7] Goldhirsch, I., *Chaos* **9**, 659 (1999).
[8] Jenkins, J. T. and Chevoir, F., *Dense Plane Flows of Frictional Spheres Down a Bumpy, Frictional Incline*, preprint (2001).
[9] Johnson, P. C., Nott, P. and Jackson, R., *J. Fluid Mech.* **210**, 501 (1990).
[10] Khakhar, D. V., Orpe, A. V., Andersen, P. and Ottino, J. M., *J. Fluid Mech.* **441**, 255 (2001).
[11] Lemieux, P. A. and Durian, D. J., *Phys. Rev. Lett.* **85**, 4273 (2000).
[12] Liu, C. H. *et al.*, *Science* **269**, 513 (1995).
[13] Losert, W., Bocquet, L., Lubensky, T. C. and Gollub, J. P., *Phys. Rev. Lett.* **85**, 1428 (2000).
[14] Lun, C. K. K. *et al.*, *J. Fluid Mech.* **140**, 223 (1984).
[15] Mills, P., Loggia, D. and Tixier. M., *Europhys. Lett.* **45**, 733 (1999).
[16] Mueth, D. M., Debregeas, G. F., Karczmar, G. S., Eng, P. J., Nagel, S. R. and Jaeger, H. M., *Nature* **406**, 385 (2000).
[17] Mueth, D. M., Jaeger, H. and Nagel, S. R., *Phys. Rev.* **E57**, 3164 (1998).
[18] Natarajan, V. V. R., Hunt, M. L. and Taylor, E. D., *J. Fluid Mech.* **304**, 1 (1995).
[19] Nedderman, R. M. and Laohakul, C., *Powder Technology* **25**, 91 (1980).
[20] Pouliquen, O. and Gutfraind, R., *Phys. Rev.* **E53**, 552 (1996).
[21] Pouliquen, O., *Phys. Fluids* **11**, 542 (1999).
[22] Radjai, F., Jean, M., Moreau, J. J. and Roux, S., *Phys. Rev. Lett.* **77**, 274 (1996).
[23] Rajchenbach, J., *Adv. Phys.* **49**, 229 (2000).
[24] Savage, S. B., in *Mechanics of Granular Materials: New Models and Constitutive Relations*, Jenkins, J. T. and Satake, M. eds. (Elsevier, Amsterdam, 1983).
[25] Savage, S. B., *J. Fluid Mech.* **377**, 1 (1998).
[26] Silbert, L. E. *et al.*, preprint cond-mat/0105071 (2001).
[27] Veje, C. T., Howell, D. W. and Behringer, R. P., *Phys. Rev.* **E59**, 739 (1999).

CHAPTER 17

Granular Media as a Physics Problem

S. F. EDWARDS and D. V. GRINEV

Polymers and Colloids Group, Cavendish Laboratory, University of Cambridge, Madingley Road, Cambridge CB3 0HE, UK

Although there is a vast engineering literature for granular materials, as a subject for physicists it has seen great growth in recent years. This is because, when stripped down to the fundamental problem, it is quite novel, and demands a rethink of the kind of laws familiar elsewhere in physics. We consider the transmission of stress in granular materials and investigate the simplest statically determinate problem.

Keywords: Stress tensor; compactivity; random packings; configuration tensor; tapping-induced compaction; volume function.

1. Introduction

There is a growing physics literature studying granular materials, with a broad ranging collection of papers and books [9, 29, 32, 37, 43]. One might think that after the vast edifice of the engineering literature [10, 13, 25, 27, 31, 36, 48, 54, 56, 64], particularly that of soil mechanics, the stage had been reached of advancing applications rather than looking to see if there are basic physical laws to be unearthed. We will argue that there are still basic laws to be found, and try to find some of them, in particular those that govern transmission of stress in static packings. We have found, in discussions with engineering colleagues, a scepticism that there are still such laws to be discovered, mainly because to discover *physics* laws one needs situations which are perfectly characterised, and the character of the problem must be at a level of simplicity that laws can be indeed expected. This means that we will demand circumstances which, though easy to describe, are not easy to create in a laboratory, indeed are such that great trouble needs to be taken to deal with cases which may seem very artificial. For example, pour sand into a box, and ask questions about the expected density, the way the sand may be expected to transmit stress. Colleagues have argued with us that one can get almost any answer to such problems according to the way the system is created. The real question, however, is to ask: Is there a way of depositing sand into a container in such a fashion that one can make physical predictions, and find what the circumstances are when the smallest amount of information is required about the sample packing? Clearly, if

First published in *Advances in Complex Systems*, Vol. 4, No. 4 (2001), pp. 451–467.

each grain is placed sequentially in the box in a fairly uniform way so that grains do not slip, and a complete inventory of the system is obtained, then the Newton equations can be solved on a computer. But can one have circumstances, as in statistical thermodynamics, where the minimum of information (e.g. the Hamiltonian and the temperature) is enough to solve for the pressure, specific heat etc.? We will argue that there is a hierarchy of situations including one as simple as statistical thermodynamics, up to the complexity of the computer assault mentioned above. What then is simplicity in a granular system?

2. Simplicity of the Model

2.1. *Shape*

Let us consider collections of different shapes (see Fig. 1). At first sight complexity of the possible analytical model could be assigned to go as $\alpha < \beta < \gamma$ but in fact one has $\alpha > \beta > \gamma$. Let us briefly characterise each system and discuss the possible obstacles and difficulties.

(i) $\boldsymbol{\alpha}$. Tube with smooth walls, containing the packing of uniformly deposited smooth monodispersive spheres. This is the most difficult problem, for this packing would definitely flow and crystallize and perhaps it is impossible to have a "jammed" packing in this case unless special boundary conditions are applied.

(ii) $\boldsymbol{\beta}$. Tube with rough walls, filled with rough particles of irregular shape. These grains may flow under certain conditions, but in majority of situations would certainly jam.

(iii) γ. Tube filled with pieces of twisted wire. This wire will entangle and not flow at all, indeed one does not need the pipe to have a jammed configuration and a stress calculation is feasible.

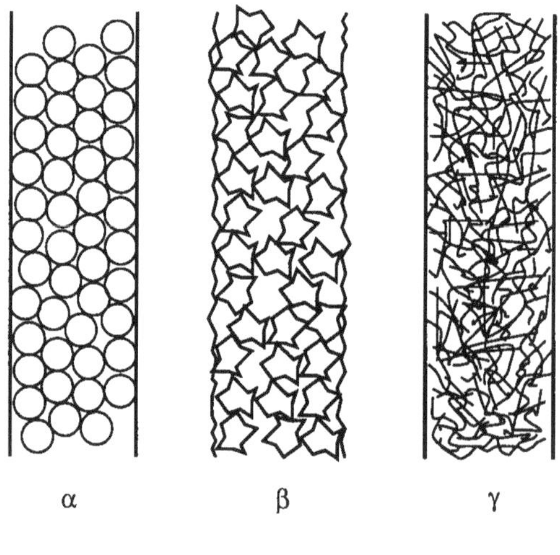

α β γ

Fig. 1. Types of granular materials.

2.2. *Friction*

Perfect friction is much easier to handle than, for example, the classical laws which relate sliding resistance to normal force [12]. By confining oneself to really rough surfaces, the contact normals are omitted from calculation and only contact points matter. Rolling takes on a special significance, for it cannot ever be forbidden, but a system which is at rest, stays at rest unless the boundary forces change dramatically.

2.3. *Elastic effects*

If the grains are infinitely hard so that there is no strain possible, equations enormously simplify at one level (no constitutive equations to bring in new information, no Hertzian deformation at contacts) but fundamental, new laws are needed to replace constitutive equations.

2.4. *Packing microstructure*

If one pours a powder into a container, it will have many voids which are reduced by tapping the container. This process, we will argue, can reach a best possible state to study, and a simplicity quest is achieved by realising that freshly poured packings, particularly in awkward geometries, are not the place to start studying. This leads us naturally into the next section.

3. The Statistical Geometry of Packings

It is well established in the literature that the empirical concept of random close packing (RCP) is relatively well founded. In a close packed system each grain touches other grains and the system is at rest. The density of such close packed systems of identical spheres can have a maximum value of 0.637 and a minimum value of 0.555 [50, 55]. The concept of RCP and its validity have recently been questioned and the notion of maximally random jammed state introduced [59]. Assuming there is no crystallinity (or that some slight shape changes forbid it), the maximum obviously exists; the minimum comes when grains begin to have the freedom to move. We believe that the region between maximum and minimum close packing (which does not need the grains to be spherical) is quite fundamentally simpler than any other range. One would not get this impression from the literature. On the contrary, there is a large body of work studying the case when there is a high density, but nevertheless low enough for the grains to have kinetic energy and collide with one another to form a well-defined statistical mechanics problem. This problem traces its history back to Boltzmann via Chapman and Enskog. Although there are many brilliant papers on this problem, we do not see it as the truly basic case. If one has a bucket of sand and pours it, kinetic energy is really irrelevant; the grains roll over one another. Equally, stirring sand, or a variety of other configuration regime a treatment for which the classical concept of a collision has no

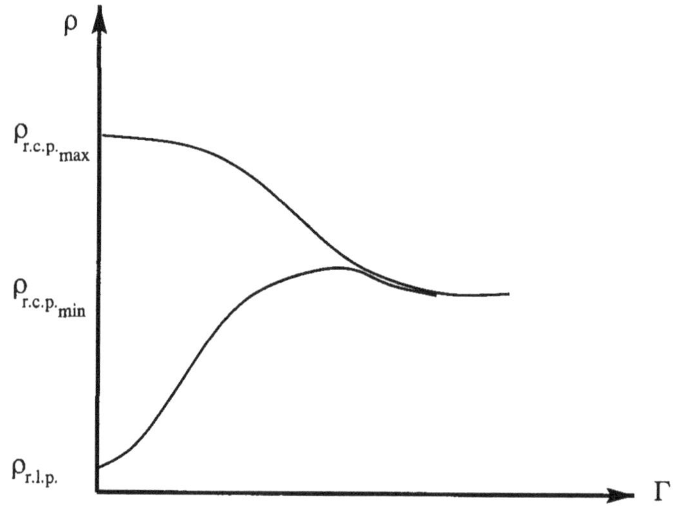

Fig. 2. Dependence of the packing density on the history of vibration intensities.

standing. Once one has decided to study the problem where grains are in continuous multiple contact, it is clear that the flow problem is much more complex than the static problem, which already offers splendid challenges in principle, even though it may not have the excitement of dynamics. A crucial experiment by the Chicago group has transformed the status of the problem [49].

They pour a powder into a vertical tube. It starts with a low density because of initial voids left, etc. It is then tapped a fixed large number of times with a certain force. The density settles and is measured by the height. The experiment is repeated with the force increased and the density increases until it plateaus at the minimum close packed density. Now the experiment continues by decreasing the magnitude of the force, when the density continues to rise until it reaches the maximum close packed density (see Fig. 2). The upper curve is *reversible.* The situation is just as in statistical mechanics. Suppose one starts with a paramagnet with spins in a non-equilibrium distribution and increases the temperature. This brings system into thermal equilibrium and changes the temperature now lead to the reversible condition. Thus, the powder now has reversible properties and all the points on the upper curve are accessible. We guess this is an analogue of ergodicity, for the assumption that in this range $\rho_{\text{r.c.p. max}} > \rho > \rho_{\text{r.c.p. min}}$ all configurations are accessible, subject only to the system having some density, i.e. volume [22, 23]. This seems to the present authors by far the simplest assumption and really no different to ergodicity in ordinary statistical mechanics. Simulation evidence does indeed support the hypothesis [5]. Therefore, just as all thermal configurations are equally probable subject to $E = H$ and entropy

$$S = k \log \int \delta\big(E - H(p,q)\big) \mathcal{D}p \mathcal{D}q\,, \tag{3.1}$$

for powders in our range

$$S = \log \int \delta(V - W)\Theta \mathcal{D}\mathcal{R}, \tag{3.2}$$

where W is the function of the position and shape of the grains, which gives the volume V, and Θ is the condition that the grains are trapped by their neighbours, i.e. $\rho_{\text{r.c.p. max}} > \rho > \rho_{\text{r.c.p. min}}$. One can then get the analogue of

$$T = \frac{\partial E}{\partial S}, \tag{3.3}$$

as temperature

$$X = \frac{\partial V}{\partial S}, \tag{3.4}$$

as compactivity and go to a canonical ensemble

$$F = E - TS, \tag{3.5}$$

becoming

$$Y = V - XS. \tag{3.6}$$

Applications such as theory of mixing can flow from this in analogy with statistical mechanics. This point of view has aroused some scepticism in the community because of course this is a very special situation. For example, if one has a box of vertical sides one can shake it and get, we believe, to the condition above. If one pours a sand pile, one cannot shake it into ergodicity for it will flatten out. One can still be optimistic and say the pouring of sand down the sides of the pile has some resemblance to shaking, and if one has very elliptical grains, a sketch of what one would expect intuitively shows the well-known dip in the pressure under the dome, and indeed the kind of law which we note later can be derived, had already been used very effectively on an empirical basis. But clearly the way a pile is poured will affect the arches proposed by Trollope long ago, and that poses a problem which is not the simplest class we seek. We would like to offer the readers a test of whether their intuition fits with the compactivity concept. Consider a pipe containing two powders separated by a membrane and suppose they are blocked off in such a way that they fill their part of the pipe (see Fig. 3). Now, have a tapping procedure Chicago style, but horizontal. Surely there must be a volume v_A occupied by A and v_B occupied by B and if the tapping strength is altered these volumes will change. A repeat experiment, however, will always set you back to v_A, v_B for the same tap history. There must be *something* the same on each side of the central membrane. Our argument is that this is an intense variable and the A, B configuration is the analogue of two materials whose energy is different, but which have the same temperature. So just as $T_A = T_B$ we argue for $X_A = X_B$. It is very hard to think of any other interpretation of such an experiment. This section has looked for situations where the analogue of thermodynamic formulae obtain. Now, we have to consider the detail of microscopic behaviour which underlies the macroscopic studies above.

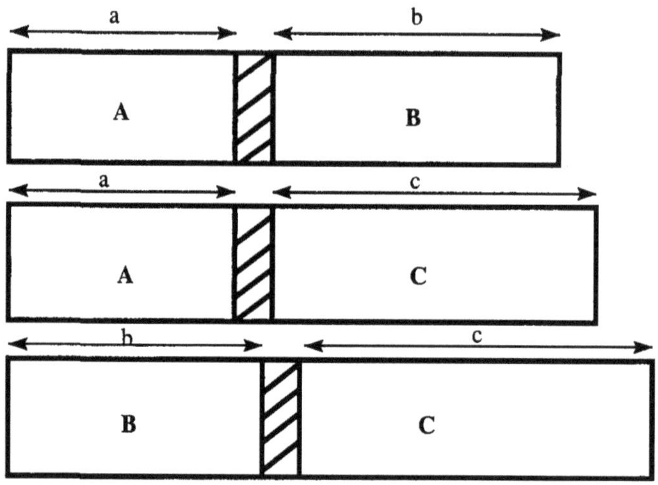

Fig. 3. "Zeroth Law" experiment: $X_A = X_B$ and $X_A = X_C$, hence $X_C = X_B$.

4. Streses and Forces

The simplest statically determinate problem of stress transmission in a static granular material is that where grains are considered to be perfectly hard, perfectly rough and each grain α has a coordination number $z^\alpha = d+1$ (a more general case exists when an average coordination number is $\bar{z} = d+1$). A problem is said to be statically determinate if the state of stress can be determined without knowledge of the displacement or motion [28]. Newton's laws of force and torque balance for every grain give us the system of $Nd(d+1)/2$ equations for $\left(\sum_{\alpha=1}^{N} z^\alpha d\right)/2$ interparticle forces $\{\mathbf{f}^{\alpha\beta}\}$ (see Fig. 5). Thus, if grain α exhibits a force $\mathbf{f}^{\alpha\beta}$ on grain β,

$$f_i^{\alpha\beta} + f_i^{\beta\alpha} = 0\,, \tag{4.1}$$

where $i = 1, \ldots, d$ is the Cartesian index. If there is an external body force $\mathbf{g}^\alpha$ acting on grain α:

$$\sum_\beta f_i^{\alpha\beta} + g_i^\alpha = 0\,. \tag{4.2}$$

The external torque balance is given by

$$\sum_\beta \epsilon_{ikl} f_k^{\alpha\beta} r_l^{\alpha\beta} + c_i^\alpha = 0\,, \tag{4.3}$$

where c_i^α is the external body torque which we take to be zero and where $\mathbf{r}^{\alpha\beta}$ is the vector joining the centroid of contact $\mathbf{R}^\alpha$ with the contact point $\mathbf{R}^{\alpha\beta *}$. The centroid of contact points of grain α is defined as

$$\mathbf{R}^\alpha = \frac{\sum_\beta \mathbf{R}^{\alpha\beta *}}{z^\alpha}\,, \tag{4.4}$$

where the summation sign $\sum_\beta$ means the sum over all nearest neighbours of the grain α which form a contact with the latter (see Fig. 4). The distance

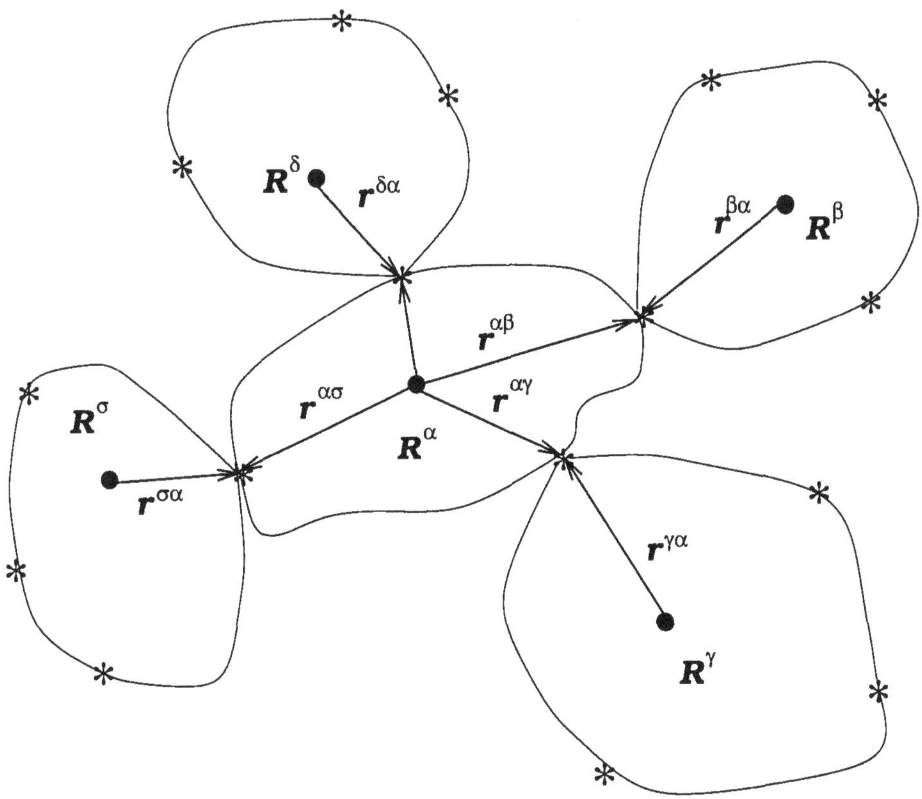

Fig. 4. Cross-section of the first coordination shell of the reference particle α.

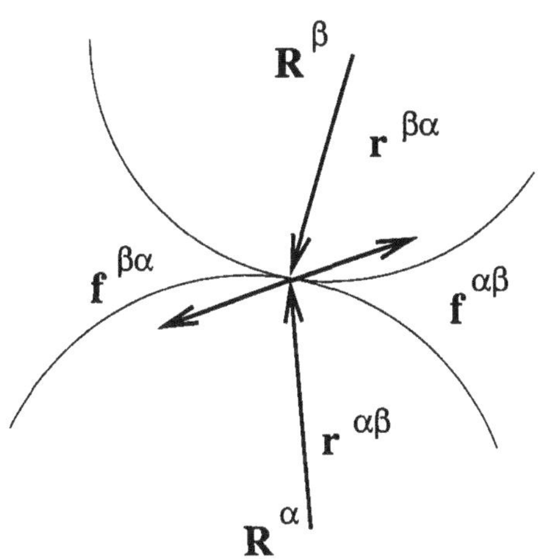

Fig. 5. Intergranular forces and the local geometry of two grains in contact.

between particles α and β is defined as the distance between their centroids of contacts:

$$\mathbf{R}^{\alpha\beta} = \mathbf{R}^{\beta} - \mathbf{R}^{\alpha} = \mathbf{r}^{\alpha\beta} - \mathbf{r}^{\beta\alpha} . \tag{4.5}$$

The system of equations of intergranular force and torque balance (4.2) and (4.3) can only be solved in d dimensions if the average number of contacts is $d+1$ per grain, i.e. four in three dimensions. Under these conditions the system of Newton's equations is determinate because the number of unknowns $\{\mathbf{f}^{\alpha\beta}\}$ is equal to the number of equations. It is clear that given the set of contact points $\{\mathbf{R}^{\alpha\beta *}\}$ it is not possible to solve the system of Newton's equations (4.2) and (4.3) analytically because of the present structural disorder and the sheer number of grains involved. Nevertheless, one can address the statistical properties of the set of possible solutions of (4.2) and (4.3) ask the following question: Given the geometric specification of the contact network, what is the probability of finding the set of intergranular forces with modulae and orientations $\{\mathbf{f}^{\alpha\beta}\}$ that satisfy the system of Newton's equations of balance? It is easy to see from (4.2) and (4.3) that the answer to this question also depends on the distribution of external forces $\mathbf{g}^{\alpha}$ and torques $\mathbf{c}^{\alpha}$. One often finds in the literature [7, 26] that contact points are counted and a fractional coordination number is recorded. However, we believe that in the "ergodic" condition the coordination number will be $d+1$ on average and indeed to make progress the simplest thing is to take exactly $d+1$ contacts per grain. The reader can now say, but what of the simulation evidence [6, 42, 62], let alone the experiments of J. D. Bernal and co-workers [7, 8], and the well-known Kepler's conjecture regarding the face-centred cubic lattice [30]? There are two responses: Firstly, one needs exact contact when dealing with perfect hardness and the chance of this in mathematical language is a set of measure zero. For perfect spheres in $d=3$, 12 point contacts are possible, but for the slightest imperfections, force will travel through the $d+1$ contacts and leave no force in the near misses. Finally and quite pragmatically, we have a theory, and if the reader would like to produce a simple theory with higher coordination number due to say, Hertzian contacts, good luck! Now let us turn to stress. Eventually, we can expect to derive a version of the force balance equation

$$\nabla_j \sigma_{ij} + \rho g_i = 0 , \tag{4.6}$$

and these equations are derived in Ref. 18 and 19. Here, there are d equations for $d(d+1)/2$ components of σ_{ij}, so there are $d(d-1)/2$, i.e. one in $d=2$ and three in $d=3$, equations to be found from the geometric configuration via Newton's laws of force and torque balance. The first example of the stress-geometry equation lies in the "fixed principal axes" hypothesis [63]:

$$P_{ijkl}\sigma_{kl} + \mathcal{O}(\nabla\sigma) = 0 . \tag{4.7}$$

The forms of P are given in the references. They are simple in anisotropic conditions where the FPA form is

$$\sigma_{11} - \sigma_{22} = 2\sigma_{12} \tan\phi , \tag{4.8}$$

where ϕ is the angle characterising anisotropy of the packing. But for isotropic conditions ϕ is random and all that can be obtained is the probability distribution of finding $\sigma_{11} - \sigma_{22}$ given σ_{12} [18, 19]:

$$P\{\sigma_{11} - \sigma_{22}|\sigma_{12}\} = \frac{2}{\pi} \frac{|\sigma_{12}|}{(\sigma_{11} - \sigma_{22})^2 + (\sigma_{12})^2}, \tag{4.9}$$

and vice versa. This distribution can then be introduced for corresponding components of the macroscopic stress tensor subject to the absence of mesoscopic correlations in the packing. Given the system of equations

$$\nabla_j \sigma_{ij} = g_i + \mathcal{O}(\nabla^3 \sigma), \tag{4.10}$$

$$P_{ijkl}\sigma_{kl} + \mathcal{O}(\nabla^2 \sigma) = 0, \tag{4.11}$$

one can construct the Green function which relates stress to external forces (including boundary conditions), i.e. if one thinks of σ_{ij} as a $d(d+1)/2$ component vector object λ and γ as $\mathbf{g}$ of Eq. (4.10) and 0 of Eq. (4.11). Then Eqs. (4.10) and (4.11) can be written

$$\mathcal{G}^{-1}\lambda = \gamma, \tag{4.12}$$

which gives

$$\lambda = \mathcal{G}\gamma, \tag{4.13}$$

and hence

$$\sigma = \tilde{\mathcal{G}} g, \tag{4.14}$$

where $\tilde{\mathcal{G}}$ is $\mathcal{G}$ in the previous notation. Thus, if we can get the coefficients P we can, for example, solve the problem of applying forces to a powder and measuring their transmission. The psychological problem is that this is a rather dull experiment compared to the weird patterns that powders can exhibit and their equally weird flows. Such experiments, though elegant and challenging are not, to our mind, really fundamental. It is also the case that an intricate understanding of a simple phenomenon is more fundamental than an oversimplified model of a complex phenomenon. It is interesting to make an analogy with polymer physics here. One can describe polymer viscosity by collecting comparatively simple expressions for a constitutive equation which separately describe some aspect of viscoelasticity, and as more phenomena are explored, add more and more terms. Alternatively, one can create a molecular model from which a constitutive equation can be produced. Such an equation can, even for a simple model, be an intricate function of strain, and the modeller has staked all on its success. But if it is successful, the rewards are complete. If it is not the failure is total. However, this, to our mind, is the real way to progress.

5. How Can We Describe a Powder?

The "missing" stress-geometry equation in the simplest version in 2D has the form

$$\begin{vmatrix} S_{11}^{\alpha} & H_{11}^{\alpha} - G_{11}^{\alpha S} & H_{11}^{\alpha} - F_{11}^{\alpha} \\ S_{12}^{\alpha} & H_{12}^{\alpha} - G_{12}^{\alpha S} & H_{12}^{\alpha} - F_{12}^{\alpha} \\ S_{22}^{\alpha} & H_{22}^{\alpha} - G_{22}^{\alpha S} & H_{22}^{\alpha} - F_{22}^{\alpha} \end{vmatrix} = 0\,, \tag{5.1}$$

where the set of tensors which describe connectivity of the contact network is introduced as

$$F_{ij}^{\alpha} = \sum_{\beta} R_i^{\alpha\beta} R_j^{\alpha\beta}\,, \tag{5.2}$$

and $G_{ij}^{\alpha S}$ is the symmetric part of the tensor G_{ij}^{α}:

$$G_{ij}^{\alpha} = \sum_{\beta} R_i^{\alpha\beta} Q_j^{\alpha\beta}\,, \tag{5.3}$$

and H_{ij} is given by the following expression:

$$H_{ij}^{\alpha} = \sum_{\beta} Q_i^{\alpha\beta} Q_j^{\alpha\beta}\,, \tag{5.4}$$

where $\mathbf{R}^{\alpha\beta}$ and $\mathbf{Q}^{\alpha\beta}$ are shown in the diagram (see Fig. 6).

This suggests that simple volume averages F_{ij} and H_{ij} (and also G_{ij}^{α} whose mean is zero) of Eqs. (5.2) and (5.4) have the status in powders that the magnetic field has in magnets or the director matrices in nematics. For 2D homogeneous

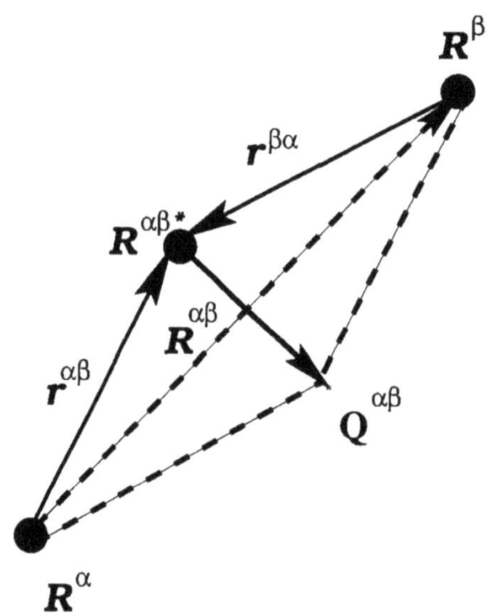

Fig. 6. Segment of the contact network between nearest neighbours α and β.

packings of spheres and fairly uniform systems of approximately spherical shapes we have

$$\langle F^\alpha_{ij}\rangle \approx \begin{pmatrix} \text{l.} & \text{v.s.} \\ \text{v.s.} & \text{l.} \end{pmatrix}, \qquad \langle H^\alpha_{ij}\rangle \approx \begin{pmatrix} \text{v.s.} & \text{v.s.} \\ \text{v.s.} & \text{v.s.} \end{pmatrix}, \qquad \langle G^\alpha_{ij}\rangle = 0\,, \tag{5.5}$$

where l. means a large tensor component and v.s. a very small one with respect to each other. For strongly elongated grains, say in the x direction, we have

$$\langle F^\alpha_{ij}\rangle \approx \begin{pmatrix} \text{l.} & \text{v.s.} \\ \text{v.s.} & \text{v.s.} \end{pmatrix}, \quad \langle H^\alpha_{ij}\rangle \approx \begin{pmatrix} \text{v.s.} & \text{v.s.} \\ \text{v.s.} & \text{l.} \end{pmatrix}, \quad \langle G^\alpha_{ij}\rangle = 0\,. \tag{5.6}$$

This analysis raises a series of questions. Firstly, we have to know how many tensors are needed to describe a powder: Recall that in crystals there is a hierarchy:

(i) Central two-body forces,
(ii) Multipole two-body forces,
(iii) Three and higher body forces.

One can often be content with the first option, but there are times when one needs the more complex cases. Secondly, given the local characteristics of the powder, we have to be able to calculate its volume V by expressing the volume function W in terms of the grain variables. The contact points are the total specification for static packings, thus the set of $\{\mathbf{R}^{\alpha\beta *}\}$ must define W. What is it? Let us pose a mathematical problem: we have a set of grains each of which has $d+1$ contacts with neighbours at specified locations $\{\mathbf{R}^{\alpha\beta *}\}$. What is $W\{\mathbf{R}^{\alpha\beta *}\}$? The crudest version is

$$W = 2\sum_{\alpha}^{N} \sqrt{\det F^\alpha_{ij}}\,, \tag{5.7}$$

but clearly there are higher order terms like $F^\alpha_{ik}U^{\alpha\beta}_{kl}F^\beta_{li}$, $G^\alpha_{ik}T^{\alpha\beta}_{kl}G^\beta_{li}$ and $H^\alpha_{ik}Y^{\alpha\beta}_{kl}H^\beta_{li}$. There should be a systematic way to write $W\{\mathbf{R}^{\alpha\beta *}\}$, perhaps in terms of first and higher order coordination shells. There are unpleasant topological constraints given the non-overlap of grains. Such topological problems are present in polymer physics and although simpler than the granular problem, have no exact mathematical solution yet (e.g. the knot problem). For polymers, however, the tube model does produce an elegant solution, so we must hope that sensible intuitive pictures will work for granular materials. Let us suppose then that a form for W in terms of the configuration tensors is available, there still remains the problem (even in the simplest case of W being a function of just $\{F^\alpha_{ij}\}$) of evaluating

$$e^{-\frac{Y}{\lambda X}} = \int e^{-\frac{W\{F\}}{\lambda X}} \prod_{\alpha}^{N} \delta\left(F^\alpha_{ij} - \sum_{\beta} R^{\alpha\beta}_i R^{\alpha\beta}_j\right) P(\{\mathbf{R}^{\alpha\beta *}\})\mathcal{D}\mathbf{R}^{\alpha\beta *}\,, \tag{5.8}$$

where $P(\{\mathbf{R}^{\alpha\beta *}\})$ is the probability of finding the set of contact points $\{\mathbf{R}^{\alpha\beta *}\}$, and the integration over $\{\mathbf{R}^{\alpha\beta *}\}$ amounts to obtaining the Jacobian required to change integration over $\{\mathbf{R}^{\alpha\beta *}\}$ to integration over $\{F^\alpha_{ij}\}$. Simple approaches just give Gaussian (obviously!), but the general attack is *not* obvious. Our third question is: How would the grains move? This paper has not attempted dynamics but it is worth posing a few final questions. When one deals with single component liquids, the equation of hydrodynamics involve, ρ, $\mathbf{v}$ and T which is intuitively clear, but also emerges from Hilbert's analysis of the linearized Boltzmann equation, where the condition for the solution requires honouring the invariant of number, momentum and energy. If nematic or magnetic degrees of freedom are present, extra variables have to be added. It is natural to expect the equation for ρ, $\mathbf{v}$, X, then $\bar{F}_{ij}(\mathbf{r})$ etc., where $\bar{F}_{ij}(\mathbf{r})$ are the mean values of $F_{ij}(\mathbf{r})$. Following the classical Boltzmann route we will have to analyze the microscopic motion. Most of the literature considers the case when the grains have enough space to have kinetic energy of consequence and a "granular temperature" is introduced, with notable success for that regime. Most granular systems are not like this, however, and pouring a powder for example is dominated by sliding and rolling. The mathematics of rolling has been developed by Gibbs and Appel [51], but many-body rolling is not well developed, and we do not know how to develop the Gibbs–Appel equations for many-body systems. A remarkable new development has come from Ball and Blumenfeld [1] who suggest that the onset of many-body rolling is heralded by the failure of the "missing" equation. In addition to its intrinsic value this suggestion gives a physical meaning to the new equations.

6. The Consequences of Being Amorphous

At the grain level, particularly with grains of mixed sizes and shapes, one cannot expect uniformity. At macroscopic dimensions, physical laws of the standard mathematical type, i.e. differential equations, can be expected, albeit of novel types. The amorphous nature of the granular system poses the problem of how to handle the microscopic equations into the macroscopic. For example one knows that the random walk problem turns into Fick's equation macroscopically, a very well behaved form. So when we have a set of equations for the intergranular forces $\{f_i^{\alpha\beta}\}$, and a set of contact points $\{R_i^{\alpha\beta *}\}$ where they act, so that relative to the contact centroid we define a force moment tensor

$$S_{ij}^{\alpha} = \frac{1}{2}\left(\sum_{\beta} f_i^{\alpha\beta} r_j^{\alpha\beta} + f_j^{\alpha\beta} r_i^{\alpha\beta}\right), \tag{6.1}$$

which satisfies the set of Nd stress-force equations

$$\sum_{\beta} S_{ij}^{\alpha} M_{jl}^{\alpha} R_l^{\alpha\beta} - \sum_{\beta} S_{ij}^{\beta} M_{jl}^{\beta} R_l^{\beta\alpha} = g_i^{\alpha}, \tag{6.2}$$

and is constrained by the set of Nd vector point fields [20, 21], ϕ_i^{α}

$$S_{ij}^{\alpha} = \frac{1}{2}\sum_{\beta}\left(\phi_i^{\beta} r_j^{\alpha\beta} + \phi_j^{\beta} r_i^{\alpha\beta}\right). \tag{6.3}$$

After eliminating $\{\phi^\alpha\}$ in Eq. (6.3) one will obtain a set of coupled $\frac{Nd(d-1)}{2}$ constraints on $\{S^\alpha_{ij}\}$. This can be accomplished analytically by writing diagonal and off-diagonal components of N tensors $\{S^\alpha_{ij}\}$ and vector fields $\{\phi^\alpha\}$ as N-dimensional vectors and inverting matrices corresponding to the diagonal elements of $\{S^\alpha_{ij}\}$ and substituting the results into the equation for the off-diagonal components. These linear constraints on $\{S^\alpha_{ij}\}$ can then be appropriately decoupled and averaged into $\frac{d(d-1)}{2}$ stress-geometry equations [20, 21, 24]. We fully expect an average

$$\sigma_{ij}(\mathbf{r}) = \left\langle \sum_{\alpha=1}^{N} S^\alpha_{ij}\delta(\mathbf{r}-\mathbf{R}^\alpha) \right\rangle , \tag{6.4}$$

to exist and be a smooth, useful function. But how well behaved is S^α_{ij}? Computer simulations and photo-elastic experiments [14–17, 34, 35, 38, 39, 40, 41, 44, 52, 53, 57, 58, 60, 61] suggest a percolation structure in which the eigenvalues of S^α_{ij} are dominated by one eigenvalue and stress percolates through the powder. However, a completely ordered system such as as a honeycomb lattice is soluble [3, 4] and is completely uniform, and each sphere behaves in the same way towards its four neighbours. Are the simulations artifact, if regarded as models of perfectly hard grains? Consider some two-dimensional picture of uniform spheres under pressure (see Fig. 7). The figure is of a honeycomb lattice and the force between the adjacent circles will be identical. Now compress the system to give the following configuration (see Fig. 8) so the bulk is now a cubic lattice (see Fig. 9). The dots mark contacts, three per sphere, and we assume the lattice is slightly distorted (see Fig. 10). The major force lines are vertical, and there is a small force exerted perpendicular to the vertical line which keeps the system stable in the same way that a small perpendicular force will delay the Euler instability in a strut. Figure 7 represents the

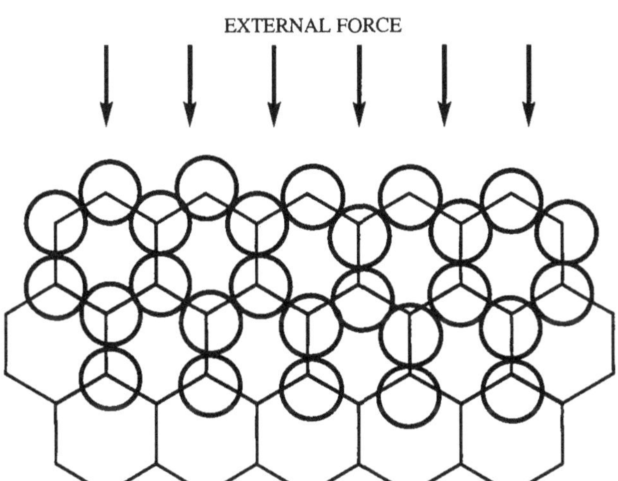

Fig. 7. Honeycomb lattice under compression.

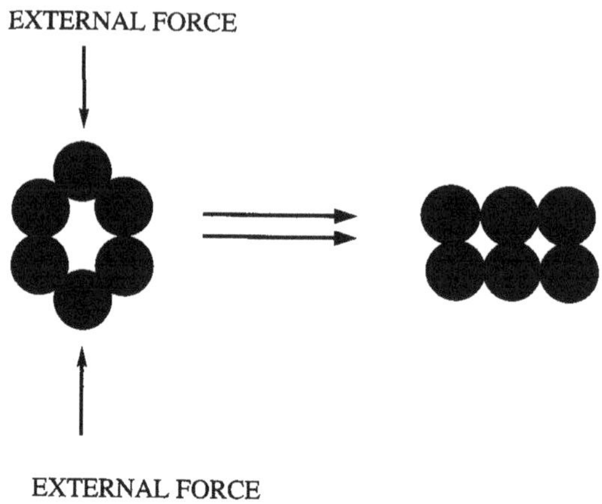

Fig. 8. Transformation of the honeycomb lattice into the cubic one under uniaxial compression.

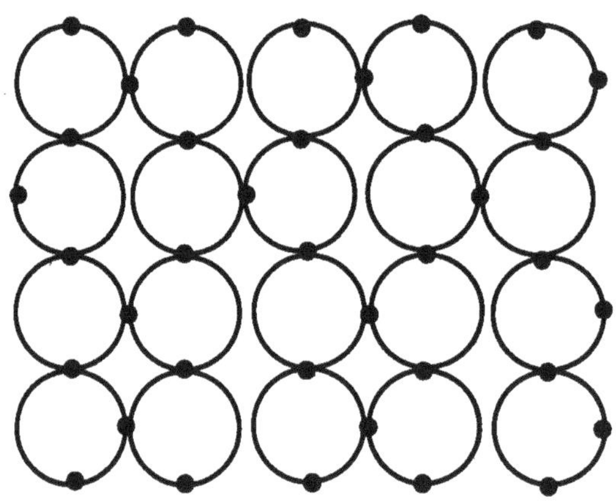

Fig. 9. Slightly distorted 2D cubic lattice with three contacts per sphere.

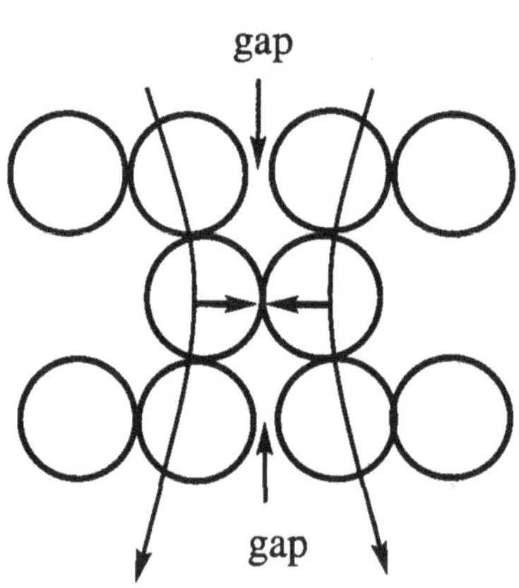

Fig. 10. The gaps between neighboring spheres of the distorted lattice.

minimum density, Fig. 9 an intermediate density but the maximum density will be closer to a hexagonal close packing than the simple cubic. Nevertheless one can, in an intuitive way, argue that there will be configurations which can sustain large forces along some line joining contacts, stabilised by smaller roughly perpendicular forces, with occasional bifurcations. One can start a general mathematical description of this [11] or try to follow the path directly, i.e. if the strong force is called $\mathbf{F}$ and runs along a path $R(s)$ such that

$$\frac{\partial R(s)}{\partial s} = \frac{\mathbf{F}}{|\mathbf{F}|}, \tag{6.5}$$

where $\left(\frac{\partial R(s)}{\partial s}\right)^2 = 1$, one can expect

$$\frac{\partial \mathbf{F}}{\partial s} = \mathbf{g} + \delta\mathbf{F}, \tag{6.6}$$

where $\mathbf{g}$ is the external force and $\delta\mathbf{F}$ the weaker "perpendicular" force which, roughly speaking, is random (see Fig. 11).

This leads one into fascinating new mathematics on how to relate this equation, already smoothed from the distribution of angles and grain sizes, to the pictures of stress lines.

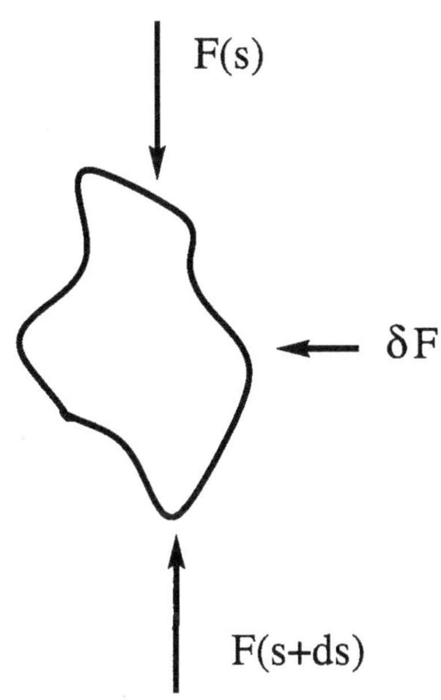

Fig. 11. The major force $F(s)$ lines are vertical, and there is a small force $\delta\mathbf{F}$ exerted perpendicular to the vertical line which keeps the system stable.

The physical picture of force transmitted in chains is consistent with an experimental observation [38, 47] that the probability of normal contact force acting on a grain contact decreases exponentially for large forces as

$$P(f) \propto \begin{cases} \left(\frac{f}{\langle f \rangle}\right)^{\alpha} & \text{if } f < \langle f \rangle\,, \\ e^{\beta\left(1-\frac{f}{\langle f \rangle}\right)} & \text{if } f > \langle f \rangle\,, \end{cases}$$

where $\langle f \rangle$ is an average contact force, α and β are constants. In order to derive this probability distribution we offer a simple theory. In this approach we consider the integral equation for the probability distribution of contact forces governed by Eq. (4.2), i.e. for the sake of feasibility of an analytical advance we ignore Eqs. (4.1) and (4.3). Forces g, h impinge on a grain which then exerts force f on a neighbor. Let f be in the x direction and use the symbols g and h for the x components of the vector forces. The probablity distribution of contact force forces is given by an intgeral equation:

$$P(f) = \int_0^{\infty} d\vec{y} \int_0^{\infty} d\vec{h}\, \delta(\vec{f} - \vec{h} - \vec{y}) P(\vec{y}) P(\vec{h})\,, \tag{6.7}$$

which becomes

$$P(f) = \int_0^{\infty} dy \int_0^{\infty} dh \int_0^1 d\mu \int_0^1 d\lambda\, P(g\mu) P(\lambda h)\,. \tag{6.8}$$

This has the solution

$$P(f) = \frac{f}{p} e^{-\frac{f}{p}}\,, \tag{6.9}$$

where $p = \frac{\langle f \rangle}{A}$ is the mean pressure and A is the area of contact, and the distribution is exponential for large f. In 3D the solution is a Bessel function with essentially the same properties. But now doubt sets in: why can the y and z components, i.e. the torque, be ignored; does the shape distribution alter the equation; what about second neighbours, etc.?

Do we know enough to proceed with this work with confidence? We leave this question to the reader.

7. Acknowledgments

We wish to acknowledge the financial support of the Leverhulme Foundation (S. F. Edwards), the ROPA grant from EPSRC (UK) and the Research Fellowship from Wolfson College (D. V. Grinev). The authors thank Prof. R. C. Ball and Dr. R. Blumenfeld for many stimulating discussions.

References

[1] Ball, R. C. and Blumenfeld, R., private communication.
[2] Ball, R. C. and Blumenfeld, R., ArXiv cond-matt/0008127, submitted to *Phys. Rev. Lett.*

[3] Ball, R. C. and Grinev, D. V., *Physica* **A292**, 167 (2001).
[4] Ball, R. C., in *Structure and Dynamics of Materials in the Mesoscopic Domain*, Lal, M., Mashelkar, R. A., Kulkarni, B. D. and Naik, V. M., eds. (Imperial College Press, London, 1999), p. 326.
[5] Barrat, A., Kurchan, J., Loreto, V. and Sellitto, M., *Phys. Rev.* **E63**, 051301 (2001).
[6] Bennett, C. H., *J. Appl. Phys.* **43**, 2727 (1972).
[7] Bernal, J. D. and Mason, J., *Nature* **188**, 910 (1960).
[8] Bernal, J. D., *Nature* **183**, 141 (1959).
[9] Bideau, D. and Hansen, A., eds., *Disorder and Granular Media* (Elsevier, Amsterdam, 1993).
[10] Bose, A., *Advances in Particulate Materials* (Boston, Butterworth–Heinemann, 1995).
[11] Bouchaud, J.-P., Claudin, P., Levine, D. and Otto, M., *Europhys. J. Phys.* **E4**, 451 (2001).
[12] Bowden, F. P. and Tabor, D., *The Friction and Lubrication of Solids* (Clarendon Press, Oxford, 1986).
[13] Briscoe, B. J. and Adams, M. J., *Tribology in Particulate Technology* (Adam Hilger, Bristol, 1987).
[14] Dantu, P., in *Proc. IV Int. Conf. Soil. Lech. Found. Eng*, **1** (Butterworths, London, 1957), p. 133.
[15] Dantu, P., *Ann. des Ponts et Chaussés*, **4** (1967).
[16] Dantu, P., *Geotechnique* **18**, 50 (1968).
[17] Drescher, A. and De Josselin de Jong, G., *J. Mech. Phys. Solids* **20**, 337 (1972).
[18] Edwards, S. F. and Grinev, D. V., *Chaos* **9**, 551 (1999).
[19] Edwards, S. F. and Grinev, D. V., *Phys. Rev. Lett.* **82**, 5397 (1999).
[20] Edwards, S. F. and Grinev, D. V., *Physica* **A294**, 451 (2001).
[21] Edwards, S. F. and Grinev, D. V., *Physica* **A302**, 147 (2001).
[22] Edwards, S. F. and Oakeshott, R. B. S., *Physica* **A157**, 1080 (1989).
[23] Edwards, S. F. and Grinev, D. V., *Phys. Rev.* **E58**, 4758 (1998).
[24] Edwards, S. F. and Grinev, D. V., submitted.
[25] Feda, J., *Mechanics of Particulate Materials: The Principles* (Elsevier, Amsterdam, 1982).
[26] Finney, J. L., *Proc. Roy. Soc. London* **319**, 479 (1970).
[27] Fleck, N. A. and Cocks, A. C. F., eds., *Mechanics of Granular and Porous Materials* (Kluwer, Dodrecht, 1997).
[28] Flügge, S., *Encyclopedia of Physics* **v.III/1**: *Principles of Classical Mechanics and Field Theory* (Springer-Verlag, Berlin, 1960), p. 580.
[29] Guazzelli, E. and Oger, L., eds., *Mobile Particulate Systems* (Kluwer, Dordrecht, 1995).
[30] Hales, T. C., Sarnak, P. and Pugh, M. C., *PNAS USA* **97**, 12963 (2000).
[31] Harr, M. E., *Mechnics of Particulate Media: A Probabilistic Approach* (McGraw-Hil, New York, 1977).
[32] Herrmann, H. J., Hovi, J.-P. and Luding, S., eds., *Physics of Dry Granular Media* (Kluwer, Dordrecht, 1998).
[33] Higgins, A. and Edwards, S. F., *Physica* **A189**, 127 (1992).
[34] Howell, D. W., Behringer, R. P. and Veje, C. T., *Chaos* **9**, 559 (1999).
[35] Jia, X., Caroli, C. and Velicky, B., *Phys. Rev. Lett.* **82**, 1863 (1999).
[36] Johnson, K. L., *Contact Mechanics* (Cambridge University Press, Cambridge, 1987).
[37] Liu, A. and Nagel, S. R., eds., *Jamming and Rheology: Constrained Dynamics on Microscopic Scales* (Taylor & Francis London, 2001).

[38] Liu, C. H., *et al.*, *Science* **269**, 513 (1995).
[39] Liu, C. and Nagel, S. R., *Phys. Rev.* **B48**, 646 (1993).
[40] Liu, C. and Nagel, S. R., *Phys. Rev. Lett.* **68**, 2301 (1992).
[41] Liu, C., *Phys. Rev.* **B50**, 782 (1994).
[42] Matheson, A. J., *J. Phys.* **C7**, 2569 (1974).
[43] Mehta, A., ed., *Granular Matter: An Interdisciplinary Approach* (Springer-Verlag, New York, 1993); for review see e.g. Jaeger, H. M., Nagel, S. R. and Behringer, R. P., *Rev. Mod. Phys.* **68**, 1259 (1996).
[44] Miller, B., O'Hern, C. and Behringer, R. P., *Phys. Rev. Lett.* **77**, 3110 (1996).
[45] Monasson, R. and Pouliquen, O., *Physica* **A236**, 395 (1997).
[46] Mounfield, C. C. and Edwards, S. F., *Physica* **A210**, 279 (1994).
[47] Mueth, D. M., Jaeger, H. M. and Nagel, S. R., *Phys. Rev.* **E57**, 3164 (1998).
[48] Nedderman, R. M., *Statics and Kinematics of Granular Materials* (Cambridge University Press, Cambridge, 1992).
[49] Nowak, E. R., Knight, J. B., Povinelli, M. L., Jaeger, H. M. and Nagel, S. R., *Powder Technology* **94**, 79 (1997).
[50] Onoda, G. Y. and Liniger, E. G., *Phys. Rev. Lett.* **64**, 2727 (1990).
[51] Pars, L. A., *A Treatise on Analytical Dynamics* (Heinemann, London, 1968).
[52] Radjai, F. *et al.*, *Phys. Rev. Lett.* **77**, 274 (1996).
[53] Radjai, F. *et al.*, *Phys. Rev. Lett.* **80**, 61 (1998).
[54] Schofield, A. N. and Roth, C. P., *Critical State Soil Mechanics* (McGraw Hill, 1968).
[55] Scott, G. D. and Kilgour, D. M., *Br. J. Appl. Phys.* **2**, 863 (1969).
[56] Sokolovskii, V. V., *Statics of Granular Materials* (Pergamon, Oxford, 1965).
[57] Thornton, C., in *Tribology in Particulate Technology*, Briscoe, B. J. and Adams, M. J., eds. (Adam Hilger, Bristol, 1987).
[58] Thornton, C., *KONA Powder and Particle* **15**, 81 (1997).
[59] Torquato, S., Truskett, T. M. and Debenedetti, P. G., *Phys. Rev. Lett.* **84**, 2064 (2000); *Phys. Rev.* **E62**, 993 (2000).
[60] Travers, T., Ammi, M., Bideau, D., Gervois, A., Troadec, J. P. and Messager, J. C., *Europhys. Lett.* **4**, 329 (1987).
[61] Travers, T., Bideau, D., Gervois, A., Troadec, J. P. and Messager, J. C., *J. Phys A: Math. Gen.* **19**, L1033 (1986).
[62] Visscher, W. M. and Bolsterli, M., *Nature* **239**, 504 (1972).
[63] Wittmer, J. P., Claudin, P. and Cates, M. E., *J. de Physique I (France)* **7** , 39 (1997).
[64] Wood, D. M., *Soil Behaviour and Critical State Soil Mechanics* (Cambridge University Press, Cambridge, 1990).

CHAPTER 18

Radial and Axial Segregation of Granular Mixtures in the Rotating-Drum Geometry

SANJAY PURI

School of Physical Sciences, Jawaharlal Nehru University, New Delhi - 110067, India

HISAO HAYAKAWA

Graduate School of Human and Environmental Studies, Kyoto University, Sakyo-ku, Kyoto 606-01, Japan

We formulate a phenomenological model for the dynamical evolution of granular mixtures in a rotating drum. In particular, we present a dynamical basis for an "effective surface tension," which penalizes sharp interfaces between regions enriched in different components of the mixture. We also present representative results which demonstrate that our model captures many experimental features for this problem.

Keywords: Granular mixtures; rotating-drum geometry; radial and axial segregation.

1. Introduction

There has been much recent interest in the physics of granular materials or powders [3, 12, 15, 18]. In part, this interest is due to the great technological relevance of these materials and their long-standing interest to the engineering community. Typically, granular systems consist of "thermodynamic" assemblies of mesoscopic particles with typical sizes ranging from 10 μm–1 cm, which dissipate energy on collision. Granular materials exhibit properties intermediate to those of solids and fluids, i.e. they can withstand stress like a solid, but can flow like a fluid.

Many studies have focused on the static properties of granular systems, e.g. heap formation, stress distribution in granular piles, etc. However, even more fascinating phenomenology is seen in granular materials which are driven by an external force, e.g. vertical or horizontal vibration [19], pouring [23], etc. Essentially, the driven system settles into a nonequilibrium steady state where the loss of energy due to collisions is balanced by the continuous input of energy due to external driving.

First published in *Advances in Complex Systems*, Vol. 4, No. 4 (2001), pp. 469–479.

This steady-state behavior is characterized by complex pattern formation, which has been of considerable research interest [3, 12, 15, 18].

In this paper, we focus on granular materials in a rotating drum. This system has been the subject of many experimental and numerical studies. Before we proceed to specific details, it is relevant to elucidate the principles which guide our study and its possible limitations. We will attempt to formulate coarse-grained or phenomenological models which capture "universal" features of granular materials. The formulation of coarse-grained models has a long tradition in conventional statistical physics [9, 14]. Of course, in that context, the microscopic scales (e.g. length $\ell \sim 10^{-9}$ m and time $\tau \sim 10^{-13}$ sec) are many orders of magnitude smaller than the scales of experimental phenomenology. However, in the case of granular systems, the separation of "microscopic" (e.g. $\ell \sim 10^{-5}$–10^{-2} m, $\tau \sim 10^{-4}$ sec) and "macroscopic" scales is not that clear. An obvious consequence is that the degree of universality in granular systems is less than that in conventional statistical physics, i.e. there are important properties at the macro-level which originate from specific properties of individual grains, e.g. size, shape, polydispersity, chemical composition, etc. Nevertheless, it is an experimental fact that there are also important properties at the macro-level which are common to diverse granular materials. Our present study will attempt to capture these universal features in the context of the rotating-drum geometry. There have been analogous studies in the context of other granular systems, e.g. vibration on a plate [28]; formulation of dynamical equations for granular fluids [5]; Ginzburg–Landau theory for inhomogeneous cooling [32], etc.

This paper summarizes and provides a fresh perspective on our recent modeling of granular mixtures in a rotating drum [24]. In particular, we provide a dynamical basis for an "effective surface tension" between domains enriched in different components of the mixture. This paper is organized as follows. Section 2 discusses the broad features of the experimental and numerical phenomenology for this problem. Section 3 presents our phenomenological model and results therefrom. Section 4 concludes this paper with a brief summary and discussion of our results.

2. Experimental and Numerical Phenomenology

2.1. *Homogeneous systems*

Let us first consider a single-component granular material, e.g. sand (S) or glass (G), in a horizontal drum of radius R, which is rotated about its longitudinal axis with angular velocity ω. In the absence of rotation ($\omega = 0$), the natural repose angle of the granular material is $\theta_0 = \tan^{-1}\mu$, where μ is the Coulombic friction coefficient. For small angular velocities ($\omega < \omega_1$), the material sloshes around in the bottom of the drum. For large angular velocities ($\omega > \omega_2$), the material is centrifuged and distributed uniformly throughout the drum. We are interested in the intermediate regime ($\omega_1 < \omega < \omega_2$), where a steady-state S-shaped surface profile is established through the accretion/depletion of particles from the "solid bulk" onto a "fluid surface layer" [26, 34]. The nature of the flowing layer determines the form of the S-shaped surface profile. For example, it is accentuated when the flowing layer has an approximately uniform thickness. On the other hand, the surface may be almost flat when the flowing layer has a nonuniform thickness.

The S-shaped surface profile is generic to a large class of systems and occurs because the maximum current of particles is transported in the central region of the flowing layer, which must therefore have the maximum local slope. On the other hand, the current goes to zero at the wings of the drum, and the local slope variable assumes its natural repose value. Following Zik *et al.* [34], we have obtained the steady-state slope profile as the solution of a cubic equation [24]:

$$\frac{p_0^3}{3\eta\rho g^2}(1+s^2)(s-\mu) = \frac{\rho\omega}{2}(R^2-d^2-x^2)\,, \tag{2.1}$$

where $s(x)$ is the local slope as a function of x, the radial coordinate along the average surface layer. In Eq. (2.1), p_0 is the pressure at the bottom of the flowing layer, η is the fluid viscosity, ρ the fluid density, g is the acceleration due to gravity, and d is the vertical distance from the drum center to the average surface profile. In the special case that the drum is half-filled, we have $d = 0$.

Equation (2.1) has only one real solution:

$$s(x) = \frac{\mu}{3} + \left\{\frac{\mu}{3} + \frac{\mu^3}{27} + \frac{A}{2} + \left[\frac{(1+\mu^2)^2}{27} + \frac{A^2}{4} + \frac{A}{3}\mu + \frac{A}{27}\mu^3\right]^{1/2}\right\}^{1/3}$$
$$+ \left\{\frac{\mu}{3} + \frac{\mu^3}{27} + \frac{A}{2} - \left[\frac{(1+\mu^2)^2}{27} + \frac{A^2}{4} + \frac{A}{3}\mu + \frac{A}{27}\mu^3\right]^{1/2}\right\}^{1/3}, \tag{2.2}$$

where we have introduced the compact notation $A = 3\eta\rho^2 g^2\omega(R^2-d^2-x^2)/2p_0^3$.

Before we proceed, some remarks are in order. Firstly, we should stress that Eq. (2.1) is obtained under rather simplistic conditions, e.g. the granular fluid is assumed to be Newtonian. Thus, Eq. (2.1) and its solution should (at best) be considered as a guide to good phenomenology. Secondly, the S-shaped profile in Eq. (2.2) is symmetric under $x \to -x$. However, in experiments, there is a difference between the dynamic friction coefficient (μ_d) and the static friction coefficient ($\mu_s > \mu_d$), which are relevant at the bottom ($x = -\sqrt{R^2-d^2}$) and top ($x = \sqrt{R^2-d^2}$) of the surface profile, respectively. Equation (2.2) is easily generalized to account for the difference in μ_d and μ_s. Thirdly, an S-shaped profile is implicit in Eq. (2.1), so it is clearly not valid in the frequency regimes $\omega < \omega_1$ and $\omega > \omega_2$.

Next, we consider a two-component granular mixture of rough and less-rough components (say, S and G) in a rotating drum. Again, the experimentally interesting regime is one where a steady-state surface profile exists. For the homogeneous mixture, the natural repose angle (at $\omega = 0$) is determined by $\theta_0 = \tan^{-1}(\mu_S\phi_S + \mu_G\phi_G)$, where μ_S, μ_G are the Coulombic friction coefficients of S and G, respectively; and ϕ_S, $\phi_G(=1-\phi_S)$ are the appropriate number fractions. The appropriate

S-shaped surface profile of the rotated homogeneous mixture is obtained from Eq. (2.2) with

$$\begin{aligned} \mu &= \mu_S\phi_S + \mu_G\phi_G \equiv \frac{\mu_S + \mu_G}{2} + \frac{\mu_S - \mu_G}{2}\psi\,, \\ \eta &= \eta_S\phi_S + \eta_G\phi_G \equiv \frac{\eta_S + \eta_G}{2} + \frac{\eta_S - \eta_G}{2}\psi\,. \end{aligned} \tag{2.3}$$

In Eq. (2.3), we have introduced the order parameter ψ such that $\phi_S = (1+\psi)/2$, $\phi_G = (1-\psi)/2$. For simplicity, we assume that the densities of the two components are matched — otherwise, one could also consider the appropriate generalization of ρ for a two-component mixture.

2.2. *Development of inhomogeneities*

The two-component granular mixture does not stay homogeneous on rotation, but rather exhibits complex segregation behavior. There have been many experiments [6–8, 10, 11, 13, 16, 20, 22, 30, 34] and numerical simulations [2, 17, 27, 31, 33] which have investigated the dynamical evolution of an initially homogeneous mixture. The rich phenomenology can be summarized as follows:

(a) There is a rapid radial segregation on the time-scale of a few drum rotations [10, 11, 13, 30, 31, 33], with the less-rough material (G) accumulating at the walls of the drum; and the rough material (S) accumulating in the central region. For example, this is clearly seen in the cellular automata (CA) simulations of Ueda *et al.* [31] (see Fig. 2(a) of Ref. 31).
(b) The radially-segregated profile set up in (a) may be destabilized and is then followed by a slower axial segregation [7, 13, 30, 31, 33, 34], where the system phase-separates into alternating bands rich in S and G (see Fig. 2(b) of Ref. 31). These bands coarsen rather slowly in time, as seen in Fig. 1 of Ref. 34; and Fig. 2 of Ref. 30.
(c) A close inspection of the experimental pictures (e.g. Fig. 1 of Ref. 34; or Fig. 2 of Ref. 30) suggests that the bands in (b) are nonuniform in the radial direction.

Furthermore, every experiment does not necessarily show all the above features. For example, some experiments do not see the rapid radial-segregation regime, whereas other experiments may exhibit only radial segregation. Furthermore, the degree of radial nonuniformity of band-size may differ from one experiment to another. We are interested in formulating a phenomenological model which replicates the behaviors discussed in (a)–(c) above.

3. Phenomenological Modeling and Results

Our phenomenological model is formulated in terms of an order-parameter field or composition variable $\psi(\mathbf{r}, t) = (1 + \phi_S(\mathbf{r}, t))/2$, which depends on space ($\mathbf{r}$) and time (t). The order-parameter values $\psi = 1, -1$ correspond to pure sand (S) and glass (G), respectively. We consider the two-dimensional reference frame (x, y), corresponding to the average surface profile of the rotated homogeneous mixture. The coordinate $x \in [-\sqrt{R^2 - d^2}, \sqrt{R^2 - d^2}]$ points along the radial direction, and $y \in [-\infty, \infty]$ points along the drum axis. We make the simplifying assumption that the system is invariant in the third direction, which is reasonable for rectangular drums and for the axially-segregated state in cylindrical drums.

The second important variable is the local slope $s(\mathbf{r}, t)$. Our model consists of coupled dynamical equations for $\psi(\mathbf{r}, t)$ and $s(\mathbf{r}, t)$. Let us first formulate the evolution equation for $\psi(\mathbf{r}, t)$. Gradients in the local slope give rise to a current of the composition variable, which we model as [24]

$$\mathbf{J}_{\text{sl}} = M_0(1 - \psi^2)\boldsymbol{\nabla}[s - s_{\text{G}}(x)]\,. \tag{3.1}$$

In Eq. (3.1), the mobility depends on the local order parameter — there is no evolution of the order parameter in a region which is purely S or G. The mobility is symmetric in the two components, as regions with pure S and G correspond to $\psi = +1$ and -1, respectively. Furthermore, the constant factor M_0 is proportional to the rotation frequency ω, which sets the time-scale of evolution dynamics. The function $s_{\text{G}}(x)$ on the right-hand side (RHS) of Eq. (3.1) refers to the S-shaped surface profile for pure G.

There is also a diffusive mixing of S and G due to random collisions. Thus, the order-parameter evolution is modeled as

$$\frac{\partial \psi(\mathbf{r}, t)}{\partial t} = -\boldsymbol{\nabla} \cdot \{(1 - \psi^2)\boldsymbol{\nabla}[M_0(s - s_{\text{G}}(x)) - D \tanh^{-1} \psi]\}\,, \tag{3.2}$$

where we have introduced the diffusion constant D (proportional to ω), and used the identity $\nabla^2 \psi = \boldsymbol{\nabla} \cdot \{(1 - \psi^2)\boldsymbol{\nabla}[\tanh^{-1} \psi]\}$.

Next, we consider the evolution of the slope variable $s(\mathbf{r}, t)$. Let us first focus on the case of a homogeneous mixture with composition ψ. We expect the slope variable to be rapidly relaxed to the local steady-state solution $f(\psi, x)$, which is determined by (say) Eq. (2.2) with appropriate generalizations of μ and ψ. Furthermore, there are fluctuations in the local slope which are relaxed diffusively. Therefore, the appropriate dynamical equation is

$$\tau \frac{\partial s(\mathbf{r}, t)}{\partial t} = f(\psi, x) - s + \xi^2 \frac{\partial^2 s}{\partial y^2}\,, \tag{3.3}$$

where we only consider diffusive fluctuations (with length-scale ξ) in the y-direction (drum axis), as fluctuations in the radial direction are suppressed by the strong relaxation. As the local slope is established by fluid flows with time-scales much faster than the diffusive time-scales which drive segregation, we consider $\tau \to 0$. In the hydrodynamic limit, where the order parameter is slowly modulated, we have the "static" solution of Eq. (3.3) as

$$\begin{aligned} s(\mathbf{r}, t) &= \left[1 - \xi^2 \frac{\partial^2}{\partial y^2}\right]^{-1} f(\psi, x) \\ &\simeq f(\psi, x) + \xi^2 \frac{\partial^2 f(\psi, x)}{\partial y^2} + \text{higher order terms}\,. \end{aligned} \tag{3.4}$$

In this approximation, the time-dependence of the local slope variable arises from the time-dependence of ψ. Then, we have a closed equation for the evolution of the order parameter as

$$\frac{\partial\psi(\mathbf{r},t)}{\partial t} = -\nabla\cdot\left\{(1-\psi^2)\nabla\left[M_0(f(\psi,x)-s_G(x))+M_0\xi^2\frac{\partial^2 f(\psi,x)}{\partial y^2}-D\tanh^{-1}\psi\right]\right\}. \tag{3.5}$$

In our earlier work, we used a different argument to arrive at an analogous model. In Ref. 24, we introduced an "interfacial penalty" term to stabilize sharp interfaces between S-rich and G-rich regions. However, such a term is reminiscent of microscopic surface tension and its physical origin is unclear in the context of granular physics. The present arguments provide a dynamical basis for an "effective surface tension" in the current context. We should also point out that recent experimental results of Turner *et al.* [29] are supportive of the existence of an "effective surface tension" in granular mixtures. Furthermore, Wakou *et al.* [32] have derived a time-dependent Ginzburg–Landau (TDGL) model for granular gases, where "surface tension" arises from the viscous term for granular fluids.

It is convenient to obtain $f(\psi,x)$ as a Taylor-expansion in ψ and ω. As we stated earlier, Eq. (2.2) was obtained under simplistic assumptions. Therefore, we invoke symmetry considerations to obtain a Taylor-expansion for $f(\psi,x)$ [24]. However, the expansion coefficients can be identified in terms of microscopic parameters by comparing with the functional form in Eq. (2.2). The resultant model (after rescaling into dimensionless units) is as follows [24]:

$$\frac{\partial\psi(\mathbf{r},t)}{\partial t} = -\nabla\cdot\left\{(1-\psi^2)\nabla\left[a(r^2-x^2)+[b+a(r^2-x^2)]\psi-\tanh^{-1}\psi+\frac{\partial^2\psi}{\partial y^2}\right]\right\}, \tag{3.6}$$

where a and b are parameters ($a\propto\omega$, b constant), and r is the dimensionless equivalent of $\sqrt{R^2-d^2}$. Equation (3.6) must be supplemented with boundary conditions at the drum edges. We impose the physical condition that there should be no radial current at the drum boundaries:

$$\frac{\partial}{\partial x}\left[a(r^2-x^2)+[b+a(r^2-x^2)]\psi-\tanh^{-1}\psi+\frac{\partial^2\psi}{\partial y^2}\right]\Bigg|_{x=\pm r}=0\,. \tag{3.7}$$

Equations (3.6) and (3.7) constitute our phenomenological model for granular materials in a rotating drum. As discussed earlier, an analogous model was obtained in Ref. 24 by directly invoking an "interfacial penalty" term to stabilize interfaces. We now present representative results for the evolution of an initially homogeneous granular mixture in a rotating drum. For detailed analytical and numerical results, we refer the reader to our extended works [24, 25].

The inhomogeneous (ψ-independent) term on the RHS of Eq. (3.6) rapidly establishes a radially-segregated profile, replicating the regime discussed in part (a) of Sec. 2.2. We have obtained approximate analytical solutions for this radially-segregated profile by considering zero-current solutions of the model in Eqs. (3.6)

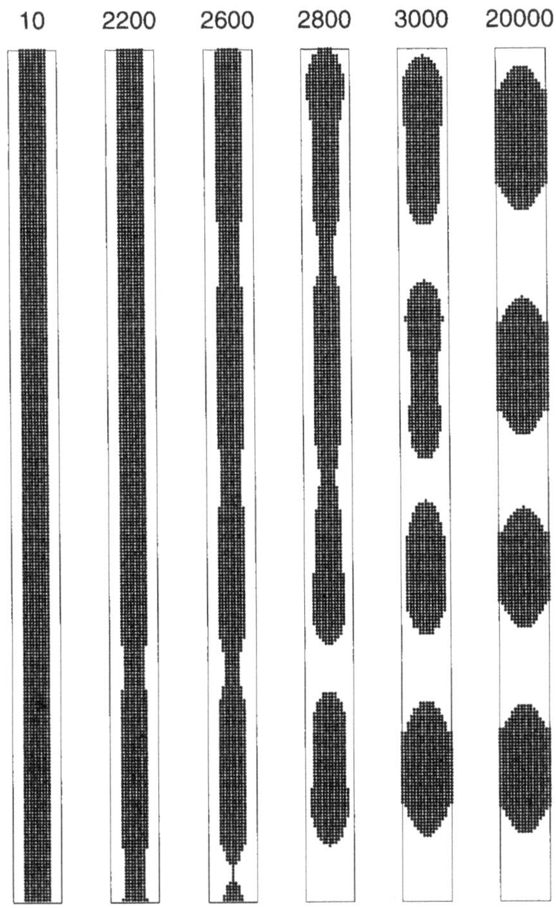

Fig. 1. Evolution of the order parameter for an initially homogeneous mixture of equal amounts of S and G. These results were obtained from a numerical solution of our phenomenological model in Eqs. (3.6) and (3.7) with parameter values $b = 0.95$ and $a = 0.04$. We used a simple Euler-discretization scheme with isotropic Laplacians and mesh sizes $\Delta t = 0.01$ and $\Delta x = 0.5$. The lattice size is $[-r, r] \times [0, L]$ with $r = 4$, $L = 128$. The boundary conditions in Eq. (3.7) were applied at $x = \pm r$, and periodic boundary conditions were applied in the other direction. Regions rich in S ($\psi > 0$) are marked in black, and regions rich in G ($\psi < 0$) are unmarked. The frames are shown at dimensionless times $t = 10$ (radially-segregated state); $t = 2200$, 2600, 2800, 3000 (destabilization of radially-segregated state); and $t = 20000$ (coarsening of axially-segregated state with radially nonuniform bands).

and (3.7). Depending on the parameter values, we observe both weak and strong radial-segregation profiles. On much longer time-scales, the radially-segregated profile may be destabilized by a Mullins–Sekerka instability [25], and then crosses over to an axially-segregated state, as discussed in part (b) of Sec. 2.2. Figures 1 and 2 depict both the radial- and axial-segregation regimes, obtained from numerical solution of our model in Eqs. (3.6) and (3.7). In the axially-segregated state, the central region of the drum is still relatively enriched in the rough component S,

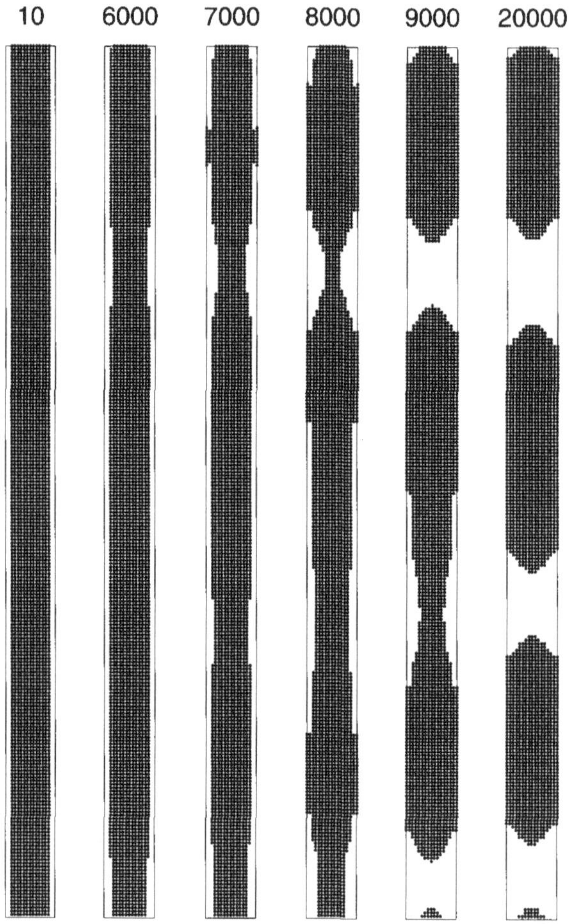

Fig. 2. Analogous to Fig. 1, but the composition of the mixture is 65% S and 35% G.

whereas the drum edges are still relatively rich in the less-rough component G. This is in conformity with the magnetic-resonance-imaging (MRI) experiments of Nakagawa *et al.* [20].

The coarsening of bands in the radially-segregated state is analogous to domain growth in the one-dimensional Cahn–Hilliard (CH) model [4]. The appropriate growth law is $L_y(t) \sim \ln(t/t_0)$, where t_0 sets the time-scale. A similar argument is due to Aranson *et al.* [1], who have also formulated a continuum model for granular mixtures in a rotating-drum geometry. Our numerical results in Fig. 3 are consistent with this prediction. We do not know of any experiment which has conclusively ascertained the time-dependence of band growth in the axial-segregation regime. We urge experimentalists to investigate this important characteristic of the evolving morphology, as it would serve as an important check on the validity of our phenomenological modeling.

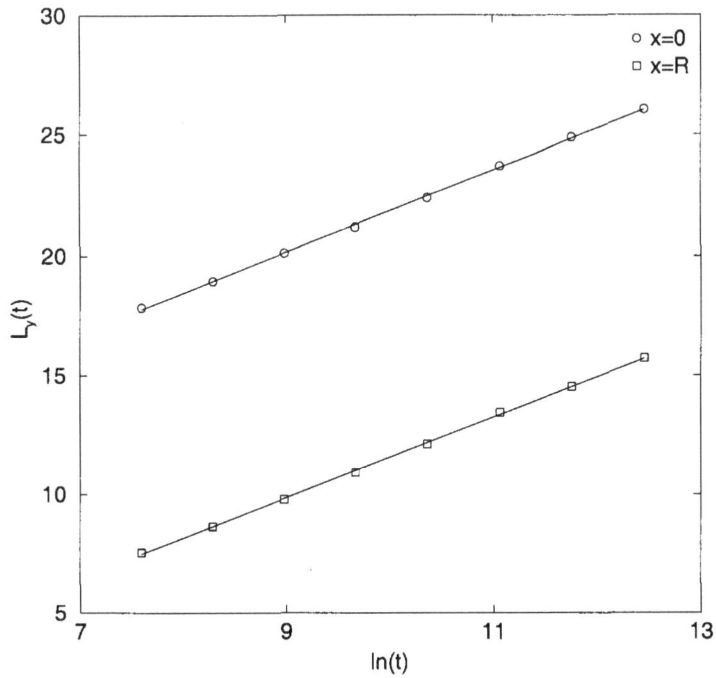

Fig. 3. Time-dependence of $L_y(t)$, the width of bands in the axially-segregated state, for the evolution depicted in Fig. 1. We present results for $L_y(t)$ versus $\ln t$ at the drum center ($x = 0$) and drum edge ($x = r$) — denoted by the specified symbols. The length scales are calculated as averages of domain-size distributions obtained from ten independent runs with lattice sizes $r = 4$ and $L = 2048$. (From Ref. 26.)

4. Summary and Discussion

Let us conclude this paper with a summary and discussion of our results. We have formulated a phenomenological model for the segregation dynamics of granular mixtures in a rotating drum. Our model considers the coupled dynamics of the composition field and the local slope field. Furthermore, we assume that the local slope is in instantaneous equilibrium with the local composition. This simplifies our model to a single dynamical equation for the composition variable. An important feature of this paper is the dynamical interpretation of an "effective surface tension," which was implicit in our earlier modeling [24]. Recent experimental results [29] also support the existence of an "effective surface tension" in the context of granular mixtures.

We have obtained both analytical and numerical results from the model presented in this paper and an earlier variant [24]. These are in good agreement with available experimental results, which are primarily at the qualitative level. We also make specific quantitative predictions, e.g. the band size is expected to grow logarithmically in the axial-segregation regime. It would be relevant for experimentalists to make more quantitative measurements and unambiguously ascertain these growth laws. Given the phenomenological nature of our modeling, the comparison of time- and length-scales between our models and experiments is difficult. Typically, comparisons must be confined to scaling behaviors and functional dependences, which in any case are of primary interest to physicists.

There are various possible directions for future study. Firstly, within the framework of our present models, a wide range of physical effects are accessible, e.g. traveling waves in axial segregation [7]; effects of asymmetry in the S-shaped surface profile; inclusion of gravitational effects, etc. Our present study is far from being exhaustive in this regard! Secondly, and perhaps more important, our model neglects interesting physical effects in the third direction — a more realistic model must account for these. For example, the nontrivial behavior of granular velocity fields below the flowing surface layer is only now being clarified experimentally [21]. Nevertheless, our study demonstrates how a few simple dynamical ingredients can recover many experimental features in the context of granular mixtures in the rotating-drum geometry. We are hopeful of extending this approach to other problems involving the dynamical properties of granular materials.

Acknowledgments

The authors would like to thank E. Fukushima, H. J. Herrmann, D. Khakhar, R. Kobayashi, S. Lipson, A. Mehta, M. Nakagawa and Y. Shiwa for many useful discussions and critical inputs.

References

[1] Aranson, I. S., Tsimring, L. S. and Vinokur, V. M., *Phys. Rev.* **E60**, 1975 (1999).
[2] Awazu, A., *Phys. Rev. Lett.* **84**, 4585 (2000).
[3] Behringer R. P. and Jenkins, J., *Powders and Grains* 97: *Proceedings of the Third International Conference on Powders and Grains* (Balkema, Rotterdam, 1997); Kishino, T. (ed.), *Powders and Grains* 2001: *Proceedings of the Fourth International Conference on Powders and Grains* (Swets and Zeitlinger, Lisse, 2001).
[4] Bray, A. J., *Adv. Phys.* **43**, 357 (1994).
[5] Brey, J. J., Moreno, F. and Dufty, J. W., *Phys. Rev.* **E54**, 445 (1996); Brey, J. J., Ruiz-Montero, M. J. and Moreno, F., *Phys. Rev.* **E63**, 061305 (2001).
[6] Cantelaube, F. and Bideau, D., *Europhys. Lett.* **30**, 133 (1995).
[7] Choo, K., Molteno, T. C. A. and Morris, S. W., *Phys. Rev. Lett.* **79**, 2975 (1997); Choo, K., Baker, M. W., Molteno, T. C. A. and Morris, S. W., *Phys. Rev.* **E58**, 6115 (1998).
[8] Clement, E., Rajchenbach, J. and Duran, J., *Europhys. Lett.* **30**, 7 (1995).
[9] Cross, M. C. and Hohenberg, P. C., *Rev. Mod. Phys.* **65**, 851 (1993).
[10] Das Gupta, S., Khakhar D. V. and Bhatia, S. K., *Chem. Eng. Sci.* **46**, 1531 (1991); *Powder Technol.* **67**, 145 (1991).
[11] Donald, M. B. and Roseman, B., *British Chem. Eng.* **7**, 749, 823 (1962).
[12] Duran, J., *Sands, Powders and Grains*: *An Introduction to the Physics of Granular Materials* (Springer-Verlag, New York, 2000).

[13] Hill, K. M. and Kakalios, J., *Phys. Rev.* **E49**, 3610 (1994); *Phys. Rev.* **E52**, 4393 (1995); Hill, K. M., Caprihan, A. and Kakalios, J., *Phys. Rev. Lett.* **78**, 50 (1997).
[14] Hohenberg, P. C. and Halperin, B. I., *Rev. Mod. Phys.* **49**, 435 (1977).
[15] Jaeger, H. M., Nagel, S. R. and Behringer, R. P., *Rev. Mod. Phys.* **68**, 1259 (1996).
[16] Khakhar, D. V., McCarthy, J. J., Shinbrot, T. and Ottino, J. M., *Phys. Fluids* **9**, 31 (1997); Khakhar, D. V., McCarthy, J. J. and Ottino, J. M., *Phys. Fluids* **9**, 3600 (1997).
[17] Lai, P.-Y., Jia, L.-C. and Chan, C. K., *Phys. Rev. Lett.* **79**, 4994 (1997).
[18] Mehta, A., *Granular Matter: An Interdisciplinary Approach* (Springer-Verlag, New York, 1994).
[19] Melo, F., Umbanhowar P. B. and Swinney, H. L., *Phys. Rev. Lett.* **75**, 3838 (1995); Umbanhowar, P. B., Melo, F. and Swinney, H. L., *Nature* **382**, 793 (1996).
[20] Nakagawa, M., *Chem. Engg. Sci.* **49**, 2544 (1994); Nakagawa, M., Altobelli, S. A., Caprihan, A. and Fukushima, E., *Chem. Engg. Sci.* **52**, 4423 (1997).
[21] Nakagawa, M., private communication.
[22] Oyama, Y., *Bull. Inst. Phys. Chem. Res. Jpn. Rep.* **18**, 600 (1939).
[23] Peng, G. and Herrmann, H. J., *Phys. Rev.* **E49**, R1796 (1994); **E51**, 1745 (1995); Moriyama, O., Kuroiwa, N., Matsushita, M. and Hayakawa, H., *Phys. Rev. Lett.* **80**, 2833 (1998).
[24] Puri, S. and Hayakawa, H., *Physica* **A290**, 218 (2001).
[25] Puri, S. and Hayakawa, H., *Physica* **A270**, 115 (1999).
[26] Rajchenbach, J., *Phys. Rev. Lett.* **65**, 2221 (1990).
[27] Ristow, G. H., *Europhys. Lett.* **34**, 263 (1996).
[28] Rothman, D. H., *Phys. Rev.* **E57**, R1239 (1998).
[29] Turner, J. L., Nakagawa, M. and Lusk, M. T., in *Powders and Grains 2001*, Kishino, T., ed. (Swets and Zeitlinger, Lisse, 2001), p. 541.
[30] Ueda, K. and Kobayashi, R., Technical Report of IEICE, NC 96-95, 1 (1997).
[31] Ueda, K., Kobayashi, R. and Yanagita, T., in *Statistical Physics: Proceedings of Second Tohwa University International Meeting* (World Scientific, Singapore, 1998), p. 141.
[32] van Noije, T. P. C. and Ernst, M. H., *Phys. Rev.* **E61**, 1765 (2000); Wakou, J., Brito, R. and Ernst, M. H., cond-mat 0103086.
[33] Yanagita, T., *Phys. Rev. Lett.* **82**, 3488 (1999).
[34] Zik, O., Levine, D., Lipson, S. G., Shtrikman, S. and Stavans, J., *Phys. Rev. Lett.* **73**, 644 (1994); see also Thomae, S., in *Nonlinearities in Complex Systems*, Puri, S. and Dattagupta, S., eds. (Narosa, Delhi, 1997).

CHAPTER 19

Applications of Synchrotron X-Ray Microtomography to Mesoscale Materials

G. T. SEIDLER,* L. J. ATKINS and E. A. BEHNE

Physics Department, University of Washington,
Seattle, Washington 98195-1560, USA
**seidler@phys.washington.edu*

U. NOOMNARM

Physics Department, University of Florida,
Gainesville, Florida 32611, USA

S. A. KOEHLER

Division of Engineering and Applied Sciences, Harvard University,
Cambridge, Massachusetts 02138, USA

R. R. GUSTAFSON and W. T. MCKEAN

Forest Resources, University of Washington,
Seattle, Washington 98195-2100, USA

We discuss the application of synchrotron X-ray microtomography (XMT) to granular matter, foams, crumpled membranes, and paper. XMT provides rapid, high-resolution, fully three-dimensional characterization of each of these classes of material. In some cases, subsequent three-dimensional image processing allows the virtual reconstruction of the disordered material as a specified assemblage of idealized basic structural units. This allows measurement of otherwise inaccessible correlation functions and can also be used as the starting point for data-initiated simulations.

Keywords: Granular matter; foam; froth; crumpled membrane; paper; virtual reconstruction; X-ray microtomography; X-ray microcomputed tomography.

1. Introduction

The purpose of this paper is to survey several applications of synchrotron X-ray microtomography (XMT) to mesoscale disordered materials. In our view the utility of XMT in studies of mesoscale materials occurs in two distinct categories. First, XMT provides the rapid measurement of the full three-dimensional (3D) structure

*Corresponding author.

First published in *Advances in Complex Systems*, Vol. 4, No. 4 (2001), pp. 481–490.

of mesoscale materials. Second, these measurements frequently have sufficient spatial resolution and signal-to-noise ratios to allow "virtual reconstruction" (VR), i.e. the representation of the real material into computer memory as some specified assemblage of idealized basic structural units. Although the simple knowledge of the spatial distribution of mass in a sample allows calculation of many important structural parameters, VR of a material opens a far wider range of possible investigations through system-specific high-order correlations [1–3, 8, 33] and data-initiated simulations [14, 15].

This paper follows a review format. First, in Sec. 2 we present experimental details. Second, in Secs. 3–6 we discuss the application of XMT and VR to four different classes of mesoscale materials: granular matter, foams, crumpled membranes, and paper. In each of these sections, we briefly motivate interest in the selected material, summarize prior work using other techniques, present and discuss our XMT measurements, and discuss the applicability of VR. Finally, we present our conclusions in Sec. 7.

2. Experimental Details

All experiments were performed at the Pacific Northwest Consortium Collaborative Access Team (PNC-CAT) beamlines (20-ID, 20-BM) at the Advanced Photon Source. The area detector of our tomography apparatus follows the general considerations of Koch *et al.* [16]. Our preliminary apparatus (UWTOM0) has been described elsewhere [33]; this apparatus was used in the measurements reported in Secs. 3 and 5 at a spatial resolution of 3.5 μm. We have recently made several upgrades to this apparatus (henceforth, UWTOM1). The new camera uses a 12-bit thermoelectrically-cooled interline-transfer CCD (Roper Scientific). For low resolution studies (10 μm or larger effective pixel size or larger) an 0.5 mm thick uniformly doped YAG:Ce scintillator was used. At higher resolutions a different YAG scintillator with only a 1 μm thick Ce-doped layer is used to provide a better match to the depth of focus of the optical portion of the detector. The sample is mounted on translation stages allowing for three-axis motion with respect to the tomography rotation stage. The rotation stage of UWTOM1 consists of an air bearing spindle (Professional Instruments, Blockhead-4R) actuated with a stepper motor and worm-gear drive. The rotation stage is supported by a cradle which provides a tilt-axis parallel to the X-ray beam direction. A second tilt-axis perpendicular to both the X-ray beam direction and the tomography rotation axis is supplied by the experimental table. Tomogram reconstruction used either a filtered backprojection or a grid-reconstruction algorithm.

3. Granular Matter

Recent experimental work in granular statics includes studies of the full stress and strain fields in two-dimensional (2D) systems [12] and of the stress field at the container boundaries of three-dimensional (3D) systems [23]. Two key differences exist between experimental capabilities in 2D and in 3D. First, while quantitative studies of the full stress field are possible in 2D [12], there is presently no method for even qualitatively imaging the 3D stress field in dry granular matter. Second, the determination of granule-level packing and corresponding strains on evolution of the structure is regularly practiced in 2D systems by digital video recording and appropriate software for VR [12, 30], but the 3D problem is considerably more challenging, especially for dry granular matter. Mechanical disassembly has been historically significant in understanding the physics of liquids [4] and glasses [32] but is too tedious for regular application. Industrial computed tomography scanners [17] and, more importantly, both medical and laboratory magnetic resonance imaging (MRI) [7, 22, 25] have proven useful in 3D imaging of granular matter. However, both of these techniques are somewhat limited for studies of the granule-scale packing.

Working independently, groups at the University of Chicago [22] and the University of Washington [33] have recently performed granule-level studies using XMT. In Ref. 22, large granules ($D \sim 3$ mm) in a 3D Couette geometry were studied with MRI, XMT and surface optical tracking. Although the XMT study was limited to a section only about one granule in height, combining all information led to a new characterization of the relationship between shear rate and fabric in the Couette geometry. In Ref. 33, several of the present authors demonstrated that XMT provides rapid imaging of granular structure to sufficiently small length scales that reliable VR of the complete granular packing is possible, even for moderately fine powders. Here, we present a sample of our work on VR of fine powders [1, 33] and discuss the range of now-practical studies on granule-level structure.

The sample is a granular bed consisting of mean diameter $D = 63$ μm glass spheres ($\geq$95% have polydispersivity less than ± 4 μm, MOSCI, USA) in a vertical 1.5 mm inner-diameter glass cylinder. The granular bed was approximately 4 cm tall. Its structure was measured approximately at mid-height both before and after compaction by 10^4 cycles of vertical agitation with a peak acceleration of 22.5 m/s^2; the sample was observed to compact without convection. In Fig. 1(a) we show a typical horizontal XMT slice through the 3D structure of the sample after compaction. The data shown in the figure is one of several hundred adjacent slices which form the full 3D data set. Note that the quality of the XMT reconstruction rules out any significant internal reorganization during data collection.

A sphere-identifying Hough transform was used to achieve VR [33]. The Hough transform is a high-level object-recognition algorithm which is known to be least-squares optimal for many problems [5]. In this case it allows determination of the center locations of each granule with a precision of 1% of the mean granule diameter. In Fig. 1(b) we show the projection down the vertical cylinder axis of the location of all 7318 granules identified in the region of interest of the compacted sample. A void-space test demonstrates no empty spaces sufficiently large to accommodate another granule with diameter D.

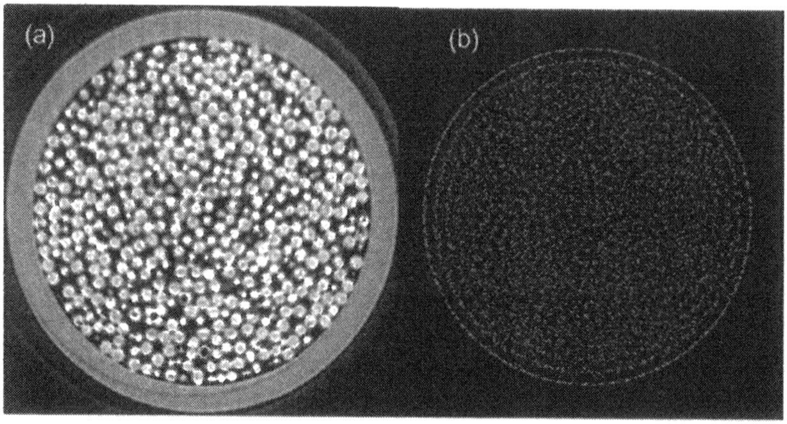

Fig. 1. (a) A typical horizontal XMT slice through a granular bed. (b) The projection down the cylinder axis of the all granule centers in the 3D tomogram. See the text for details.

A detailed comparison of the VR of the sample before and after compaction shows several interesting features, including evidence for significant internal reorganization of the sample during the deposition process and also a strong dependence of two-bond orientational order on the angle between the pair of bonds and their mutual displacement vector [1]. The combination of XMT with VR for granular materials will have many future applications. Our own studies will next focus on the fabric in granular beds with anisotropic particles and the granule-level structure of sediments.

4. Foams

Liquid foams and low-Z solid foams are near-ideal materials for XMT. First, they are two-phase materials with a large density difference between the two phases. Second, their low average density allows studies of statistically large samples at convenient photon energies. Third, their cellular nature is amenable to VR; in fact this has already been performed from optical measurements of small 3D samples of a liquid foam [21]. Below, we address liquid foams and then solid foams. In both cases, we find that XMT provides sufficiently high resolution for VR.

4.1. *Liquid foams*

Liquid foams are metastable soft condensed matter systems that are in mechanical equilibrium [28, 36] but are far from thermodynamic equilibrium. The large surface area of a foam is energetically expensive, and the foam evolves towards states with lower surface area through coarsening. The way a foam partitions space into bubbles and the time evolution of these bubbles are each very complex process, which have relevance to the grain growth of crystals in metals [34], the formation of solid foams [9], the preparation of foodstuffs [27], and the structure of the cosmos [27].

Although there has been substantial progress in understanding the evolution of two-dimensional liquid foams [24, 34], the situation is far more challenging for fully 3D foams. Optical imaging [21], diffusing light spectroscopy [6], and MRI [11] studies have each proven useful for certain questions, but none of these techniques allow for direct observation of the detailed geometry of statistically large 3D foam samples.

To address this situation, we have performed a preliminary XMT study of 3D coarsening of a liquid foam using UWTOM1 at beamline 20 BM. The sample was a commercial cosmetic foam [10] injected into a 50 mm tall thin-walled polypropylene cylinder having an average 6.0 mm inner-diameter. The cylinder was then sealed at both ends to prevent evaporation; the region of interest is approximately at mid-height. First, with a photon energy of 8 keV, we measured absorption XMT when the sample had aged 22 hours after preparation. The total measurement time was 60 min, the spatial resolution was 13 μm, the sample to detector distance was 1 cm, and the total foam volume imaged was 74 mm^3 consisting of an estimated 8000 bubbles. In Fig. 2(a) we show a typical XMT slice through the 3D data set. Vertices and Plateau borders are apparent in the figure, however the bubbles' faces were in general too thin to be resolved. To the best of our knowledge, this is the first XMT study of a liquid foam.

A total of 70 hours aging after preparation, the liquid fraction of the sample had decreased to a point where the X-ray absorption of the Plateau borders was insufficient to perform absorption XMT. Consequently, we increased the photon energy to 15 keV and the sample-to-detector distance to 12 cm and used phase-contrast XMT. In Fig. 2(b) we show a typical XMT slice through the 3D data set. The measurement time for the same volume as above was only 30 min. A comparison of Figs. 2(a) and (b) shows a significant increase in both the average bubble size and the bubble polydispersivity, in addition to the expected large decrease in liquid

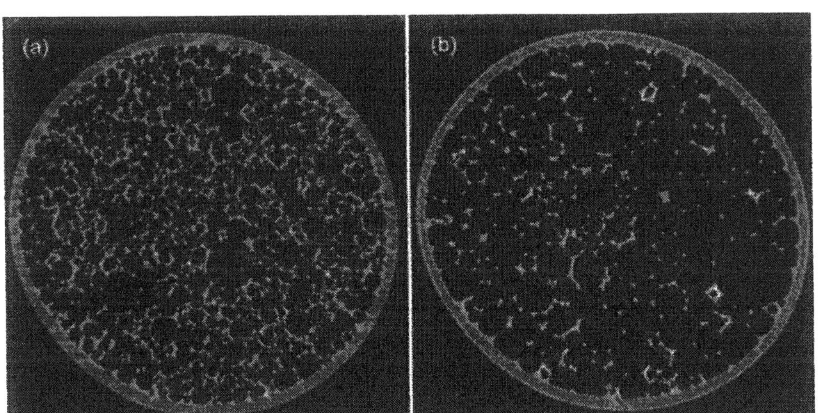

Fig. 2. (a) An XMT slice through a liquid foam 22 hours after preparation. (b) The same sample a total of 70 hours after preparation. The average i.d. of the cylindrical container is 6 mm.

fraction. A more complete analysis of these samples is in progress and will be reported elsewhere [2].

Having established that large samples of liquid foams can be effectively imaged with 3D XMT, we expect that many previously difficult issues can now be studied with improved detail. Future studies could investigate the validity of the decoration theorem [36], the interrelationships between and distribution functions of the number of faces, number of edges, and volumes of bubbles [19], and the functional dependence of the coarsening rate on the real foam topology [38].

4.2. *Solid foams*

Synthetic solid foams are used extensively in insulation, in shock absorbing, as low-density structural materials, and as filters [9]. Despite this widespread industrial relevance, fully 3D investigations of the micromechanics of solid foams are rare [15, 35]. From the physicist's standpoint, we propose that elastic solid foams may provide an interesting environment in which to investigate stress-propagation in disordered materials. Unlike granular matter where inter-particle friction richly complicates the microscopic constitutive relationships [18], complete knowledge of the 3D strain field in an elastic solid foam should allow reliable determination of the spatial distribution of elastic energy through finite element modeling. Initial work along these lines for closed-cell foams has recently been published by Kinney *et al.* [15]. It will be especially interesting to investigate the sensitivity of the deviations from continuum elasticity models to local foam topology.

Using UWTOM1 at beamline 20 BM, we recently characterized the microstructure of several reticulated solid foams. We present a tomographic slice and a 3D rendering of the ordinary reticulated polyurethane foam [3] under zero strain in Figs. 3(a) and (b), respectively. The polyhedral foam topology is well resolved. More detailed discussions of the XMT of reticulated solid foams and their VR will be presented elsewhere. [3]

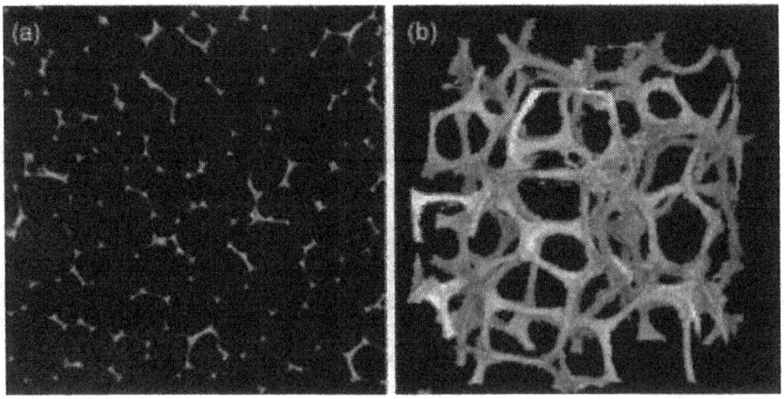

Fig. 3. (a) A 2 mm × 2 mm subregion of a typical XMT slice through a reticulated polyurethane foam with average pore size 0.25 mm. (b) A 3D rendering of a subvolume of the same sample.

5. Crumpled Membranes

Several aspects of the physics of tethered 2D manifolds, i.e. non-intersecting membranes, have attracted recent theoretical [13, 20] and experimental interest [29]. In many cases the relevant structural correlation function to describe crumpling is not the mass-mass pair correlation function, but instead the joint probability density function $P(l,d)$ for path lengths l along the manifold relative to distances d in the embedding into Euclidean space [13]. Hence, a strong interaction between theory and experiment will require real-space structural information. A complete analysis of the structure of a crumpled membrane would necessarily involve the "virtual un-crumpling" (henceforth VUC) of the membrane, i.e. determination of the one-to-one mapping from the original flat manifold to the final crumpled state.

We have performed a preliminary XMT study of a crumpled membrane to evaluate the difficulty of a complete VUC study. The measurements were taken at the 20-ID beamline with the prototype UWTOM0 apparatus at an energy of 20 keV. The sample is a 6 cm^2 area of standard aluminum foil (Reynolds, USA) manually crumpled into a spheroid. We show an XMT cut through the largest cross-section of the sample in Fig. 4(a). The variation in the apparent width of the meandering membrane is due to the variation of the angle between the membrane and the plane of the figure. Interestingly, the sample shows a strong tendency for lamination of different subsections of the membrane and for the formation of tubules.

Two particular challenges must be overcome in order to perform a VUC study. First, the large ratio of the sample size to the thickness of the membrane will likely require the sample to be rotated several times at different sample translations perpendicular to the X-ray beam. Second, special effort may be needed to resolve the structure of the membrane near places where the membrane touches itself. If this cannot be addressed by software, then one might use a tri-layer membrane consisting of two low density sheets coating a higher density interior region. The higher density regions would then never touch, resolving the problem.

6. Paper

Recent theoretical work on paper [26] has included simulations, scaling descriptions of sedimentation mechanisms, and general discussions of statistical measures for characterizing fiber packing geometry. Although extensive work on paper microstructure exists in the forest resources and paper sciences literature, very little of this provides the fully 3D information necessary for comparison with statistical mechanical theories of paper formation. For example, there has been only sporadic attempts to use confocal microscopy to study paper [37] and the first two brief XMT studies of paper have just recently been published [31].

Paper manufacture involves several steps whose physics is of current interest, including sedimentation, hydrodynamically-induced segregation, network-formation, and non-equilibrium mixing. Paper is a lamellate assemblage of fibers that generally

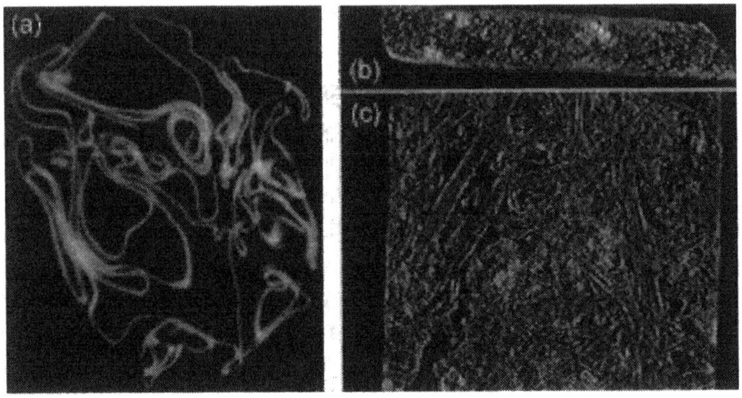

Fig. 4. (a) An XMT slice through a crumpled metal foil. (b) and (c) Two perpendicular XMT slices through a piece of commercial paper used for four-color printing.

come from wood. The fibers are typically 1–4 mm long and 10–50 μm wide. Prior to papermaking the fibers are pulped to remove lignin and mechanically refined to make them pliable and to fibrillate their surfaces. To make paper the fibers are dispersed in water at a solids fraction of about 0.1% fibers. The pulp slurry is pumped into a pressurized headbox that discharges a uniform jet of the slurry on to a moving forming fabric; the deposit is forcibly drained by suction through the porous fabric. The pulp jetting out of the headbox partially orients the fibers in the sheet along the direction the papermachine is running. The sheet is then conveyed through a series of subsequent steps for additional drying, pressing, and smoothing. It is during these steps that the originally open-celled fibers collapse into ribbons and also form additional bonds at contacts with neighboring fibers.

XMT measurements were performed at beamline 20-ID with UWTOM1. The photon energy was 11.5 keV and the spatial resolution was 1.7 μm. In Figs. 4(b) and (c) we show XMT slices perpendicular (Fig. 4(b)) and parallel (Fig. 4(c)) to the face of a sheet of a commercial coated paper. The width across the figures is 0.7 mm. Many important structural characteristics are apparent, including the irregular penetration of the polymer coating into the sheet, the strong anisotropy of the fiber orientations with the ribbon-like fibers laying in the plane of the sheet, the irregular mixture of larger (soft wood) and small (hard wood) fibers, and the presence of isolated voids. VR of paper, i.e. the numerical decomposition into the constituent fibers, will be difficult for a dense paper such as in the figure but should be possible for bulkier paper products where the density of contacts between fibers is smaller.

7. Conclusions

We report synchrotron X-ray microtomography (XMT) measurements for several materials of contemporary interest in the statistical physics community, specifically granular matter, foams, crumpled membranes, and paper. We find that XMT, especially when linked with virtual reconstruction (VR), can be a powerful tool for understanding the microstructure and micromechanics of these materials. The availability of XMT user facilities and the low-cost of modern computing power makes the combination of XMT and VR readily accessible to the experimental community.

Acknowledgments

The PNC-CAT is supported by the U.S. Department of Energy, Basic Energy Sciences, under Contract No. DE-FG03-97ER45629, by the University of Washington, and by grants from the NSERC of Canada. Use of the Advanced Photon Source was supported by the U.S. Department of Energy, Basic Energy Sciences, Office of Science, under Contract No. W-31-109-Eng-38. G. T. Seidler acknowledges support from the Research Corporation as a Cottrell Scholar. U. Noomnarm acknowledges support from the U.S. National Science Foundation under the Research Experiences for Undergraduates program, contract No. PHY-0097551.

References

[1] Atkins, L. J., Martinez, G., Seeley, L. H. and Seidler, G. T., Compaction and three-dimensional structure of a granular bed, to be submitted to *Phys. Rev. E.*
[2] Behne, E. A., Koehler, S. A., Noomnarm, U. and Seidler, G. T., X-ray microtomography of liquid foams, in preparation.
[3] Behne, E. A., Wells, D., Atkins, L. J., Noomnarm, U. and Seidler, G. T., Three-dimensional structure of a reticulated polyurethane foam, in preparation.
[4] Bernal, J. D., *Proc. Roy. Soc. London* **A280**, 299 (1964).
[5] Davies, E. R., *Machine Vision*, 2nd edn. (Academic Press, 1997).
[6] Durian, D. J., Weitz, D. A. and Pine, D. J., *J. Phys. Condens. Matter* **2**, 433 (1990).
[7] Ehrichs, E. E. *et al.*, *Science* **267**, 1632 (1995); Hill, K. M., Caprihan, A. and Kakalios, J., *Phys. Rev. Lett.* **78**, 50 (1997); Yamane, K. *et al.*, *Phys. Fluids* **10**, 1419 (1998); Fukushima, E., *Ann. Rev. Fluid Mech.* **31**, 95 (1999).
[8] For example, see: Lindquist, W. B. *et al.*, *J. Geophys. Res. – Sol. Earth* **105**, 21509 (2000); Lindquist, W. B. and Venkatarangan, A., *Phys. Chem. Earth* **A24**, 593 (1999); Salome, M. *et al.*, *Med. Phys.* **26**, 2194 (1999); Rintoul, M. D. *et al.*, *Phys. Rev.* **E54**, 2663 (1996); Spanne, P. *et al.*, *Phys. Rev. Lett.* **73**, 2001 (1994).
[9] Gibson, L. J. and Ashby, M. F., *Cellular Solids: Structure and Properties* (Cambridge University Press, 1997).
[10] Gillette Foamy Regular (The Gillette Company).
[11] Gonatas, C. P. *et al.*, *Phys. Rev. Lett.* **75**, 573 (1995).
[12] Howell, D., Behringer, R. P. and Veje, C., *Phys. Rev. Lett.* **82**, 5241 (1999).
[13] Kantor, Y., Kardar, M. and Nelson, D. R., *Phys. Rev. Lett.* **57**, 791 (1986); Aronovitz, J. A. and Lubensky, T. C., *Phys. Rev. Lett.* **60**, 2634–2637 (1988); Paczuski, M., Kardar, M. and Nelson, D. R., *Phys. Rev. Lett.* **60**, 2638 (1988); Guitter, E. *et al.*, *Phys. Rev. Lett.* **61**, 2949 (1988); Radsihovsky, L. and Toner, J., *Phys. Rev. Lett.* **75**, 4752 (1995).
[14] Kinney, J. H. and Ladd, A. J. C., *J. Bone Miner. Res.* **13**, 839 (1998); von Eisenhart-Rothe, R. *et al.*, *Anat. and Embryology* **195**, 279 (1997).
[15] Kinney, J. H. *et al.*, *J. Appl. Polym. Sci.* **80**, 1746 (2001).

[16] Koch, A. *et al.*, *J. Opt. Soc. Am.* **15**, 1940–1951 (1998).
[17] Lee, X. and Dass, W. C., in *Powders and Grains 93*, Thornton, C., ed. (Balkema, Rotterdam, 1993), p. 17.
[18] Levine, D., in *Jamming and Rheology*, Liu, A. J. and Nagel, S. R., eds. (Taylor and Francis, 2001), p. 9.
[19] Lewis, F. T., *Anat. Rec.* **38**, 241 (1928); Lambert, C. J. and Weaire, D., *Metallography* **14**, 307 (1983); Coxeter, H. S. M., *Ill. J. Math* **2**, 746 (1958).
[20] Lobkovsky, A. *et al.*, *Science* **270**, 1482 (1995); Lobkovsky, A. E. and Witten, T. A., *Phys. Rev.* **E55**, 1577 (1997); Lobkovsky, A. E., *Phys. Rev.* **E53**, 3750 (1996); Kramer, E. M., *J. Math. Phys.* **38**, 830 (1997); Moldovan, D. and Golubovic, L., *Phys. Rev. Lett.* **82**, 2884 (1999); Kramer, E. M. and Witten, T. A., *Phys. Rev. Lett.* **78**, 1303 (1997).
[21] Monnereau, C., Prunet-Foch, B. and Vignes-Adler, M., *Phys. Rev.* **E63**, 061402 (2001).
[22] Mueth, D. M. *et al.*, *Nature* **406**, 385 (2000).
[23] Mueth, D. M., Jaeger, H. M. and Nagel, S. R., *Phys. Rev.* **57**, 3164 (1998); Reydellet, G. and Clement, E., *Phys. Rev. Lett.* **86**, 3308 (2001).
[24] Mullins, W. W., *J. Appl. Phys.* **59**, 1341 (1986); Glazier, J. A., Gross, S. P. and Stavans, J., *Phys. Rev.* **A36**, 306 (1987); Glazier, J. A. and Weaire, D., *J. Phys. Condens. Matter* **4**, 1867 (1992); Stavans, J., *Rep. Prog. Phys.* **56**, 733 (1993); Miri, M. F. and Rivier, N., *Europhys. Lett.* **54**, 112 (2001).
[25] Nakagawa, M., Waggoner, R. A. and Fukushima, E., in *An Introduction to the Mechanics of Granular Materials*, Oda, M. and Iwashita, K., eds. (Balkema, Rotterdam, 1999), p. 240.
[26] Niskanen, K. J. and Alava, M. J., *Phys. Rev. Lett.* **73**, 3475 (1994); Provatas, N., Alanissila, T. and Alava, M. J., *Phys. Rev. Lett.* **75**, 3556 (1995); Gates, D. J. and Wescott, M., *J. Stat. Phys.* **94**, 31 (1999); Provatas, N. *et al.*, *Colloids and Surfaces* **A165**, 209 (2000); Alava, M., *Sci. Comp. World* **56**, 30 (2001).
[27] Perkowitz, S., *Universal Foam* (Walker & Co., New York, 2000).
[28] Plateau, J. A. F., *Statique Experimentale et Theorique des Liquides Soumis aux Seules Forces Moleculaire* (Gauthier-Villars, Trubner et cie, Paris, 1873).
[29] Plouraboue, F. and Roux, S., *Physica* **A227**, 173 (1996); Houle, P. A. and Sethna, J. P., *Phys. Rev.* **E54**, 278 (1996); Kramer, E. M. and Lobkovsky, A. E., *Phys. Rev.* **E53**, 1465 (1996).
[30] Rouyer, F. and Menon, N., *Phys. Rev. Lett.* **85**, 3676 (2000).
[31] Samuelson, E. J. *et al.*, *J. Pulp and Pap. Sci.* **27**, 50 (2001); Goel, A. *et al.*, *TAPPI J.* **84**, 72 (2001).
[32] See Cargill, C. S. III, in *Solid State Physics*, Seitz, F. and Turnbull, D., eds. (Academic Press, New York, 1975), Vol. 30, p. 227.
[33] Seidler, G. T. *et al.*, *Phys. Rev.* **E62**, 8175 (2000).
[34] Smith, C. S., in *Metal Interfaces*, Herring, C., ed. (American Society for Metals, Cleveland, OH, 1952), p. 65.
[35] Sugimura, Y. *et al.*, *Acta Met.* **45**, 5245 (1997); Bastawros, A. F., Bart-Smith, H. and Evans, A. G., *J Mech. Phys. Solids* **48**, 301 (2000).
[36] Weaire, D. and Hutzler, S., *The Physics of Foams* (Oxford University Press, Oxford, 2000).
[37] Xu, L., Parker, I. and Osborne, C., *Appita J.* **50**, 325 (1997).
[38] von Neumann, J., in *Metal Interfaces*, Herring, C., ed. (American Society for Metals, Cleveland, OH, 1952), p. 108; Monnereau, C., Pittet, N. and Weaire, D., *Europhys. Lett.* **52**, 361 (2000); Hilgenfeldt, S. *et al.*, *Phys. Rev. Lett.* **86**, 2685 (2001).

CHAPTER 20

Nonlinear Elasticity and Thermodynamics of Granular Materials

HERNÁN A. MAKSE
Levich Institute and Physics Department, City College of New York, New York, New York 10031, USA
makse@mailaps.org

The elastic properties of granular materials can be enormously nonlinear as compared with the properties of non-porous materials. Experiments on isotropic compression of a granular assembly of spheres show that the shear μ, and bulk k, moduli vary with the confining pressure faster than the 1/3 power law predicted by Hertz–Mindlin elasticity theory. Moreover, the ratio between the experimental bulk and shear moduli is found to be constant but with a value larger than the theoretical prediction. Numerical simulations resolve the question as to whether the problem lies with the treatment of the grain-grain contact or with the elastic framework. We find that the problem lies principally with the latter; the *affine* assumption (which underlies the elastic formulation) is found to be valid for k but to breakdown seriously for μ. This explains why the experimental and numerical values of $\mu(p)$ are much smaller than the elastic predictions. In this paper we review recent progress on the understanding of this problem based on microscopic simulations, elasticity theory and more innovative ideas such as jamming, fragility and thermodynamics of granular materials.

Keywords: Granular matter; nonlinear elasticity; thermodynamics; effective temperature.

1. Introduction

The study of elastic properties of granular materials [14, 15] has been treated by many researchers since the pioneering work of Mindlin in the 1950's [6–8, 10, 13, 16, 24, 27, 34, 35]. However, a basic understanding of the physics of nonlinear granular elasticity is currently lacking. Here, nonlinear effects refer to the non-linear dependence on pressure of the linear elastic moduli — which is calculated using small-amplitude strains — and they do not refer to the nonlinear mechanical response to large strain perturbations.

In a typical experiment, a set of glass beads is confined at a hydrostatic pressure p and the compressional sound speed v_p and the shear sound speed v_s are measured as a function of pressure (see for instance Domenico [8], Yin [35] and Refs. 13, 16 and 27). The P-wave and S-wave speeds are related to the elastic constants of the

First published in *Advances in Complex Systems*, Vol. 4, No. 4 (2001), pp. 491–501.

material: $v_p = \sqrt{(k + 4/3\mu)/\rho}$ and $v_s = \sqrt{\mu/\rho}$, where k and μ are the bulk and shear moduli of the system and ρ is the density of the system.

Important insight into this problem comes first from the Hertzian contact theory to model the intergrain forces [17]. In this case, nonlinearity arises from the increase with the external pressure of the contact area between two spherical grains. Other nonlinearities arise due to the fact that the number of contacts per grain increases as the confining pressure increases.

Conventional theories to describe this problem are based on the theory of elasticity of continuum media developed for non-porous materials [20]. Inherent to an elastic formulation is a uniform strain at all scales and that the displacement field of the grains is affine with the macroscopic deformation. A common approximation used in this theory is to consider the disordered medium as an effective medium that exerts a mean-field force on a single representative grain. This approximation is usually referred to as Effective Medium Theory (EMT) [10, 34].

It is found experimentally that the shear and bulk moduli of an assembly of spherical grains vary with the confining pressure p faster than the power law predicted by conventional elasticity theory (EMT) based on Hertz–Mindlin contact mechanics [10, 27, 34] (see Fig. 1). EMT predicts that both moduli vary as $k_e \sim \mu_e \sim p^{1/3}$, and that the ratio k_e/μ_e is a constant independent of pressure, independent of coordination number, and dependent only on Poisson's ratio of the individual grains.[a] The value of k/μ found experimentally, though constant, is larger than the EMT prediction.

The origin of the above discrepancies has not been clear. These discrepancies between theory and experiment could be due to the breakdown of the Hertz–Mindlin force law at each grain contact or they could be associated with the breakdown of the continuum theory of elasticity. De Gennes [7] proposed that a thin shell of oxide layer would give a faster growth with pressure of the elastic moduli of the system, which may explain the behavior for metallic beads. Goddard [13] proposed that sharp angularities of the grains (for instance sand grains) may modify the contact law force between grains, giving rise to a different pressure dependence than that predicted by Hertzian theory which is valid for spherical grains. Other authors [6, 13] have suggested that the increasing number of contacts with pressure may be the reason for the discrepancies in the pressure dependence of the moduli. Jenkins *et al.* [6] measured the elastic moduli using numerical simulations and concluded that EMT does not correctly describe the shear modulus but it describes fairly well the bulk modulus. Other experimental work done by Liu and Nagel [24] and Jia *et al.* [16] concentrated on the role played by force chains in sound propagation. Interesting approaches based on original ideas by Nesterenko for propagating waves in one-dimensional elastic chains has also been applied to wave propagation in granular media [30].

[a] The subscript denotes that these values are calculated considering granular materials as purely elastic solids.

There is a clear need for a microscopic study in order to explore the above discrepancies. In collaboration with D. Johnson and L. Schwartz from Schlumberger and N. Gland from Ecole Normale Supérieure we have performed molecular dynamics (MD) simulations for a system of unconsolidated spherical glass beads. We were able to calculate the elastic properties of a disordered array of Hertz–Mindlin spherical grains [27].

We found agreement between our simulations and existing experimental data, thus confirming the validity of the Hertz–Mindlin contact theory to glass bead aggregates. Our study quantifies two problems with the conventional elastic continuum approach: (1) If the calculated increase of the average coordination number with increasing pressure is taken into account, the modified EMT gives an accurate description of the bulk modulus $k(p)$, but it gives only the correct trend of the pressure dependence of the shear modulus $\mu(p)$. (2) Most importantly, we showed that the affine assumption is approximately valid for the bulk modulus and seriously in error for the shear modulus; this fact is why the EMT prediction of k_e/μ_e differs significantly from the experimental value. We conclude that in order to develop a better understanding of the problem, one must abandon the purely elastic framework and consider granular matter as a full viscoelastic body. Viscoelastic effects can account for the discrepancy in the shear modulus in comparison with the elastic prediction; the corrections increase dramatically in the case of loose materials or for weak intergrain shear forces. The complex relaxation mechanism is related to the structural disorder of the packing structure, such as heterogeneities in the contact force network or "force chains" (stress paths of grains carrying most of the forces in the material [9, 25, 28]), and the nonaffine motion of grains.

Here, we briefly describe our efforts to understand the nonlinear elasticity of granular materials. We will review also recent approaches to this problem based on innovative approaches based on thermodynamics of granular matter, as well as concepts such as jamming and fragility.

2. Theory of Granular Elasticity

In the microscopic model two spherical grains with radii R_1 and R_2 interact via a normal Hertz force $F_n = \frac{2}{3}k_n R^{1/2}\xi^{3/2}$ and a transverse Mindlin force $\Delta F_t = k_t(R\xi)^{1/2}\Delta s$ [17]. Here $R = 2R_1R_2/(R_1+R_2)$, the normal overlap is $\xi = (1/2)[(R_1+R_2) - |\mathbf{x}_1 - \mathbf{x}_2|] > 0$, and $\mathbf{x}_1$, $\mathbf{x}_2$ are the positions of the grain centers. The normal force acts only in compression, $F_n = 0$ when $\xi < 0$. The variable s is defined such that the relative shear displacement between the two grain centers is $2s$. The prefactors $k_n = 4\mu_g/(1-\nu_g)$ and $k_t = 8\mu_g/(2-\nu_g)$ are defined in terms of the shear modulus μ_g and Poisson's ratio ν_g of the material from which the grains are made (typically $\mu_g = 29$ GPa and $\nu_g = 0.2$, for spherical glass beads). Internal dissipation is included in two ways: (1) via a viscous damping term proportional to the relative normal and tangential velocities, and (2) via sliding friction; when F_t exceeds the Coulomb threshold $\mu_f F_n$ the grains slide and $F_t = \mu_f F_n$, where μ_f

is the kinematic friction coefficient between the spheres (typically $\mu_f = 0.3$). We usually assume a distribution of grain radii in which $R_1 = 0.105$ mm for half the grains and $R_2 = 0.075$ mm for the other half.

The basic idea of elastic theories relevant to our study is that the macroscopic work done in deforming the system is set equal to the sum of the work done of each grain-grain contact and that the latter is replaced by a suitable average [10, 34].

In the case of an isotropic deformable solid, the energy of the assembly as a function of the strain ϵ_{ij} is $U = \mu_e(\epsilon_{ij} - \frac{1}{3}\delta_{ij}\epsilon_{ll})^2 + k_e\epsilon_{ij}^2$ [20], and the work done by the imposed strain is calculated in terms of the intergrain forces and displacements. There are two assumptions inherent to the elastic EMT: (1) All the spheres are statistically the same, and it is assumed that there is an isotropic distribution of contacts around a given sphere. (2) An affine approximation is used, i.e. the spheres at position X_j are moved a distance δu_i in a time interval δt according to the macroscopic strain rate $\dot{\epsilon}_{ij}$ by

$$\delta u_i = \dot{\epsilon}_{ij} X_j \delta t \,. \tag{2.1}$$

The grains are always at equilibrium due to the assumption of isotropic distribution of contacts and further relaxation is not required. This sort of mean field theory is analogous to a simple average of the non-linear spring constants.

The EMT predictions for the bulk and shear modulus are

$$k_e(p) = \frac{k_n}{12\pi}(\phi Z)^{2/3}\left(\frac{6\pi p}{k_n}\right)^{1/3}, \tag{2.2}$$

$$\mu_e(p) = \frac{k_n + \frac{3}{2}k_t}{20\pi}(\phi Z)^{2/3}\left(\frac{6\pi p}{k_n}\right)^{1/3}, \tag{2.3}$$

where ϕ is the volume fraction and Z the average coordination number of the spheres.

As discussed above, the $p^{1/3}$ dependence of the moduli is not corroborated by experiments. Another way of seeing the breakdown of the elastic theory is to focus on the ratio k/μ. According to Eqs. (2.2) and (2.3), $k_e/\mu_e = 5/3(2-\nu_g)/(5-4\nu_g)$, independent of pressure, a value which depends only on Poisson's ratio of the bead material. The experiments give $k/\mu \approx 1.1$–1.3. Our simulations give $k/\mu \approx 1.05 \pm 0.1$ [27] in good agreement with experiments. EMT predicts $k_e/\mu_e = 0.71$, if we take $\nu_g = 0.2$ for Poisson's ratio of glass. (The EMT prediction is quite insensitive to variations of ν_g; $k_e/\mu_e = 0.71 \pm 0.04$ for $\nu_g = 0.2 \pm 0.1$.) Conversely, a value $\nu_g \simeq 1.2$ would be needed in order to fit the experimental k/μ, clearly violating the upper thermodynamic limit on $\nu_g \leq 1/2$ [20].

3. Numerical Results

We perform MD simulations of systems of the order of 10,000 particles [27]. A model used in current granular simulations is the Discrete Element Method developed by Cundall and Strack [5, 32] for a system of soft spheres [17]. Our calculations begin with a numerical protocol designed to mimic the experimental procedure used to prepare dense packed granular materials. Generating a mechanically stable packing is not a trivial task. The simulations begin with a gas of spherical particles located at random positions in a periodically repeated cubic cell of side L. At the outset, a series of strain-controlled isotropic compressions and expansions are applied until a volume fraction slightly below the critical density is reached. The system is then compressed and extended slowly until a specified value of the pressure and volume fraction — above the random close packing (RCP) fraction — is achieved at static equilibrium.

Consider now, the calculation of the elastic moduli of the system as function of pressure. Beginning with the equilibrium state describe above, we apply a small perturbation to the system and measure the resulting response. The shear modulus is calculated in two ways, from a pure shear test, $\mu = (1/2)\Delta\sigma_{12}/\Delta\epsilon_{12}$, and also from a biaxial test, $\mu = (\Delta\sigma_{22} - \Delta\sigma_{11})/2(\Delta\epsilon_{22} - \Delta\epsilon_{11})$. The bulk modulus is obtained from a uniaxial compression test, $k + 4/3\mu = \Delta\sigma_{11}/\Delta\epsilon_{11}$. Here, the stress σ_{ij} is determined from the measured forces on the grains, and the strain ϵ_{ij} is determined from the imposed dimensions of the unit cell.

In Fig. 1 our calculated values of the elastic moduli as a function of pressure are compared with EMT and with experimental data. From Fig. 1 we see that our experimental and numerical results are in reasonably good agreement. Also shown are measured data from Domenico [8] and Yin [35]. Clearly, the experimental data are somewhat scattered. This scatter reflects the difficulty of the measurements, especially at the lowest pressures where there is significant signal loss. Nevertheless, our calculated results pass through the collection of available data. Also shown in Fig. 1 are the EMT predictions (2.2) and (2.3). The EMT curves are obtained

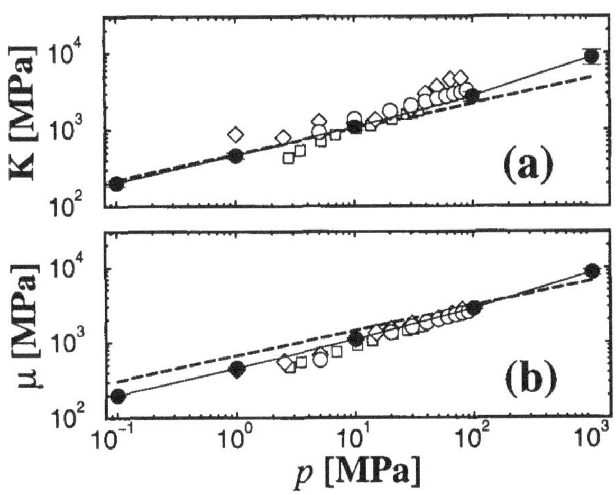

Fig. 1. Pressure dependence of the elastic moduli from MD (•), experiments [Domenico (square) [8], Yin (diamond) [35], and our experiments (circle)], and theory (dashed line): (a) bulk modulus, and (b) shear modulus.

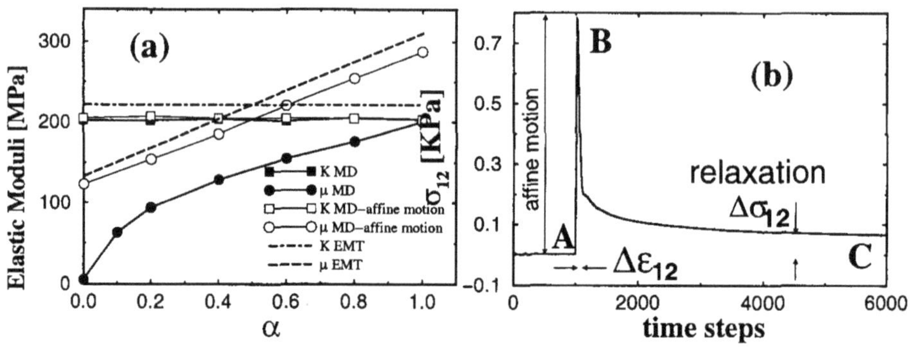

Fig. 2. (a) $k(\alpha)$ and $\mu(\alpha)$ versus α for a fixed $p = 100$ KPa. (b) Relaxation of the shear stress (B $\to$ C) after an affine motion (A $\to$ B) in the calculation of the shear modulus.

using the same parameters as in the simulations; we also set $Z = 6$ and $\phi = 0.64$, independent of pressure. At low pressures we see that k is well described by EMT. At larger pressures, however, the experimental and numerical values of k grow faster than the $p^{1/3}$ law predicted by EMT. The situation with the shear modulus is even less satisfactory. EMT overestimates $\mu(p)$ at low pressures but, again, underestimates the increase in $\mu(p)$ with pressure.

To understand why μ is over-predicted by EMT we must examine the role of transverse forces and rotations in the relaxation of the grains. (These effects do not play any role in the calculation of the bulk modulus.) Suppose we redefine the transverse force by introducing a multiplicative coefficient α, viz: $\Delta f_t = \alpha k_t (R\xi)^{1/2} \Delta s$; with $\alpha = 1$ we recover our previous results. To quantify the role of the transverse force on the elastic moduli, we calculate $k(\alpha)$ and $\mu(\alpha)$ at a given pressure $p = 100$ KPa (Fig. 2(a)). This pressure is low enough that the changing number of contacts is not an issue. Surprisingly, the shear modulus becomes negligibly small as $\alpha \to 0$. As expected, k is independent of the strength of the transverse force. To compare with the theory we also plot the prediction of the EMT, Eq. (2.3) in which k_t is rescaled by αk_t. We see that the EMT fails to take into account the vanishing of the shear modulus as $\alpha \to 0$. However, it accurately predicts the value of the bulk modulus, which is independent of α.

Figure 2(b) shows the typical viscoelastic response of the granular medium (this behavior seems not to be related to aging as we obtain similar results after repeated application of the infinitesimal perturbation). In the MD calculation of the shear modulus an affine perturbation is first applied to the system. The shear stress increases (from A to B in Fig. 2(b)) and the grains are far from equilibrium since the system is disordered. The grains then relax towards equilibrium (from B to C), and we measure the resulting change in stress from which the modulus is calculated. To better understand the approximations involved in the EMT, suppose we repeat the MD calculations taking into account only the affine motion of the grains and ignoring the subsequent relaxation. The resulting values of the moduli are plotted in Fig. 2(a) as open symbols and we see that the moduli calculated this way are very close to the EMT predictions. Thus, the difference between the MD and EMT results for the shear modulus lies in the nonaffine relaxation of the grains; this difference being largest when there is no transverse force. By contrast, grain relaxation after an applied *isotropic* affine perturbation is not particularly significant and the EMT predictions for the bulk modulus are quite accurate.

The surprisingly small values of μ found as $\alpha \to 0$ can be understood as a melting of the system that occurs when the system is close to the RCP fraction. This fluid like behavior (when $k_t = 0$) is closely related to the melting transition seen in compressed emulsions [19] and foams [11]. At the RCP fraction the system behaves like a fluid with no resistance to shear, and might indicate that the system is in a fragile state.

Our results can be analyzed in terms of jamming [22, 31] theory. It has been proposed that jamming — and in some particular cases fragility [3] — are universal features of many complex systems with a common slow relaxation dynamics such as glasses, foams and granular matter [22, 31]. We notice that granular systems can jam at RCP [2] but also at other states of lower density known as random loose packing (RLP) states [33]. It is near the jamming transition that we expect to find the most interesting viscoelastic properties.

In the limit of weak compression, the stress vanishes continuously as $\sigma \sim (\phi - \phi_c)^\beta$, where ϕ_c is the critical density of the jamming transition [23, 28]. Jamming can occur at RLP or RCP according to the existence or not of transverse forces between the grains, respectively. At the jamming transition, the coordination number approaches a minimal value Z_c (=4 for friction and 6 for frictionless grains) also as a power law. Our result $Z_c = 6$ agrees with experimental analysis of Bernal packings for close contacts between spheres fixed by means of wax, and our own analysis of Finney packings [2] using the actual sphere center coordinates of 7934 steel balls. However, no similar experimental study exists for RLP which could be able to confirm $Z_c = 4$. A critical slowing down — the time to equilibrate the system increases near ϕ_c — and the emergence of shear rigidity is also found at criticality.

Concomitant with the approach to the jamming transition we find a change in the probability distribution and in the degree of spatial correlation of the interparticle forces; localized force chains with an exponential probability distribution — as observed in recent studies [25] — are found at low confining stress giving way to a considerably less localized and homogeneous arrangement of the forces with a Gaussian probability distribution at higher stress away from the critical density [28].

4. Thermodynamics Formulation of Granular Materials

In equilibrium statistical mechanics of fluids microscopic information on the fluctuations of the particle motion equilibrated at temperature T is related to macroscopic dissipative variables, such as the viscosity, via the fluctuation-dissipation theorem

(FDT), that in one of its forms — the Stokes–Einstein relation — relates the random diffusion of thermally fluctuating positions of particles, and the response function to an external force [21]. Even though the diffusivity of the tracer D and mobility χ are strongly dependent on size and shape of the tracer, they turn out to be always related by the Einstein relation $D/\chi = T$, where T is the temperature of the liquid.

One may wonder if this approach could be applied to an out-of-equilibrium system such as granular matter. If so, we could have a theoretical framework for understanding macroscopic quantities such as complex shear modulus from detailed microscopic variables.

At first sight, investigations of a thermodynamics relation such as FDT for granular materials of size larger than 1 μm might seem to be condemned to fail; granular materials are athermal systems characterized by the fact that energy is supplied by external driving (via shear or tapping) and dissipated by inelastic interactions and slippage between the grains — a very different "heat exchange mechanism" from that of a thermal bath in a molecular or colloidal system. However, recent theoretical and numerical work has yielded mounting evidence of an underlying thermodynamics valid for dense granular media. This train of ideas started more than a decade ago by S. Edwards and collaborators [12]. Edwards proposed a thermodynamics and statistical ensemble valid for dense slowly moving granular materials and glasses, and through it, thermodynamic notions of entropy and temperature.

The thermodynamic construction of Edwards is based on two postulates [12]: (1) While in the Gibbs construction for equilibrium statistical mechanics, one assumes that the physical quantities are obtained as average over all possible configurations at a given energy, Edwards' ensemble consists of only the jammed configurations (static) at the appropriate energy and volume. (2) The strong 'ergodic' hypothesis is that all jammed configurations of given volume and energy can be taken to have equal statistical weights. This formulation leads to an entropy $S_{\rm Edw}(E, V)$ as the logarithm of the number of jammed configurations, and the corresponding temperature

$$T_{\rm Edw}^{-1} = \frac{\partial S_{\rm Edw}}{\partial E}, \tag{4.1}$$

and compactivity $X_{\rm Edw}^{-1} = \partial S_{\rm Edw}/\partial V$.

Recent theoretical support to this formulation came from glass theory [1, 18]; theoretical and numerical work on schematic models has yielded evidence supporting this thermodynamic picture — and this has spurred a renewed interest in Edwards' thermodynamics. Indeed, recent analytic schemes for out-of-equilibrium glassy dynamics have suggested the existence of an FDT thermodynamical temperature (unrelated to the room temperature) governing the slow components of fluctuations and responses of all observables in the glassy state [4, 18]. Furthermore, this fluctuation-dissipation temperature has been shown to be directly related to Edwards' ideas for a statistical ensemble for granular matter [12] — within schematic models [1] and in MD models of granular matter [26].

We have recently performed in collaboration with J. Kurchan, with a realistic numerical model, a diffusion-mobility numerical experiment [26]. Our results support the existence of a notion of temperature for long-scale displacements (structural rearrangements) independently of the viscous friction dissipation between grains or even the presence of sliding friction (Coulomb's law).

We performed MD simulations for a binary system of large and small spheres in a periodic 3D cell. We applied a gentle simple shear flow at constant volume and measured the spontaneous fluctuations $\langle \Delta x^2 \rangle$ and force-induced displacements $\langle \Delta x \rangle / f$, where f is an small external force, for two types of tracers with different sizes. Our data was consistent with an extended Einstein relation:

$$\langle \Delta x^2 \rangle = 2T_{\rm FDT} \frac{\langle \Delta x \rangle}{f} , \tag{4.2}$$

valid for both particles *with the same* $T_{\rm FDT}$ for widely separated time scales. This suggested that $T_{\rm FDT}$ can be considered to be the temperature of the slow modes.

We then performed an explicit computation to show that the temperature arising from Edwards' thermodynamic description and the dynamic one measured from Einstein's relation coincide (for details see Ref. 26). This last step cannot be performed in the laboratory, so the numerical simulation provides the missing link between thermodynamic ideas and diffusion-mobility checks. In order to calculate $T_{\rm Edw}$ and compare with $T_{\rm FDT}$, we counted the number of jammed configurations at a given energy and volume. To do this in practice, we extended the 'auxiliary model' method of Ref. 1 to the case of deformable grains. This work strongly supports the thermodynamic picture. A hypothesis currently being tested is whether the temperature arising from Edwards' formalism may provide the link between the mechanical properties and the microstructural relaxation properties of granular materials. This approach is motivated by studies of microrheology in complex fluids. In 1995 Mason and Weitz [29] developed a technique, called microrheology, using a frequency-dependent Stokes–Einstein relation to extract the complex shear modulus from the equilibrium thermal fluctuations of a tracer moving in the viscoelastic material. In the case of athermal, slowly-driven granular materials, the thermodynamic Edwards' temperature may provides the link between the microscopic quantities and the shear modulus.

5. Summary

Our MD simulations are in good agreement with the available experimental data on the pressure dependence of the elastic moduli of granular packings. They also serve to clarify the deficiencies of EMT. Grain relaxation after an infinitesimal affine strain transformation is an essential component of the shear (but not the bulk) modulus. This viscoelastic behavior is not taken into account in the EMT. In the limit $\alpha \to 0$ a packing of nearly rigid particles responds to an external isotropic load with an elastic deformation and a finite k. By contrast, such a system cannot support a shear load ($\mu \to 0$) without severe particle rearrangements. This may indicate a "fragile" state of the system [3] where inter-particle forces are organized along "force chains" (stress paths carrying most of the forces in the system) oriented along the principal stress axes. Such fragile networks support, elastically, only perturbations compatible with the structure of force chains and deform plastically otherwise. Clearly, there is a need for an improved theory; recent work on thermodynamics and jamming of granular materials may provide an alternative formulation and allow one to describe properly the response of granular materials to perturbations.

References

[1] Barrat, A., Kurchan, J., Loreto, V. and Sellitto, M., *Phys. Rev. Lett.* **85**, 5034 (2000).
[2] Bernal, J. D. and Mason, J., *Nature* **188**, 910 (1960).
[3] Cates, M. E., Wittmer, P., Bouchaud, J.-P. and Claudin, P., *Phys. Rev. Lett.* **81**, 1841 (1998).
[4] Cugliandolo, L. F., Kurchan, J. and Peliti, L., *Phys. Rev.* **E55**, 3898 (1997).
[5] Cundall, P. A. and Strack, O. D. L., *Géotechnique* **29**, 47 (1979).
[6] Cundall, P. A., Jenkins, J. T. and Ishibashi, I., in *Powder & Grains 89*, Biarez and Gourvès, eds. (Balkema, Rotterdam, 1989).
[7] de Gennes, P.-G., *Europhys. Lett.* **35**, 145 (1996).
[8] Domenico, S. N., *Geophysics* **42**, 1339 (1977).
[9] Drescher, A. and De Josselin De Jong, V. F., *J. Mech. Phys. Solids* **20**, 337 (1972).
[10] Duffy, J. and Mindlin, R. D., *J. Appl. Mech. (ASME)* **24**, 585 (1957).
[11] Durian, D. J., *Phys. Rev. Lett.* **75**, 4780 (1995).
[12] Edwards, S. F., in *Granular Matter: An Interdisciplinary Approach*, Mehta, A., ed. (Springer-Verlag, New York, 1994), p. 121.
[13] Goddard, J. D., *Proc. R. Soc. London* **A430**, 105 (1990).
[14] Guyer, R. A. and Johnson, P. A., *Phys. Today* **52**, 30 (1999).
[15] Herrmann, H. J., Hovi, J. P. and Luding, S., *Physics of Dry Granular Matter* (Kluwer, Dordrecht, 1998).
[16] Jia, X., Caroli, C. and Velicky, B., *Phys. Rev. Lett.* **82**, 1863 (1999).
[17] Johnson, K. L., *Contact Mechanics* (Cambridge University Press, Cambridge, 1985).
[18] Kurchan, J., *J. Phys.: Cond. Matter* **29**, 6611 (2000).
[19] Lacasse, M.-D., Grest, G. S., Levine, D., Mason, T. G. and Weitz, D. A., *Phys. Rev. Lett.* **76**, 3448 (1996).
[20] Landau, L. D. and Lifshitz, E. M., *Theory of Elasticity* (Pergamon, New York, 1970).
[21] Landau, L. D., Lifshitz, E. M. and Pitaevskii, L. P., *Statistical Physics* (Pergamon Press, New York, 1980).
[22] Liu, A. J. and Nagel, S. R., *Nature* **396**, 21 (1998).
[23] Liu, A. J. and O'Hern, C. S., private communication.
[24] Liu, C. H. and Nagel, S. R., *Phys. Rev. Lett.* **68**, 2301 (1992).
[25] Liu, C.-H. *et al.*, *Science* **269**, 513 (1995).
[26] Makse, H. A. and Kurchan, J., cond-mat/0107163, to appear in *Nature*.
[27] Makse, H. A., Gland, N., Johnson, D. L. and Schwartz, L. M., *Phys. Rev. Lett.* **83**, 5070 (1999).
[28] Makse, H. A., Johnson, D. L. and Schwartz, L. M., *Phys. Rev. Lett.* **84**, 4160 (2000).
[29] Mason, T. G. and Weitz, D., *Phys. Rev. Lett.* **74**, 1250 (1995).

[30] Nesterenko, V. F., *J. Phys. IV* **4**, 729 (1994); Hascoët, E., Herrmann, H. and Loreto, V., *Phys. Rev.* **E59**, 3202 (1999).
[31] O'Hern, C. S., Langer, S. A., Liu, A. J. and Nagel, S. R., *Phys Rev. Lett.* **86**, 111 (2001).
[32] Schäfer, J., Dippel, S. and Wolf, D. E., *J. Phys. I (France)* **6**, 5 (1996).
[33] Scott, G. D., *Nature* **188**, 908 (1960).
[34] Walton, K., *J. Mech. Phys. Solids* **35**, 213 (1987).
[35] Yin, H., PhD thesis, Stanford University (1993).

CHAPTER 21

Granular Flow Studies by NMR: A Chronology

E. FUKUSHIMA
New Mexico Resonance, 2301 Yale SE Suite C1,
Albuquerque, New Mexico 87106, USA
eiichi@nmr.org

Granular flow is all around us. Examples include hourglasses, avalanches, flows of grains and stress distributions in silos and sandpiles, mixing of granular matter such as pharmaceutical components, tumbler drying, segregation of heterogeneous components such as in dry food products, etc. Indeed, the majority of industries have to deal with granular flow problems and the monetary loss due to these problems is unimaginable.

A major barrier to studying the internal structure and movement of granular matter is that its opacity prevents visual observation or the use of other optical methods. Ultrasound methods are also not suited to observing granular matter because of the innumerable boundaries that reflect sound, preventing it from penetrating very far into the matter. X-ray and other radiation-beam methods can be used to obtain structures of granular matter, but they are not well suited to studies of higher order parameters such as velocity and diffusion. However, if the imaging speed becomes sufficiently high, it should be possible, in principle, to track individual (or assemblies of) particles. One promising approach uses X-rays at the Advanced Photon Source (APS), Argonne National Laboratory [10].

In this review, we outline the subject of using nuclear magnetic resonance (NMR) to study granular flows and refer the reader to the articles and reviews that are scattered in many diverse journals. A recent review was a Materials Research Society Symposium presentation in San Francisco, April 24-27, 2000 [1]. It contained a chronological bibliography, through 1999, which is extended in this review.

NMR is an excellent tool for measuring many experimental parameters in granular flow. This is due, in part, to its ability to non-invasively obtain data from deep inside the granular assembly. However, an equally important advantage of NMR is its ability to measure a multiplicity of spatially resolved parameters such as concentration, velocity and diffusion. These measurements can lead to parameters such as shear and granular temperature. (Granular temperature of a localized assembly of particles represents the kinetic energy of random motion of the assembly, a

First published in *Advances in Complex Systems*, Vol. 4, No. 4 (2001), pp. 503–507.

parameter that is difficult to measure but important to theories that are based on kinetic theory.)

NMR has been used to study the velocity profile of certain granular flows [7]. It has also tracked the evolution of axial and radial segregation [3–5]. Finally, recent experiments have shown the feasibility of using NMR to measure spatially resolved granular temperature [2]. Such measurements would shed light on the physics of granular flow by relating the dynamic data to variables such as the sample geometry, rotation speed, and the shape of the flowing layer.

Even though these parameters can now be measured, no systematic measurements have been made. The reason for the delay between the development of the methodology and its systematic application is due, solely, to logistics. For example, New Mexico Resonance, which pioneered many of these developments, is first and foremost a NMR lab rather than a granular flow lab. Therefore, it needs collaborators that are willing to devote time and manpower to making the needed measurements. This is a situation similar to that for the X-ray microimaging apparatus at APS, mentioned above [10].

A downside to using NMR for studies of granular flows and structures is that it is difficult to get useful NMR signals from solid particles. Therefore, common granular particles such as sand and coal are not directly accessible for studies by NMR. Thus, every granular flow study by NMR, so far, has used grains containing liquid phase matter that can easily be imaged by NMR. Examples include natural particles containing oils such as mustard and sesame seeds [6, 7], pharmaceutical pills containing liquids [2, 11], and other particles containing imbibed liquids [8, 9].

In summary, NMR is a versatile tool that can spatially resolve many useful parameters of granular flow for certain kinds of grains. Thus, its growing application will assist in understanding the fascinating and useful phenomenon of granular flow.

Acknowledgment

I acknowledge the support of the Engineering Science Program, Basic Energy Sciences, US Department of Energy, via grant DE-FG03-98ER14912. I would also like to thank Braunen Smith for helping in the preparation of this article.

References

[1] Altobelli, S. A., Caprihan, A., Fukushima, E. and Seymour, J. D., Nuclear magnetic resonance studies of granular flows — current status, in *The Granular State — Materials Research Soc. Symposium Proceedings*, Vol. 627, Sen, S. and Hunt, M. L., eds. (Materials Research Society, Warrendale, Pennsylvania, 2001), p. BB2.1.1.

[2] Caprihan, A. and Seymour, J. D., Correlation time and diffusion coefficient imaging: Application to a granular flow system, *J. Magn. Reson.* **144**, 96 (2000).

[3] Hill, K. M., Caprihan, A. and Kakalios, J., Bulk segregation in rotated granular materials measured by magnetic resonance imaging, *Phys. Rev. Lett.* **78**, 50 (1997).

[4] Hill, K. M., Kakalios, J., Yamane, K., Tsuji, Y. and Caprihan, A., Dynamic angle of repose as a function of mixture concentration: Results from MRI experiments and

DEM simulations, in *Powders and Grains 97 — Proceedings of the Third International Conference on Powders and Grains*, Behringer, R. P. and Jenkins, J. T., eds. (A. A. Balkema, Rotterdam, 1997), p. 483.
[5] Hill, K. M., Caprihan, A. and Kakalios, J., Axial segregation of granular media rotated in a drum mixer: Pattern evolution, *Phys. Rev.* **E56**, 4386 (1997).
[6] Mueth, D. M., Debregeas, G. F., Karczmar, G. S., Eng, P. J., Nagel, S. R. and Jaeger, H. M., Signatures of granular microstructure in dense shear flows, *Nature* **406**, 385 (2000).
[7] Nakagawa, M., Altobelli, S. A., Caprihan, A., Fukushima, E. and Jeong, E.-K., Non-invasive measurements of granular flows by magnetic resonance imaging, *Experiments in Fluids* **16**, 54 (1993).
[8] Porion, P., Sommier, N. and Evesque, P., Mixing and segregation of grains studied by NMR imaging investigation — application to a turbulence mixer: The turbula blender, in *IUTAM Symposium on Segregation in Granular Flows*, Rosato, A. D. and Blackmore, D. L., eds. (Kluwer Academic Publishers, Dordrecht, 2000), p. 141.
[9] Porion, P., Sommier, N. and Evesque, P., Dynamics of mixing and segregation processes of grains in 3D blender by NMR imaging investigation, *Europhys. Lett.* **50**, 319 (2000).
[10] Seidler, G. T., Proceedings of this conference, Trieste, 2001.
[11] Seymour, J. D., Caprihan, A., Altobelli, S. A. and Fukushima, E., Pulsed gradient spin echo nuclear magnetic resonance imaging of diffusion in granular flow, *Phys. Rev. Lett.* **84**, 266 (2000).

Chronology: Granular Flow NMR Publications, 1993 to 2001

- Nakagawa, M., Altobelli, S. A., Caprihan, A., Fukushima, E. and Jeong, E.-K., Non-invasive measurements of granular flows by magnetic resonance imaging, *Experiments in Fluids* **16**, 54 (1993).
- Nakagawa, M., Axial segregation of granular flows in a horizontal rotating cylinder, *Chem. Engng. Sci.* **49**, 2540 (1994).
- Ehrichs, E. E., Jaeger, H. M., Karczmar, G. S., Knight, J. B., Kuperman, V. Yu. and Nagel, S. R., Granular convection observed by magnetic resonance imaging, *Science* **267**, 1632 (1995).
- Kuperman, V. Yu., Ehrichs, E. E., Jaeger, H. M. and Karczmar, G. S., A new technique for differentiating between diffusion and flow in granular media using magnetic resonance imaging, *Rev. Sci. Instrum.* **66**, 4350 (1995).
- Knight, J. B., Ehrichs, E. E., Kuperman, V. Yu., Flint, J. K., Jaeger, H. M. and Nagel, S. R., Experimental study of granular convection, *Phys. Rev.* **E54**, 5726 (1996).
- Kuperman, V. Yu., Nuclear magnetic resonance measurements of diffusion in granular media, *Phys. Rev. Lett.* **77**, 1178 (1996).
- Metcalf, G. and Shattuck, M., Pattern formation during mixing and segregation of flowing granular materials, *Physica* **A233**, 709 (1996).
- Hill, K. M., Caprihan, A. and Kakalios, J., Bulk segregation in rotated granular materials measured by magnetic resonance imaging, *Phys. Rev. Lett.* **78**, 50 (1997).

- Knight, J. B., External boundaries and internal shear bands in granular convection, *Phys. Rev.* **E55**, 6016 (1997).
- Kuperman, V. Yu., NMR measurements demonstrate increased intensity of collisions near walls in a vibrating granular material, in *Powders and Grains 97 — Proceedings of the Third International Conference on Powders and Grains*, Behringer, R. P. and Jenkins, J. T., eds. (A. A. Balkema, Rotterdam, 1997), p. 393.
- Nakagawa, M., Altobelli, S. A., Caprihan, A. and Fukushima, E., NMR measurement and approximate derivation of the velocity depth-profile of granular flow in a rotating, partially filled, horizontal cylinder, in *Powders and Grains 97 — Proceedings of the Third International Conference on Powders and Grains*, Behringer, R. P. and Jenkins, J. T., eds. (A. A. Balkema, Rotterdam, 1997), p. 447.
- Cheng, H. A., Altobelli, S. A., Caprihan, A. and Fukushima, E., NMR and mechanical measurements of the collisional dissipation of granular flow in a rotating, partially filled, horizontal cylinder, in *Powders and Grains 97 — Proceedings of the Third International Conference on Powders and Grains*, Behringer, R. P. and Jenkins, J. T., eds. (A. A. Balkema, Rotterdam, 1997), p. 463.
- Hill, K. M., Kakalios, J., Yamane, K., Tsuji, Y. and Caprihan, A., Dynamic angle of repose as a function of mixture concentration: Results from MRI experiments and DEM simulations, in *Powders and Grains 97 — Proceedings of the Third International Conference on Powders and Grains*, Behringer, R. P. and Jenkins, J. T., eds. (A. A. Balkema, Rotterdam, 1997), p. 483.
- Caprihan, A., Fukushima, E., Rosato, A. D. and Kos, M., Magnetic resonance imaging of vibrating granular beds by spatial imaging, *Rev. Sci. Instrum.* **68**, 4217 (1997).
- Nakagawa, M., Altobelli, S. A., Caprihan, A. and Fukushima, E., An MRI study: Axial migration of radially segregated core of granular mixture in a horizontal rotating cylinder, *Chemical Engineering Science* **52**, 4423 (1997).
- Hill, K. M., Caprihan, A. and Kakalios, J., Axial segregation of granular media rotated in a drum mixer: Pattern evolution, *Phys. Rev.* **E56**, 4386 (1997).
- Yamane, K., Nakagawa, M., Altobelli, S. A., Tanaka, T. and Tsuji, Y., Steady particulate flows in a horizontal rotating cylinder, *Phys. Fluids* **10**, 1419 (1998).
- Nakagawa, M., Moss, J. L. and Altobelli, S. A., Segregation of granular particles in a nearly packed rotating cylinder: A new insight for axial segregation, in *Physics of Dry Granular Media*, Herrmann, H. J., Hovi, H. J. and Luding, S., eds. (Kluwer Academic, Dordrecht, 1998), p. 703.
- Dury, C., Ristow, G. H., Moss, J. L. and Nakagawa, M., Boundary effects on the angle of repose in rotating cylinders, *Phys. Rev.* **E57**, 4491 (1998).
- Caprihan, A., Seymour, J. D., Altobelli, S. A. and Fukushima, E., Velocity fluctuation measurements of granular media in a rotating cylinder by nuclear magnetic resonance imaging, in *IUTAM Symposium on Segregation in Granular

Flows, Rosato, A. D. and Blackmore, D. L., eds. (Kluwer Academic Publishers, Dordrecht, 2000), p. 153.

- Porion, P., Sommier, N. and Evesque, P., Mixing and segregation of grains studied by NMR imaging investigation — application to a turbulence mixer: The turbula blender, in *IUTAM Symposium on Segregation in Granular Flows*, Rosato, A. D. and Blackmore, D. L., eds. (Kluwer Academic Publishers, Dordrecht, 2000), p. 141.
- Mueth, D. M., Debregeas, G. F., Karczmar, G. S., Eng, P. J., Nagel, S. R. and Jaeger, H. M., Signatures of granular microstructure in dense shear flows, *Nature* **406**, 385 (2000).
- Seymour, J. D., Caprihan, A., Altobelli, S. A. and Fukushima, E., Pulsed gradient spin echo nuclear magnetic resonance imaging of diffusion in granular flow, *Phys. Rev. Lett.* **84**, 266 (2000).
- Caprihan, A. and Seymour, J. D., Correlation time and diffusion coefficient imaging: Application to a granular flow system, *J. Magn. Reson.* **144**, 96 (2000).
- Porion, P., Sommier, N. and Evesque, P., Dynamics of mixing and segregation processes of grains in 3D blender by NMR imaging investigation, *Europhys. Lett.* **50**, 319 (2000).
- Altobelli, S. A., Caprihan, A., Fukushima, E. and Seymour, J. D., Nuclear magnetic resonance studies of granular flows — current status, in *The Granular State — Materials Research Soc. Symposium Proceedings*, Vol. 627, Sen, S. and Hunt, M. L., eds. (Materials Research Society, Warrendale, Pennsylvania, 2001), p. BB2.1.1.
- Mueth, D. M., Measurements of particle dynamics in slow, dense granular couette flow, cond-mat/0103557 preprint, May 2001.
- Yang, X., Huan, C., Candela, D., Mair, R. W. and Walsworth, R. L., A dense, vibrated granular fluid in gravity: NMR measurements of grain motion in three dimensions, cond-mat/0108256 preprint, August 2001.

Reviews Mentioning Granular Flow NMR

- Jaeger, H. M., Nagel, S. R. and Behringer, R. P., Granular solids, liquids, and gases, *Rev. Mod. Phys.* **68**, 1259 (1996).
- Jaeger, H. M., Nagel, S. R. and Behringer, R. P., The physics of granular materials, *Phys. Today* **4**, 32 (1996).
- Fukushima, E., Nuclear magnetic resonance as a tool to study flow, *Annu. Rev. Fluid Mech.* **31**, 95 (1999).
- Shinbrot, T. and Muzzio, F. J., Nonequilibrium patterns in granular mixing and segregation, *Phys. Today* **3**, 25 (2000).

CHAPTER 22

The Four Avalanche Fronts: A Test Case for Granular Surface Flow Modeling

STÉPHANE DOUADY, BRUNO ANDREOTTI, PIERRE CLADÉ and ADRIAN DAERR

Laboratoire de Physique Statistique de l'E.N.S.,
24 rue Lhomond, F-75005 Paris, France

Granular surface flows have still to be fully modelled. Here, we present the four types of front that can be observed in avalanches. These strongly inhomogeneous and unsteady flows are very sensitive test cases for the different types of model. We show that, at least qualitatively for the moment, the model we propose, based on the analysis of the motion of a single grain and layers of grains, can reproduce the different characteristics of these four fronts.

Keywords: Avalanche fronts; St Venant equations; granular surface flow; velocity profile; sub-critical bifurcation; inelasticity; non-local shocks; trapping.

1. Introduction

With its ubiquity and practical importance, it is surprising that granular surface flows have not yet been described as clearly and completely as has been done long ago for liquid flows. One reason may be that the physical mechanisms at work in the flowing granular layer are different enough from the liquid case, and yet hardly described. To investigate these mechanisms a first step is to look at the motion of a single grain (Sec. 2). This already provides many clues of the essential physical mechanisms, namely inelastic collisions and trapping, leading to a strong sub-critical bifurcation between motion and rest [22].

A second step can then be to look at layers of grains flowing one above the other (Sec. 3). In this case we have the same effects as for a single grain (inelastic collisions and trapping), but we also find an essential effect: the non-locality of the collisions. This non-locality of the shocks explains the otherwise surprising linear velocity profile, and its combination with the trapping gives the depth selection of the surface flow [2].

From these observations, and other experimental ones, we propose a model [12]. In Sec. 4 we compare it with other propositions and possibilities. Then, we present the different types of fronts that can be observed in avalanches, and show how their

First published in *Advances in Complex Systems*, Vol. 4, No. 4 (2001), pp. 509–522.

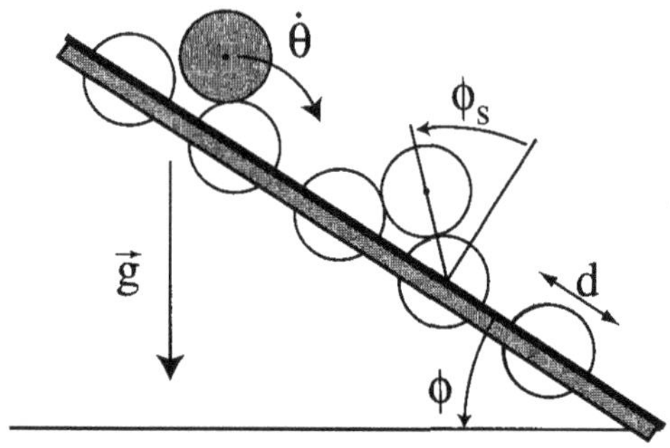

Fig. 1. Schematic set-up of the single bead experiment with notations.

different characteristics point to particular aspects that have to be present in a model trying to describe them (Sec. 5). Finally, we conclude with future work.

2. The Motion of a Single Grain

A bead (in practice, a cylinder) placed on a row of identical beads presents several states (Fig. 1). The first is static equilibrium, when the bead is trapped in the hole in-between the beads underneath. If the row of beads is tilted by an angle ϕ, this bead remains static up to the angle at which the trap disappears. The bead will then spontaneously start to roll down. This defines the starting angle ϕ_s (usually referred as the static angle).

Another possible state is to have the bead rolling down indefinitely. Experimentally, it quickly reaches a constant mean velocity [14, 24]. If the angle of the row is decreased, this periodic motion abruptly becomes impossible below a second lower angle, the stopping angle ϕ_d (usually referred as the dynamical angle).

Analysis of the motion [22] shows that the bead rolls on those underneath, without sliding or jumping (for not too large an angle). During rolling it gains kinetic energy $\dot{\theta}^2_{\mathrm{down}} = \dot{\theta}^2_{\mathrm{up}} + 2\alpha^2 \frac{g}{d} \sin\phi_s \sin\phi$, where α is related to the ratio of potential energy transferred into translation kinetic energy and the angular momentum. Then, it collides with the next bead. Even though the materials used are not completely inelastic (for instance, steel), we do not observe any rebound, but rather a perfectly inelastic normal collision. There remains some tangential velocity, and also part of the angular momentum is transformed into tangential kinetic energy. The bead thus starts rolling on the next bead, with a new angular velocity given by $\dot{\theta}_{\mathrm{after}} = \rho\dot{\theta}_{\mathrm{before}}$ where ρ (of the order of 1/2) is the coefficient describing the loss of kinetic energy in the shock.[a]

[a]This perfectly inelastic normal shock can be interpreted as the signature of an inelastic collapse; it makes a small rebound, but then an infinite number of small rebounds in a small finite time. It would be interesting to look precisely at the real shocks as carefully done by Zippelius *et al.* (this volume).

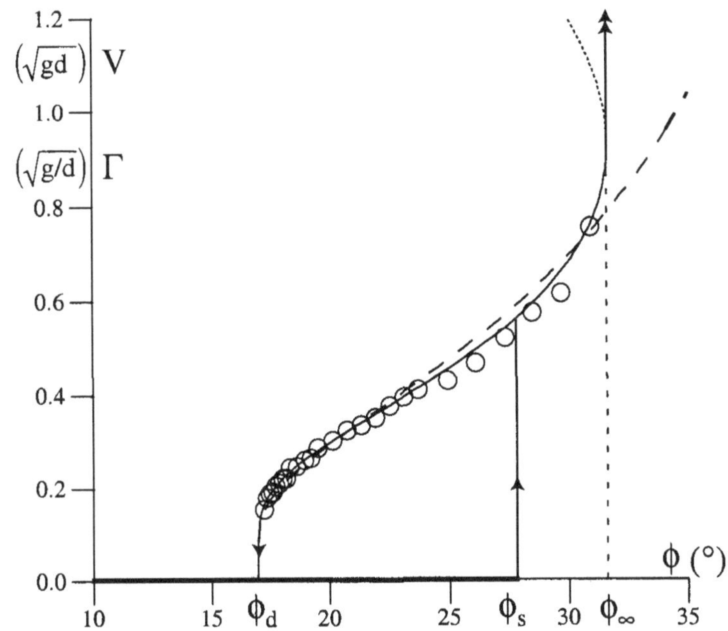

Fig. 2. Mean velocity for a single bead V. Assuming this is the relative layer velocity, it also corresponds to the velocity gradient $\Gamma = V/d$, where d is the bead diameter.

This motion quickly leads to a constant angular velocity:

$$\dot{\theta}_l = \beta\sqrt{\frac{g\sin\phi}{d}}\,, \tag{2.1}$$

with $\beta = \alpha\rho\sqrt{2\sin\phi_s/(1-\rho^2)}$. We can see that this limit angular velocity is far from the mean velocity experimentally observed (cf. Fig. 2). The main difference is that no bead motion is observed below ϕ_d, and clearly not down to a horizontal row ($\Phi = 0$). This comes simply from the fact that even if the beads have a non-zero angular velocity after one shock, it will not necessarily go down to the next dip; in-between the two dips it has to pass above the bead. When the bumpy line angle is decreased, the limit angular velocity decreases. At the same time, the height above which it has to pass (the trapping height) increases; then comes an angle for which this limit angular velocity is not enough to escape from the trap, and this explains the origin of ϕ_d. Just above this angle, the bead slows down infinitely when passing above the one underneath, so the mean bead velocity goes abruptly to zero.

If the bead motion above the trap is numerically integrated, the experimental velocity is well reproduced (Fig. 2). This constant velocity regime ends at an upper angle ϕ_∞. Slightly below this value, the bead has too large an angular velocity so that just before a collision it stops rolling, takes off and simply jumps straight forward. The next collision occurs more tangentially, the loss of normal velocity is reduced, hence a larger mean velocity. At ϕ_∞, the bead even starts to jump above the following bead. It then makes a longer way down before the next collision, transferring more potential energy into kinetic energy, and thus making the next jump longer, and so on. As a result, the bead accelerates indefinitely.

In contrast to the previous formula (Eq. (2.1)), a good fit of the mean (macroscopic) velocity is given by

$$V = \gamma\sqrt{gd} \Big/ \ln\left(\frac{\tan\phi_\infty - \tan\phi_{\rm d}}{\tan\phi - \tan\phi_{\rm d}}\right), \tag{2.2}$$

with γ of the order of 0.4. However, with this formula, the fitted value for ϕ_∞ is too large and the divergence too slow (long dashed line) compared to experiment (see Fig. 2). To be closer to measurements, we add an upper unstable branch (dotted curve).

We can also look at the transients leading to this mean velocity, i.e. to the apparent forces acting on the bead. The measurements [22] give a force of the form:

$$F = \sigma g(\sin\phi - \mu\cos\phi) - \kappa\frac{V^2}{d}. \tag{2.3}$$

The effect of the collisions thus corresponds to a Bagnold type of term, as it should be, the shocks dissipating a part of V with a frequency V/d [5]. The driving force is to a first approximation a force proportional to the gravity (but smaller than its projection along the beads row). It is striking to see that a friction force also spontaneously appears with a friction coefficient μ.[b]

3. The Motion of Layers of Grains

How can we go from the motion of one bead to the propagation of an avalanche? The next step can be to look at the motion of superposed layers of beads. Even if it is difficult to do it experimentally with rigid layers, we can still turn to numerical simulations. We thus consider layers of periodic beads as sketched in Fig. 3. We assume the same motion as previously: each bead rolls without sliding on the ones underneath.

The crucial assumption concerns the shocks, when a bead in one layer collides with the one underneath. The assumption used for liquids, by Bagnold as well, is that the exchange of momentum is local, affecting only the two layers involved in the collision. In the case of liquids, the frequency of shocks is related to thermal agitation, while with grains, the frequency of shocks comes from the motion of the grains themselves. This gives the viscous or Bagnold stress, respectively. In both cases, the characteristics are the positive curvature of the velocity profile, the fact that the velocity derivative is null at the free surface (where no stress is exerted), and to flow down to the bottom, as sketched in Figs. 4(b) and (c).

[b]By setting this force to zero, we should recover the mean velocity. This is correct for the central range; this fit does not extend close to ϕ_∞, and the observed cut-off is sharper ($\mu < \tan\phi_{\rm d}$).

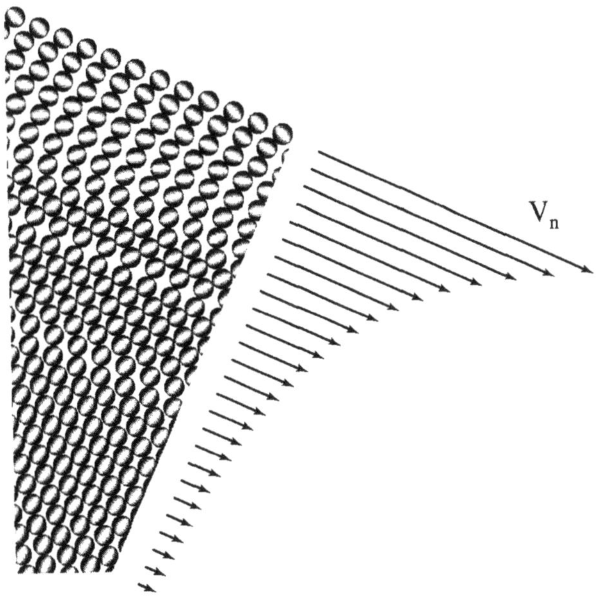

Fig. 3. A sketch of the flowing layers at one moment of their motion, with the resulting mean velocity profile.

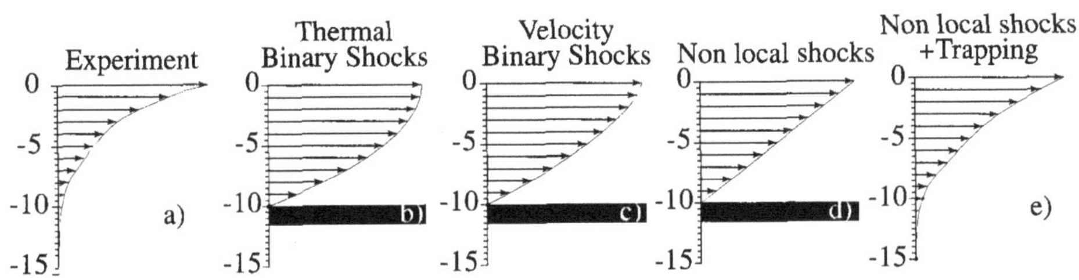

Fig. 4. Types of velocity profiles: (a) as experimentally observed using PIV, (b) with thermally induced binary shocks (viscous fluid), (c) with binary shocks induced by the motion itself (Bagnold's profile), (d) with non-local shocks in connected grains (this layer model), and (e) as obtained with this layer model.

Here, the main problem is precisely to model a surface flow, i.e. only a thin flow of grains on a remaining static layer (Fig. 4(a)). In our model, the beads are always rolling above one another, so they are always in contact (cf. Fig. 3). So when there is a shock, it is not just one bead colliding with another bead, but a small column of beads colliding with another column of beads, the latter going down to the bottom. In this case the transfer of momentum is complex and depends on the angular position of the beads along these columns.

Simulating this model numerically gives the velocity profiles of Fig. 3 [2]. What we observe is precisely a flow only on the upper part for a small tilt angle. The flow depth increases with the inclination angle, eventually reaching the bottom above a given one. The velocity profile looks very close to those observed experimentally (Figs. 4(a) and (e) [6, 23]): roughly a linear velocity profile, followed by a slow creeping tail. This slow deformation tail has been carefully studied in Ref. 16. Komatsu *et al.* have shown that it is an exponential with a characteristic length of five beads, independent of what is flowing above. This is very reminiscent of the exponential deformation observed in shear experiments (see Losert or Nagel, this volume). In other words, it seems that we just observe a flowing layer with a linear velocity profile which shears a static pile underneath.

If we look at the velocity gradient of this linear velocity profile, we find that it simply corresponds to the velocity of a single bead above a static layer [2] — roughly as if each layer of grain were rolling on a static underneath layer, with purely normal inelastic collisions [23]. This apparent inelasticity comes from the non-locality of the shocks: the momentum is transmitted down through the column underneath, so it looks as if it was dissipated, exactly as the momentum of a single bead is transmitted through the static bead, the plane, the table and the lab, down to the earth.

To further understand the selection of flow depth, we have to take into account the fact that one layer, to go above the layer underneath and fall into the next dip, has to lift up all the layers lying on it. In other words, the amplitude of the potential trapping increases linearly with depth. Now the shocks occurring above give only part of their momentum to this layer as some is still transmitted further down. The momentum transmitted thus increases with depth, but not as quickly as the trapping. So for a given angle, the condition that the shocks give enough momentum to pass the trapping selects a particular flowing depth.[c]

This roughly linear velocity profile observed in surface flows seems to be in contradiction with experiments and numerical simulations of granular flows on rough planes, showing rather a Bagnold type of velocity profile [1, 4, 21]. A possible reduction of this discrepancy is to point out two differences: the chute experiments are done for angles largely above the ones at which surface flows are considered; the boundary condition (hard rough bottom) is very different from the soft plastically deforming pile. From both these differences an interpretation is then that chute flows are slightly more diluted than surface flows, so that the contact chains could then be finite in length, and we may recover a Bagnold type of velocity profile. A way to check this interpretation would be to look at the characteristic length in the stress that can be derived from the velocity profile. This length should not be the grain diameter as in the Bagnold expression (see Hasley, this volume), but rather the typical chain length.

The presence of these contact chains giving rise to non-local shocks is precisely the idea underlying present models built to recover the characteristics of the chute

[c]The fact that the transferred momentum is constantly fluctuating, and that the beads in the sheared exponential tail wriggles before suddenly jumping over from time to time is very reminiscent of the model of thermal activation proposed by Pouliquen (this volume).

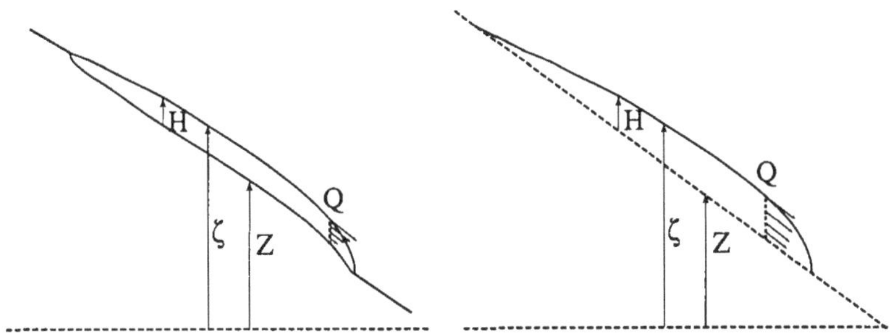

Fig. 5. Sketch and notations for a surface flow over a thick static pile (left) and a flow down to a fixed bottom (right).

velocity profiles [15, 18]. This is also the introduction of a non-local length scale (simply the depth of the flow) in front of the Bagnold shear stress that allows Khakhar (this volume) to recover not only the linear velocity profile but also the good scaling.

4. Modeling

Looking at a single bead and layers of beads helps to understand the hysteresis, the apparent forces, and the particular velocity profile. Now a macroscopic modeling of the whole surface flow can be done using the formalism first introduced by St-Venant [11], integrating the equations of motion vertically. In the case of a flowing layer, as shown in Fig. 5, we obtain the general equations from mass and momentum conservation:

$$\partial_t \zeta + \partial_x Q = 0\,, \tag{4.1}$$

$$\partial_t Q + \partial_x E = \frac{1}{\rho} F\,. \tag{4.2}$$

All the terms are vertical integrals; the free surface elevation ζ is the integral of 1, the flux Q of the velocity, the overall energy E of the velocity square and F of the resulting forces. These two equations are exact and quite general. All the particularities and approximations come from the various possible expressions of the different terms. In the case of a surface flow as in Fig. 5 (left), the unknowns are: the free surface ζ, the flowing depth H and the mean velocity of the flowing layer U (simply related to the flux by $Q = UH$). We see that we have only two equations for three unknowns, so that at least one relation is missing.

4.1. *Fixed bottom*

A first solution is to consider flows down to a fixed bottom Z (Fig. 5, right). Then, the free surface is directly linked to the flowing height, $\zeta = Z + H$, and the two previous equations are enough to find the two remaining unknowns H and U. The first equation (4.1) gives the flowing depth H. The second (4.2) gives the mean velocity U, using $\partial_t Q = H\partial_t U + U\partial_t H$ and that $\partial_t H$ is given by the first equation.

Historically, this modeling was first used to study *flood waves* in rivers. In this case the flux Q is assumed to be known as a function of the local flowing height (derived from the local cross-sections of the river and/or flow calibrations). Then, only the first mass conservation equation (4.1) is enough to solve the problem, with the characteristics method, for instance [26].

Another classical use of these equations is for thin fluid layers. A first case is to assume a purely inviscid fluid (the shallow water approximation). Then, the velocity profile is a constant one (plug flow), the fluid layer just sliding as a whole on the bottom (so that $E = U^2H$). The force is just derived from gravity and pressure (with at this place approximations of a thin, slowly varying layer). A second case is on the contrary to assume a very viscous fluid (the lubrification theory). The velocity profile is then half of a parabola (with curvature given by viscosity), as shown in Fig. 4, and there is a viscous force.

This fixed bottom approximation was also used in the case of granular flows. A first model by Savage and Hutter [25] is similar to the thin fluid film approximation, except that a friction force between the flowing layer and the bottom is added. Pouliquen [19, 20] has further refined this model by putting an effective force derived from his measurements of granular flows on a rough plane. This is very similar to the flood wave studies, using measurements in constant stable flows to predict non-uniform ones.

4.2. *Free bottom (thick pile)*

Few models have been written for the case of flow on an otherwise static pile. The first was proposed by BCRE [7], and later modified by BRdG [8]. As in the previous models, it assumes a plug flow (constant velocity profile), but there still remains three unknowns. So the velocity itself is assumed to be constant. The first equation (4.1) then gives an equation for the free surface ζ and the second (4.2) for the flow depth H, as now $\partial_t Q = U \partial_t H$. In other words, if the flowing layer wants to gain (lose) momentum, it simply increases (decreases) its thickness. The model was not presented in this way, so that the left part of the second equation was not recognized as deriving from a physical force. However, it corresponds roughly to a constant friction force. A model with two equations was also proposed by Mehta *et al.* [17], but it seems to be more difficult to translate into this St-Venant framework.

For our part we proposed, as well as Khakhar (this volume), a model using experimental results as those described above. So we rather assume a linear velocity profile [3]. Then, as above, we need another relation, to fix one unknown. So we assume that the velocity gradient Γ is locally constant and equal to that of a single bead in the same local conditions. We thus relate the mean velocity to the flowing depth, $U = \frac{1}{2}\Gamma H$, and $E = \frac{1}{3}\Gamma^2 H^3$. If the flowing layer wants to gain (lose) momentum, it now increases (decreases) both its mean velocity and depth, keeping its local velocity gradient unchanged. We also assume directly an hysteresis between static and dynamical friction forces with a typical transition depth (Fig. 6).

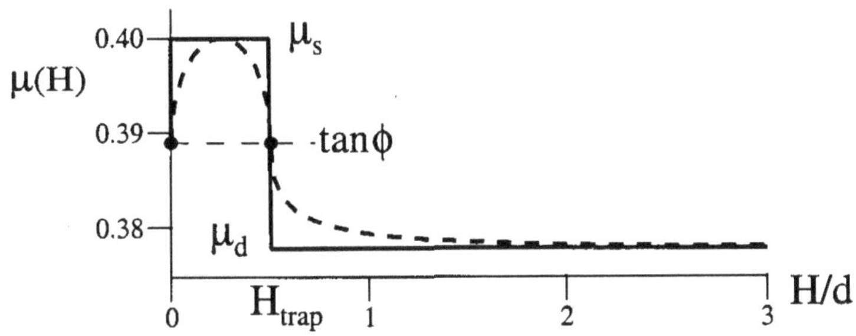

Fig. 6. Variation of the friction coefficient with flowing depth. Below a trapping height H_{trap}, of the order of one grain diameter, the friction coefficient exchanges from dynamical to static.

With these assumptions, the model summarizes as

$$\partial_t \zeta + 2U\partial_x H = 0\,, \tag{4.3}$$

$$\partial_t H + 2U\partial_x H = \frac{g}{\Gamma}(\tan\phi - \mu(H))\,, \tag{4.4}$$

with $U = 1/2\Gamma H$, Γ as shown in Fig. 2, and $\mu(H)$ as in Fig. 6, and ϕ the local surface slope. The factor two in the advective terms comes from the linear velocity profile. This model can be checked in the case of stationary flows. This is what is done by Khakhar (this volume), but also Bonamy [6] for instance, with encouraging results. But we also want to check it for more demanding cases, namely transients and unstable situations. Apart from our work on the avalanches on a rough plane ([9, 10]), we will discuss here another type of strongly inhomogeneous and unsteady flows.

5. The Four Fronts of Avalanches

When sand is poured at the top of a pile with a small enough flux, it accumulates and eventually an avalanche starts. This avalanche front is the intuitive one. At a certain place, we observe the transition from a static surface to a flowing one, and the front goes down so we call this front the "start-down" front (Fig. 7(a)). When this front reaches the end of the pile, it spreads and stops. There is then a stopping front moving upward (Fig. 7(b)). When this "stop-up" front reaches the top of the pile, the sand accumulates again and so we have successive avalanches starting down and stopping up.

If we make the symmetric of this classical experiment, i.e. a sand pile emptying at its lower end, we also observe two other fronts. The complete experiment, with an upper flat silo emptying and a lower one filling up, could be described as a "half flat hourglass" (Fig. 8). In the upper part, the sand goes out at the foot of the sand pile, and it creates an increasing step. When this step is steep enough, the upper static grains start to roll down, and we observe a "start-up" front (Fig. 7(c)). When it has reached the top end of the pile, it also flattens and stops. We then observe a

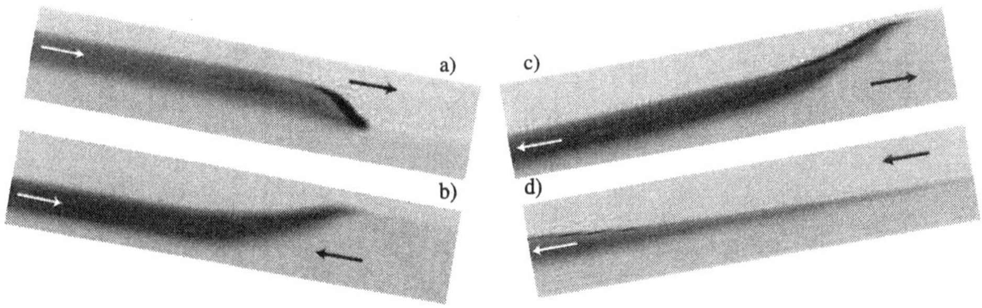

Fig. 7. The four types of fronts as observed with image differences.

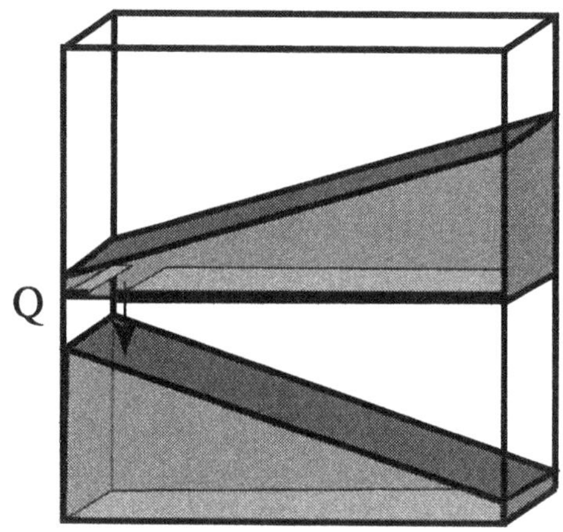

Fig. 8. Schema of the half flat hourglass experiment: the upper part is emptying into the lower part, with a fixed flux Q.

"stop-down" front (Fig. 7(d)). These four fronts are the four possible fronts either starting or stopping, either moving up or down.

These four fronts have different characteristics. The "start-down" and "start-up" fronts are sharp. We can also see that the "start-down" front sharpens with time, but tends towards a fixed shape. On the other hand, the "stop-down" and "stop-up" fronts are flat, and we can see them flattening with time. These observations are already very restrictive for possible models.

5.1. *Velocity profile*

The evolution of the front shapes depends first on the velocity profile inside the flowing layers. For a plug flow, the shapes are translated without any change at first order in time. The possible shape evolution is only second order in time, due to the effects of a force depending on the flowing height (pressure, for instance). This results in an effective diffusion of the flowing height. So it would flatten the "start-down" front, for instance.

In contrast, the effect of a non-constant velocity profile is first order in time, due to the advective term of the mass conservation equation (Eq. (4.3)). If the mean velocity U increases with flow depth H, then the "start-down" front precisely sharpens with time, and evolves toward breakdown and shock formation. The "stop-down" front will also flatten continuously with time. Thus, the observations point toward a mean velocity increasing with flow depth.[d]

5.2. *Starting fronts shapes*

If an increasing velocity profile is assumed, the mean velocity usually goes to zero for a null flowing height. Then, a simple problem for the propagation of the "start-down" front arises: its contact line is the place where the height is going to zero, so that the velocity there is null. In other words, such fronts cannot propagate at all. A first solution is to say that by breaking down it will create a shock, and then the velocity of this shock is given by conservation equations, without looking at any further detail. Another possibility is to say that the front is periodically breaking down and spreading, with a different law of motion when spreading, so that we have some kind of stick/slip motion. Yet another possibility is to set a slip velocity: a non-zero velocity limit for null height. This is what is usually done for liquids, but also in Ref. 3.

Experimentally we do not strictly speaking observe a shock at the "start-down" front, but a well-defined shape with a well-defined front angle, similar to wetting liquid fronts with a wetting angle. We also do not observe a visible stick/slip motion, but rather a continuous motion, with various velocities depending on the front height.

The case of the "start-up" fronts is different. The advective term may effectively create a shock. However, the grains spontaneously start to roll down if the angle becomes larger than ϕ_s. Thus, its slope should saturate and simply be this starting angle ϕ_s.

5.3. *"Stop-down" front propagation*

With a increasing velocity profile the "stop-down" front clearly flattens with time. But the fact that it propagates down implies a further condition: that for a small enough height (or velocity), the moving layer simply freezes. Otherwise, the front would just flatten out, but its foot should remain always at the same place.

5.4. *Simulation*

The consequence of the first observation of these four fronts is first that the mean velocity should be increasing with flow depth. The "stop-down" front also indicates

[d] At this stage, more precise investigation is required to separate a linear velocity profile from a Bagnold one, but the way the "stop-down" front flattens (as $1/t$) is, for the moment, in favor of a linear velocity profile.

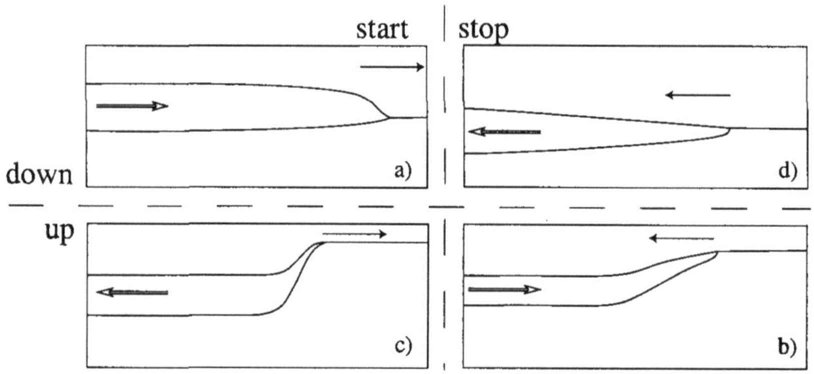

Fig. 9. The four types of fronts as obtained with numerical integration of our model. (a–d) as in Fig. 7.

that there must be a typical height or velocity under which the friction force becomes the static one. These ingredients are present in our model. Its numerical integration indeed presents the four types of fronts with the expected characteristics, as shown in Fig. 9. In particular, we observe a "start-down" front propagating with a constant shape and velocity.

In our model, this result derives from the divergence of the grain velocity at ϕ_∞. We see that the slope of the free surface is continuously increasing when going to the front foot. Using, in our model, a local velocity gradient corresponding to the velocity of one bead, when the front foot reaches the angle ϕ_∞, then locally the velocity gradient becomes infinite. Physically, it just means that the grains start locally to jump and accelerate. Such gaseous grain, ahead of the front are indeed observed. Mathematically, we then have H going to zero, while Γ goes to infinity. The mean velocity $U = 1/2\Gamma H$ can thus tend, at the front foot, toward a constant value. Thus, the front can propagate with an arbitrary velocity. A consequence of this interpretation is that the front angle should always be precisely ϕ_∞.

6. Conclusion

We have seen that looking at the motion of a single bead and layer of beads gives much information on granular surface flows. The first point is the presence of the trapping of grains in the dip between the ones underneath. The second is that the complex grain motion results on average to the hysteretic friction laws. The third is the non-locality of the shocks in dense slow flows, as all the grains are in permanent contact. The fact that for a larger angle, a more rapid flow, or a different bottom, the contact line may become finite and small, could explain the difference between the velocity profiles observed for surface flows and chute flows. These points deserve further exploration.

The St-Venant approach seems to be satisfactory in the case of stationary granular surface flows if the hypothesis of a linear velocity profile with a velocity gradient given by the velocity of one grain is used. Its results are also compatible, at least qualitatively, with more demanding cases such as the four avalanche fronts. The next work is to make a quantitative comparison. In particular, this model depends only on five parameters with well-defined physical meanings (see Fig. 2): the starting angle ϕ_s, the stopping angle ϕ_d, the jumping angle ϕ_∞, the freezing height H_{trap}, and a normalization constant γ (in front of the velocity). As these parameters can be obtained from several independent experiments, numerous cross checks are possible. Finally, a more distant goal would be to see how far these types of models can be extended, in particular in the case of geological flows. Is the inertial effect already present with the advective term derived from a linear velocity profile enough to describe these huge flows, or are further modifications needed?

References

[1] Ancey, C., Evesque, P. and Coussot, P., *J. Phys. (France)* **I6**, 725 (1996).

[2] Andreotti, B. and Douady, S., Selection of velocity profile and flowing depth in granular flows, *Phys. Rev.* **E63**, 031305 (2001).

[3] Aradian, A., Raphaël, E. and de Gennes, P.-G., Thick surface flow of granular materials: The effect of the velocity profile on the avalanches amplitude, *Phys. Rev.* **E60**, 2009 (1999).

[4] Azanza, E., Chevoir, F. and Moucheront, P., *J. Fluid Mech.* **400**, 199 (1999).

[5] Bagnolds, R. A., The shearing and dilatation of dry sand and the 'singing' mechanism, *Proc. R. Soc. London, Ser.* **A295**, 219–232 (1966).

[6] Bonamy, D., Faucherand, B., Planelles, M., Daviaud, F. and Laurent, L., Granular surface flows in a rotating drum: Experiments and continuous description, in *Powder & Grains 2001*, Kishino, Y. and Balkema, A. A., eds. (Amsterdam, 2001), p. 463; Bonamy, D., Daviaud. F. and Laurent, L., Experimental study of granular surface flows via a fast camera: A continuous description, to appear in *Phys. Fluids*.

[7] Bouchaud, J. P., Cates, M., Prakash, J. R. and Edwards, S. F., A model for the dynamics of sandpile surfaces, *J. Phys. (France)* **I4**, 1383 (1994).

[8] Boutreux, T., Raphael, E. and de Gennes, P. G., Surface flows of granular material: A modified picture for thick avalanches, *Phys. Rev.* **E58**, 4692 (1998).

[9] Daerr, A. and Douady, S., Two types of avalanche behaviour in granular media, *Nature* **399**, 6733 (1999).

[10] Daerr, A., Dynamical equilibrium of avalanches on a rough plane, *Phys. Fluids* **13**, 2115 (2001).

[11] De Barré Saint-Venant, A. J. C., Mémoire sur des formules nouvelles pour la solution des problèmes relatifs aux eaux courantes, *C. R. Acad. Sci. (Paris)* **31**, 283 (1850).

[12] Douady, S., Andreotti, B. and Daerr, A., On granular surface flow equations, *Eur. Phys. J.* **B11**, 131 (1999).

[13] Forterre, Y. and Pouliquen, O., Longitudinal vortices in granular flows, *Phys. Rev. Lett.* **86**, 5886 (2001).

[14] Jan, C. D., Shen, H. W., Ling, C. H. and Chen, C. L., *Proc. of the 9th Conf. On Eng. Mech. College Station* (Texas, 1992), p. 768.

[15] Jenkins, J. and Chevoir, F., Dense plane flow of frictional spheres down a bumpy, frictional incline, preprint (2001).

[16] Komatsu, T. S., Inagaki, S., Nakagawa, N. and Nasuno, N., Creep motion in a granular pile exhibiting steady surface flow, *Phys. Rev. Lett.* **86**, 1757 (2001).

[17] Metha, A., Luck, J. M. and Needs, R. J., Dynamics of sand piles: Physical mechanisms, coupled stochastic equations, and alternative universality classes, *Phys. Rev.* **E53**, 92 (1996).
[18] Mills, P., Loggia, D. and Tixier, M., Model for a dense stationary granular flow along an inclined wall, *Europhys. Lett.* **45**, 733 (1999).
[19] Pouliquen, O., On the shape of granular fronts down rough inclined planes, *Phys. Fluids* **11**, 1956 (1999).
[20] Pouliquen, O., Scaling laws in granular flows down rough inclined planes, *Phys. Fluids* **11**, 542 (1999).
[21] Prochnow, M., Chevoir, F. and Albertelli, M., in *13th International Conference on Rheology* (Cambridge, 2000).
[22] Quartier, L., Andreotti, B., Daerr, A. and Douady, S., On the dynamics of a grain on a model sandpile, *Phys. Rev.* **E62**, 8299 (2000).
[23] Rajchenbach, J., Continuous flows and avalanches of grains, in *Physics of Dry Granular Media*, Hermann, Hovi and Luding, eds. (Kluwer Academic, Nethterlands, 1998), p. 421.
[24] Ristow, G. H., Riguidel, F. X. and Bideau, D., *J. Phys. (France)* **I4**, 261 (1994).
[25] Savage, S. B. and Hutter, K., *J. Fluid Mech.* **199**, 177 (1989).
[26] Whitham, G. B., *Linear and Nonlinear Waves* (John Wiley & Sons, 1974).

CHAPTER 23

Random Multiplicative Response Functions in Granular Contact Networks

CRISTIAN F. MOUKARZEL
CINVESTAV del IPN, Unidad Mérida,
AP 73 Cordemex, 97310 Mérida, Yucatán, Mexico

The contact network of a frictionless polydisperse granular packing is isostatic in the limit of low applied pressure. It is argued here that, on disordered isostatic networks, displacement–displacement and stress–stress static Green functions are described by random multiplicative processes and have a truncated power-law distribution, with a cut-off that grows exponentially with distance. If the external pressure is increased sufficiently, excess contacts are created, the packing becomes hyperstatic, and the abovementioned anomalous properties disappear because Green functions now have a bounded distribution. Thus, the low-pressure, isostatic, limit is a *critical point*.

Keywords: Granular matter; isostaticity; response functions; stress distributions; multiplicative processes.

1. Introduction

Disordered packings of frictionless, cohesionless spheres, have, in any dimension d, an *isostatic* contact network when the external pressure is small with respect to the stiffness of its particles [13, 14, 16]. Because of isostaticity, the stress induced in bond b by a vertical load applied at site i, is exactly equal to the vertical displacement Δ_b^i that site i suffers when the interparticle bond b is stretched. This symmetry between force-force and displacement-displacement response functions links together two apparently unrelated phenomena observed in granular systems; (a) stress-distributions which are not Gaussian as in "usual" disordered media, but much wider, perhaps with an exponential or stretched-exponential tail [1–3, 5, 8, 9, 12, 17, 18, 21, 25] and (b) large-scale rearrangements under small perturbations [4, 12, 25]. Both phenomena are due to an "anomalous" probability distribution of Green functions on isostatic systems.

It is argued here that on generic disordered isostatic networks, Green functions are the result of a multidimensional *random multiplicative process* [19]. Several examples of linear stochastic processes with multiplicative noise have been studied recently and found to give rise to power-law distributions [6, 7, 11, 22, 23]. With

First published in *Advances in Complex Systems*, Vol. 4, No. 4 (2001), pp. 523–533.

the help of two-dimensional numerical models of isostatic packings, we show that Green functions have a truncated power-law distribution, with a cut-off that grows exponentially with distance.

When the packing is compressed the resulting network is no longer isostatic and the observed abnormal properties of granular materials disappear, giving place to a "normal" elastic behavior, i.e. a bounded distribution of Green functions. Therefore, disordered networks in the isostatic limit show a "noise induced" transition whereby the distribution of response functions acquires a critical (power-law) character.

2. Green Functions are Random Multiplicative Variables on Disordered Isostatic Networks

On an isostatic system, the stress λ_{mn} induced in a contact (mn) by a unit force acting on site i in the direction $\mathbf{e}_\alpha$, is equal to the component in the direction $\mathbf{e}_\alpha$ of the displacement induced in site i by a unit stretching of bond mn [13–16, 20]. The force-force and displacement-displacement response functions are equal on isostatic systems (but notice that they are defined in such a way that propagation of forces and displacements occurs in opposite directions). These remarks only apply for *infinitesimal* perturbations. Our considerations do not hold for response functions when rearrangements of the contact network are allowed [20].

Let us first consider a square lattice (Fig. 1(a)). Because of this symmetry in response functions, the vertical stress Green function G_y is obtained by measuring

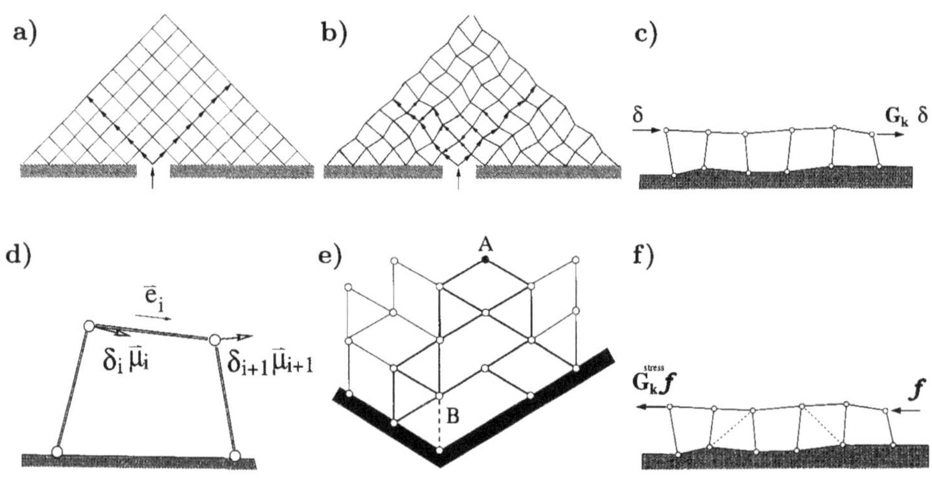

Fig. 1. (a) The response function of a regular square lattice is nonzero only along the diagonals stemming from the perturbation point. (b) If positional disorder is introduced, displacements propagate multiplicatively and diffuse inside the cone. (c) Propagation of displacements along the diagonals. (d) Site motions preserve bond lengths, thus the propagation of displacement is a random multiplicative process (see text). (e) A pantograph: When bond B (dashed) is stretched by an infinitesimal amount δ, site A (shaded) moves upwards by an amount 2δ. (f) The propagation of forces and displacements in an overconstrained network is damped because the system is rigid and opposes deformation by storing elastic energy. Thus, the response functions decay with distance.

the vertical displacement Δ_y induced on a given site by a vertical displacement of a site below it. Because our system is originally isostatic (minimally rigid), this displacement produces a deformation without any change in bond lengths. Only those sites along the diagonals (characteristics) stemming from the perturbation point will be displaced. Consequently G is some constant on the sites of the diagonals and zero everywhere else. Such a network cannot have an exponential distribution of stresses under gravitational load. Some ingredient is missing, and this is disorder.

We now introduce positional disorder by randomly displacing the sites of a square lattice by a small amount, as shown in Fig. 1(b). Because the bond angles are now random, some of the displacement diffuses inside the cone defined by the two characteristics. One notable feature is that the displacement of some sites inside this cone may now be negative, meaning that an overweight there would reduce the stress on the bottom site. Moreover, it is easy to see that the propagation of displacements δ is a *random multiplicative process*, as we now show. Considering first propagation along one of the characteristics and given that the sites below the cone (shaded sites in Fig. 1(b) do not move, this is equivalent to propagation of displacements along a one-dimensional chain of sites as depicted in Fig. 1(c). Let $\boldsymbol{\mu}_i$ be the unit vector normal to the bond supporting site i. Since displacements preserve bond lengths, site i can only move along the direction of $\boldsymbol{\mu}_i$. Let δ_i be the amplitude of its displacement. Analogously site $i+1$ moves by an amount $\delta_{i+1}\boldsymbol{\mu}_{i+1}$. Now let $\mathbf{e}_i$ be the unit vector from i to $i+1$. The condition that the length of the "horizontal" bond be unchanged reads $(\delta_i\boldsymbol{\mu}_i - \delta_{i+1}\boldsymbol{\mu}_{i+1}) \cdot \mathbf{e}_i = 0$, from which we get the recursion relation $\delta_{i+1} = \delta_i(\boldsymbol{\mu}_i \cdot \mathbf{e}_i)/(\boldsymbol{\mu}_{i+1} \cdot \mathbf{e}_i)$. Therefore, after k steps, $\delta_k = G_k\delta_0$ with

$$G_k = \prod_{i=0}^{k-1} \frac{\boldsymbol{\mu}_i \cdot \mathbf{e}_i}{\boldsymbol{\mu}_{i+1} \cdot \mathbf{e}_i} \tag{2.1}$$

showing that G is the result of a random multiplicative process [19] along the two characteristics. The multiplier $(\boldsymbol{\mu}_i \cdot \mathbf{e}_i)/(\boldsymbol{\mu}_{i+1} \cdot \mathbf{e}_i)$ can take values smaller and larger than one. Therefore, values of G_k much larger and much smaller than one will appear for large k. Notice however that, because of the left-right symmetry of the problem, after disorder average one must have $\langle G_k \rangle = 1$, but for each realization of disorder G_k may take very large (of order $e^{c_1 k}$) and very small (of order $e^{-c_2 k}$) positive values. In other words, the distribution $P(G_k)$ at each site is such that its first moment is one, but higher moments grow exponentially with distance [19].

Next consider any path inside the cone whose vertex is the displaced site. Assume for a moment that the "supporting sites" (sites not in the path but connected to it by an upwards-pointing bond) did not move. Under this approximation, propagation of displacements along any path would be a RMP of exactly the same kind as described by Eq. (2.1). The contribution of the motions of the supporting sites can be thought of in a first approximation as "additive noise" on top of a multiplicative process. It is known that random multiplicative processes with additive noise give rise to truncated power-law distributions [6, 7, 11, 22, 23]. A more accurate description would be given by a $(d-1)$-dimensional set of multiplicative variables with local coupling. These systems also display power-law distributions.

A similar reasoning can be used for isostatic networks with contact disorder. We may now think of a network whose sites are located on a regular triangular lattice, but whose bonds are an isostatic subset of the triangular lattice [13, 14, 16]. Considering now the propagation of displacements along arbitrary paths on such networks,. we again find that they can be roughly though of as a random multiplicative process with additive noise. Here the multipliers take discrete values which include zero and negative values, instead of being continuously distributed as in the previous example. The multiplication of displacements is in this case due to the existence of particular configurations of contacts (called "pantographs" — Fig. 1(e)) which multiply displacements by an integer number.

These considerations suffice to remark that the existence of random multiplicative processes in the propagation of stresses and displacements is a generic property of *isostatic* disordered networks. Numerical results to be described later show that Green functions have an approximately symmetric, truncated power-law distribution, with a cut-off that grows exponentially with distance, on isostatic networks.

Strongly compressed granular packings have a hyperstatic contact network, with more contacts than needed for minimal rigidity. The equivalence between stress–stress and displacement–displacement Green functions is no longer valid, thus one must distinguish between G^{stress} and G^{displ}. Moreover, on hyperstatic networks, displacements and stresses cannot be calculated by local propagation.

However, we can still argue in simple terms that the presence of "overconstraints" removes the anomalous properties of $P(G)$. For this purpose consider for example the one-dimensional propagation of displacements or forces along one of the characteristics on the distorted square lattice with extra bonds, as shown in Fig. 1(f). Because this system is hyperstatic, excess constraints make it rigid; its bonds store elastic energy when the network is deformed, and this introduces a dissipative term in the equations that rule, the propagation of δ_k [15]. Therefore, the average perturbation (and all its higher moments) decay with distance. In this sense, overconstraints introduce a sort of "damping term" in the (discrete) equations which rule the propagation of perturbations along the chain. A similar mechanism operates on d-dimensional hyperstatic networks.

3. Numerical Results

3.1. *Regular triangular networks with contact disorder*

We first consider a two-dimensional triangular packing of $N = L^2$ slightly polydisperse disks of radius $\approx R$ [13–16]. All particles have a unit weight. Boundary conditions are periodic in the horizontal direction. The width of the system is twice its height, therefore force lines stemming from one particle do not "interfere" with each other around the packing. Each particle has three lower neighbors on which it can eventually "rest," i.e. establish a contact with. The total number of contacts will in general depend on the "applied pressure," being $2N$ for low pressure and $3N$ for large enough pressure. The applied pressure is not an explicit parameter in our model, we only mimic its effect by controlling the density O_v of three-legged sites in the contact network.

The pile is built by adding layers *from top to bottom.* All particles in a layer have the same height. For each particle, we choose one among the three possible configurations of two supporting contacts, respectively labeled L (leftwards bond plus vertical bond), S (symmetric) and R (rightwards plus vertical). The type of contact disorder is determined by a triplet of numbers $\{p_{\rm L}, p_{\rm S}, p_{\rm R}\}$ adding up to one, which give the *a priori* probability to choose configurations L, S and R, respectively. The case $p_{\rm S} = 1$ is the square lattice. In the case $p_{\rm S} = 0$ we have that stresses propagate only vertically along independent columns, and diagonal bonds suffer zero stress (because we only have vertical gravitational forces). In both these limits there is effectively no contact disorder for the loading conditions described here. The preferred choices in this work were usually $p_{\rm L} = p_{\rm S} = p_{\rm R} = 1/3$. The procedure described above produces isostatic networks, but has *no built-in sign constraint* for stresses, and therefore tensile stresses appear. We call it the "random-stress" isostatic model (RIM). It is easy to introduce a modification in order to only obtain compressive stresses, as follows: (a) If configuration S is chosen at a given site (which happens with probability $p_{\rm S}$), and since the "incoming forces" are also compressive by construction, we can rest assured that the supporting stresses will also be compressive. (b) If configuration S is not chosen (which happens with probability $1-p_{\rm S}$), then we look at the horizontal component of the incoming forces. If it points leftwards then we choose configuration L. If it points rightwards we choose configuration R, and if it is exactly zero we choose any of those. This we call the CIM (Compressive-stress isostatic model). Once the supporting configuration (L, S or R) of a particle is chosen, we can calculate stresses on both supporting legs, because we already know the "incoming" forces due to contacts with the previous (upper) layer. Thus, when the bottom is reached and the pile is finally built, all stresses are already known. Green functions are then calculated by upwards propagation in a similar fashion.

Figure 2 shows Green function distributions for CIM networks. We can see that their distribution is a truncated power-law, with a cut-off that grows exponentially with distance. The exponent of the power-law is close to one, therefore all moments of order larger than one grow exponentially fast with height. According to our discussions in Sec. 2, both features of the distribution of Green functions, namely a truncated-power law form and exponentially growing cut-off, are due to the existence of random multiplicative processes on disordered isostatic networks.

Because of the fact that each particle has a unit gravitational force acting on it, the stress λ_b on a given bond b is given by $\lambda_b = \sum_i \Delta_i^b$. Given that Green

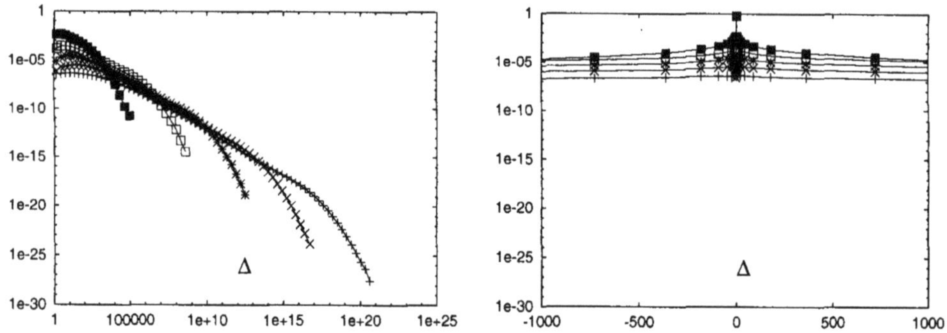

Fig. 2. Compressive Isostatic Model (CIM): Distribution of stress–stress Green functions Δ (the excess stress induced by a localized extra weight on the top) in log-log (left) and log-lin (right) scales, measured at a depth of 200 (filled squares), 400 (squares), 600 (asterisks), 800 (crosses), and 1000 (pluses) layers. The lines are only for guiding the eye. Averages were done over 10^4 to 10^5 realizations, each containing 10^6 disks with $p_R = p_L = p_S = 1/3$.

functions Δ_i^b have an exponentially broad distribution, one could naively expect stresses to behave in the same way. However, the scale of stresses grows only linearly with depth on CIM models, and not exponentially. It is therefore evident that strong correlations exist, so that Green functions of different signs (and orders of magnitude larger than the stresses themselves) cancel each other exactly at each site. The origin of these correlations is easily discovered. They are a consequence of the fact that for CIM networks we choose the local supporting configurations at each site in order to avoid negative stresses. This is equivalent to requiring that $\sum_i \Delta_i^b$ never be negative, and introduces correlations among Green functions.

When the supporting configurations are chosen entirely at random (RIM) the abovementioned correlations are absent and stresses of both signs appear (Fig. 3). The stress distribution $P(\lambda)$ is now approximately symmetric around zero — not totally symmetric because its first moment is of course still proportional to the depth as it should, but $P(\lambda) - P(-\lambda)$ is nonzero only for small stresses. The large-stress behavior is symmetric. The cut-off in $P(\lambda)$ now *grows exponentially with depth.* The second moment of $P(\lambda)$ grows exponentially with depth (not shown). The fact that the stress-scale is exponentially amplified with distance is a generic property of randomly disordered isostatic structures. As we can see, the sign constraint on stresses, and the resulting correlations induced by it, are crucial to avoiding this behavior.

3.2. *Square networks with positional disorder*

In order to show that multiplicative processes are a generic property of disordered isostatic networks, let us now briefly consider square lattices with positional disorder (Fig. 1(b)). Each disk has a unit weight and rests on two neighbors, but their

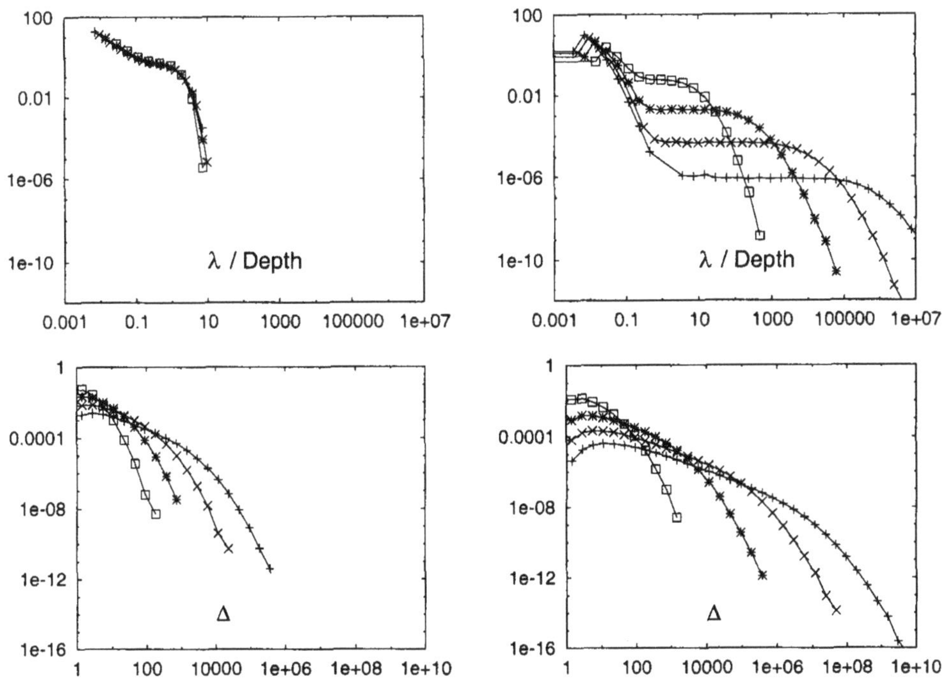

Fig. 3. Distribution of normalized stress $w = \lambda/\text{depth}$ (upper plots) and of Green functions Δ (lower plots), for isostatic packings with compressive-only stresses (CIM, left) and unrestricted stresses (RIM, right), at a depth of 50 (squares), 100 (asterisks), 150 (crosses) and 200 (pluses) layers. Only positive stresses are shown. Notice that for RIM (upper right plot), there is a stress-cut-off which grows *exponentially* with depth, instead of being proportional to depth as in the compressive-stress case (upper left).

centers are slightly displaced from the sites of a regular lattice. Our results for Green function distributions and stress distributions are shown in Fig. 4. For these systems no constraint on stresses was imposed, so that stresses of both signs appear, even when the gravitational forces act downwards. Again, the higher moments of the stress-distribution are seen to grow exponentially with depth, a result which follows from the exponentially broad distribution of response functions.

3.3. *The effect of increasing the applied pressure*

In order to simulate the effect of increasing the external pressure we introduce a variable density O_v of "overconstraints." This is done by letting each site be supported by three legs with probability O_v. This parameter "mimics" the effect of external pressure. If $O_v > 0$, stresses are no longer determined by local conditions as in the isostatic ($O_v = 0$) case, but depend on the rheology and loading conditions all over the material [13, 16]. In this case, the elastic equations must be solved and this is done iteratively by means of a Conjugate Gradient [24] method, assuming linear elastic bonds. As argued in Refs. 13, 14 and 16 and here, when the pressure

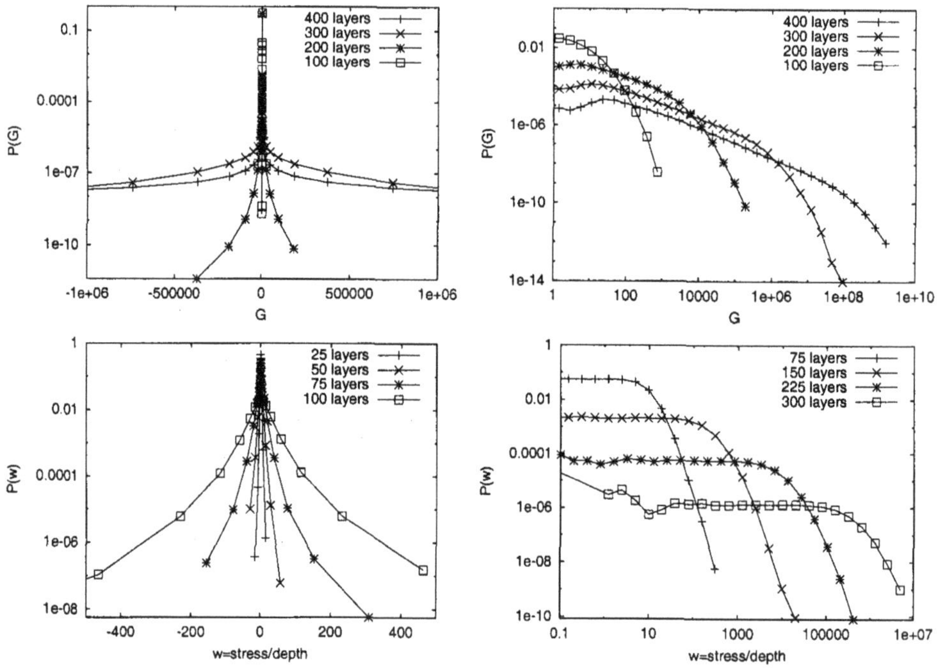

Fig. 4. Log-lin (left) and log-log (right) plots of Green function distributions $P(G)$ (upper row) and stress distributions $P(w)$ (lower row) for position-disordered square lattices, measured at several heights.

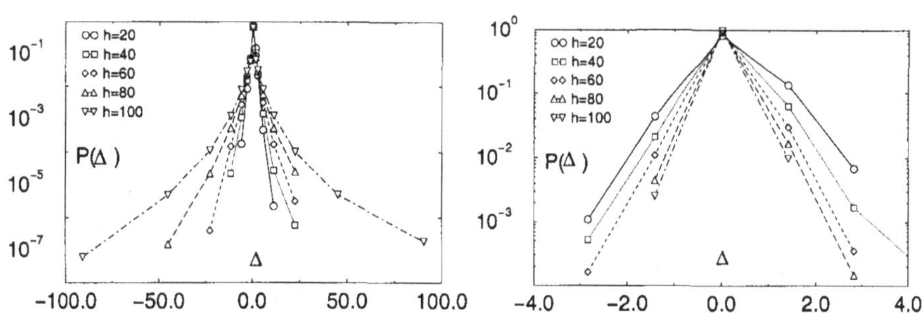

Fig. 5. On isostatic lattices (left), $P(\Delta)$, which measures the distribution of load-load and displacement-displacement Green functions, gets broader when the distance h increases. It eventually becomes a truncated power-law distribution (see Fig. 2) with an exponentially large (of order $e^{\alpha h}$) cut-off. However, a density of overconstraints $O_v = 0.02$ is enough to revert to this situation (right). In the overconstrained case, $P(\Delta)$ gets narrower with increasing distance h.

applied on a packing is increased, excess contacts appear. The resulting hyperstatic structures now damp the multiplicative effects, thus eliminating the unusual critical properties of Green functions that isostatic networks display. This is illustrated in Fig. 5, showing that in the contact-disordered numerical model with $O_v > 0$, the distribution of induced displacements Δ is bounded, and its moments decay

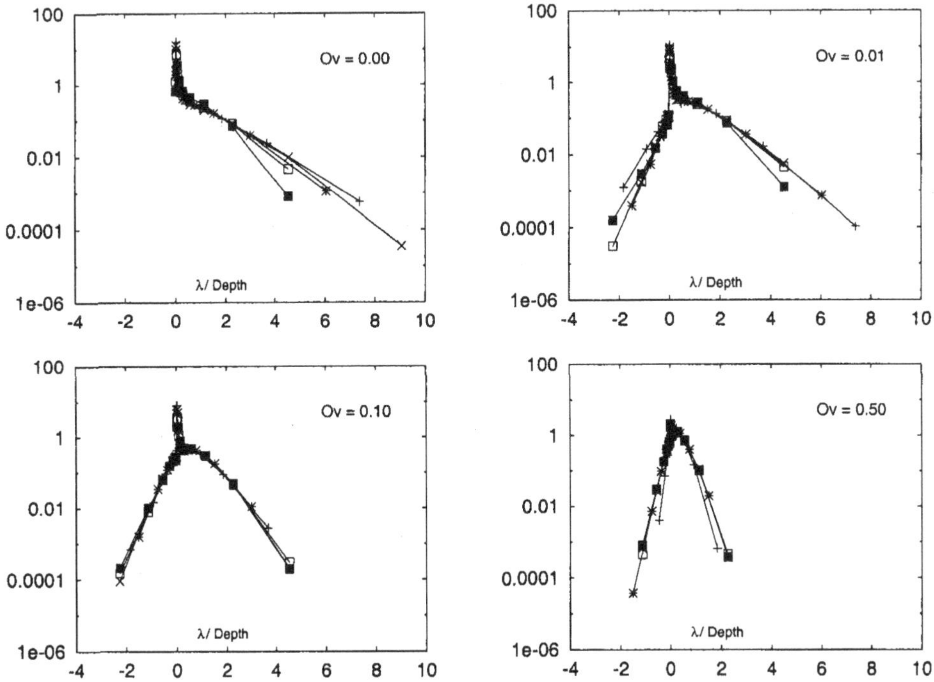

Fig. 6. Normalized stress distributions on a CIM model with increasing densities O_v of overconstraints, measured at a depth of 20 (filled squares), 40 (squares), 60 (asterisks), 80 (crosses) and 100 (pluses) layers.

with distance instead of increasing exponentially as in the isostatic case. This is consistent with the results in Fig. 6, showing how stress distributions are modified by the introduction of overconstraints. One sees that overconstraints induced by external pressure make $P(\lambda)$ become similar to a Gaussian, that is, $P(\lambda) \sim e^{-\lambda^2}$ for large λ.

4. Discussion

Rigidity considerations for the contact network of disordered frictionless packings of spheres predict the property of isostaticity in the low-pressure (or large stiffness) limit. Isostaticity means that the contact network is rigid and minimally so. The removal of *any* contact would produce the loss of rigidity. On isostatic networks, stress-stress and displacement-displacement Green functions are equal.

In the case of sequentially deposited packings we have shown that Green functions Δ of disordered isostatic networks are distributed according to a truncated power-law: $P(\Delta, h) \sim \Delta^{-\alpha}$ for $\Delta < \Delta^{\max}(h)$, where the cut-off $\Delta^{\max}(h)$ grows exponentially with distance h. The reason for this surprising behavior is that on disordered isostatic networks, response functions Δ are the result of random multiplicative processes [19]. Several instances of stochastic processes of this sort have recently been studied and found to give rise [6, 7, 11, 22, 23] to power-law distributions.

Although granular packings have by definition no tensile stresses, we also considered here isostatic structures with stresses of any sign. Our numerical results suggest the following picture: On disordered isostatic networks the distribution $P(G)$ of Green functions is a truncated power-law, and its cut-off grows exponentially with distance. If no sign constraint exists for stresses, the stress distribution $P(\lambda)$ presents surprising characteristics; all moments of the stress distribution of order larger than one grow *exponentially* with depth. In other words, the distribution of stresses is exponentially broad, but its first moment only grows linearly with depth. This is a consequence of cancellations among stresses of different signs, and is required by weight conservation.

If, on the other hand, only compressive stresses are allowed, nontrivial correlations are introduced among different Green functions at a given point. In this case the resulting stresses, which are the superpositions of Green functions, are "well behaved," i.e. its nth moment $\langle \lambda^n \rangle$ scales as depth^n.

When the external pressure is increased, excess contacts appear, the network is no longer isostatic but hyperstatic. Hyperstaticity produces a bounded distribution of Green functions, and now stress-distributions decay in a quasi-Gaussian fashion for large stresses. This transition from "anomalous" (exponential tails in $P(\lambda)$) to "normal" (Gaussian tails) elastic behavior under increasing pressure was observed in recent experiments [5] and numerical simulations [10], in which the packing fraction γ is controlled. The mechanism by which this transition happens is, I argue, that excess contacts damp the effect of multiplicative noise by introducing an extra term in the equations which rule the propagation of Green functions. This extra term is due to the fact that an overconstrained network opposes deformation, requiring a finite expenditure of elastic energy to be deformed. A discussion of specific models describing this "noise induced" transition is left for a forthcoming publication [15].

Acknowledgment

The author would like to acknowledge financial support from CONACYT.

References

[1] Brockbank, R., Huntley, J. and Ball, R., *J. Phys. II* **7**, 1521 (1997).
[2] Coppersmith, S. N., *Physica* **D107**, 183 (1997).
[3] Coppersmith, S., *et al.*, *Phys. Rev.* **E53**, 4673 (1996).
[4] Guyon, E., *et al.*, *Rep. Prog. Phys.* **53**, 373 (1990).
[5] Howell, D., Behringer, R. and Veje, C., *Chaos* **9**, 559 (1999).
[6] Kesten, H., *Acta Math.* **207** (1973).
[7] Levy M. and Solomon, S., *Int. J. Mod. Phys. C - Phys. Comput.* **7**, 595 (1996).
[8] Liu, C., *et al.*, *Science* **269**, 513 (1995).
[9] Lovoll, G., Maloy, K. and Flekkoy, E., *Phys. Rev.* **E60**, 5872 (1999).
[10] Makse, H., Johnson, D. and Schwartz, L., *Phys. Rev. Lett.* **84**, 4160 (2000).

[11] Malcai, O., Biham, O. and Solomon, S., *Phys. Rev.* **E60**, 1299 (1999).
[12] Miller, B., OHern, C. and Behringer, R., *Phys. Rev. Lett.* **77**, 3110 (1996).
[13] Moukarzel, C., Rigidity theory and applications, in *Fundamental Materials Science* (Plenum Press, New York, 1999), p. 125.
[14] Moukarzel, C. F., *Granul. Matter* **3**, 41 (2001).
[15] Moukarzel, C. F., (unpublished).
[16] Moukarzel, C. F., *Phys. Rev. Lett.* **81**, 1634 (1998).
[17] Mueth, D., Jaeger, H. and Nagel, S., *Phys. Rev.* **E57**, 3164 (1998).
[18] Radjai, F., Jean, M., Moreau, J. and Roux, S., *Phys. Rev. Lett.* **77**, 274 (1996).
[19] Redner, S., *Am. J. Phys.* **58**, 267 (1990).
[20] Roux, J. N., *Phys. Rev.* **E61**, 6802 (2000).
[21] Schollmann, S., *Phys. Rev.* **E59**, 889 (1999).
[22] Sornette, D., *Phys. Rev.* **E57**, 4811 (1998).
[23] Takayasu, H., Sato, A. and Takayasu, M., *Phys. Rev. Lett.* **79**, 966 (1997).
[24] Vetterling, W. T., Teukolsky, S. A., Press, W. H. and Flannery, B. P., *Numerical Recipes*, 2nd edn. (Cambridge University Press, New York, 1995).
[25] Wolf, D. E. and Grassberger, P., *Friction, Arching and Contact Dynamics* (World Scientific, Singapore, 1997).

INDEX

www.ingramcontent.com/pod-product-compliance
Lightning Source LLC
LaVergne TN
LVHW071222011026
849280LV00009B/46